GROWTH FACTORS AND CYTOKINES IN HEALTH AND DISEASE

A Multi-Volume Treatise

Volume 3A • 1997

SYSTEMS

GROWTH FACTORS AND CYTOKINES IN HEALTH AND DISEASE

A Multi-Volume Treatise

SYSTEMS

Editors: DEREK LEROITH
Diabetes Branch
NIDDK
National Institutes of Health
Bethesda, Maryland

CAROLYN BONDY
Developmental Endocrine Branch
NICHD
National Institutes of Health
Bethesda, Maryland

Volume 3A • 1997

JAI PRESS INC.

Greenwich, Connecticut *London, England*

55 Old Post Road No. 2
Greenwich, Connecticut 06836

JAI PRESS LTD.
38 Tavistock Street
Covent Garden
London WC2E 7PB
England

ISBN: 0-7623-0118-X

Manufactured in the United States of America

Transferred to digital printing in 2008.

CONTENTS

CONTENTS

LIST OF CONTRIBUTORS

Carlos L. Arteaga	Departments of Medicine and Cell Biology Vanderbilt University School of Medicine
Marie Beckner	Laboratory of Pathology Bethesda, Maryland
Ernesto Canalis	Department of Research Saint Francis Hospital and Medical Center
Susan Carswell	Cephalon, Inc. West Chester, Pennsylvania
Nicholas A. Cataldo	Department of Obstetrics, Gynecology, and Reproductive Sciences University of California, San Francisco
Yasmin A. Chandrasekher	Department of Gynecology and Obstetrics Division of Reproductive Endocrinology Stanford University Medical Center
David R. Clemmons	Department of Medicine University of North Carolina School of Medicine
Anne M. Delany	Departments of Research and Medicine Saint Francis Hospital and Medical Center
Solomon Epstein	Department of Medicine Albert Einstein Medical Center

Linda C. Giudice	Department of Gynecology and Obstetrics Division of Reproductive Endocrinology Stanford University Medical Center
Stefan C. Grant	Department of Medicine Memorial Sloan-Kettering Cancer Center and Cornell University Medical College
Victor K.M. Han	Departments of Pediatrics, Anatomy, and Biochemistry The Lawson Research Institute University of Western Ontario
Lee J. Helman	Pediatric Branch National Cancer Institute
David J. Hill	Departments of Medicine, Pediatrics, and Physiology The Lawson Research Institute University of Western Ontario
John E. Janik	Clinical Research Branch National Cancer Institute
Weerapan Khovidhunkit	Department of Medicine Albert Einstein Medical Center
Lance A. Liotta	Laboratory of Pathology Bethesda, Maryland
Sally A. Lynch	Memorial Sloan-Kettering Cancer Center
Patrick Mallucci	Departments of Anatomy, Developmental Biology, and Surgery University College London

Paul Martin	Departments of Anatomy, Developmental Biology, and Surgery University College London
Jane McCluskey	Departments of Anatomy, Developmental Biology, and Surgery University College London
John Mendelsohn	Department of Medicine Memorial Sloan-Kettering Cancer Center and Cornell University Medical College
Langdon L. Miller	Biologics Evaluation Section National Cancer Institute
Philip J. Morrissey	Department of Molecular Immunology Immunex Corporation
Sue Nodder	Departments of Anatomy, Developmental Biology, and Surgery University College London
Michael S. Saporito	Cephalon, Inc. West Chester, Pennsylvania
Walid Saleh	Department of Gynecology and Obstetrics Division of Reproductive Endocrinology Stanford University Medical Center
Gary A. Ulaner	Department of Gynecology and Obstetrics Division of Reproductive Endocrinology Stanford University Medical Center
Thierry van Dessel	Department of Gynecology and Obstetrics Division of Reproductive Endocrinology Stanford University Medical Center
Leonard H. Wexler	Pediatric Branch National Cancer Institute

O.W. Stephanie Yap

Department of Gynecology and Obstetrics
Division of Reproductive Endocrinology
Stanford University Medical Center

PREFACE

Advances in molecular technology in recent years have catalyzed an explosive growth of information about intercellular peptide messengers and their receptors. For example, 10 years ago the only neurotrophin characterized at the molecular level was nerve growth factor (NGF) and the only recognized neurotrophin receptor was the p75 NGF receptor. At present, the number of described neutotrophic peptides approaches 30 and the number of receptors is increasing apace. Just six years ago, the characterized interleukins numbered about three while now there are at least 16. Because many of these new peptide ligands and receptors were identified by "reverse genetic" techniques the understanding of their biological roles lags behind the knowledge of their molecular structures. Over the past few years, however, a new era of functional studies has begun because recombinant proteins have become available for clinical studies. In addition, animal models have been and are being developed using recombinant DNA techniques. Both the clinical studies and studies of transgenic and target deleted mice will allow for further physiologic elucidation of the biological roles of these messenger peptides and their receptors.

This series on *Growth Factors and Cytokines in Health and Disease* is divided into three main sections: *Growth Factors* (Volume 1), *Cytokines*

(Volume 2), and *Systems* (Volume 3). Although Volumes 1 and 2 are separate the distinction between "growth factors" and "cytokines" is probably more historical or pragmatic than indicative of differences in function. The term "growth factors" refers to a wide variety of locally of systemically produced proteins with pleiotropic actions on tissue growth and differentiation. The term "cytokines" describes a group of proteins identified primarily within the immune and hematopoietic systems, although it is likely that such a narrow view of cytokines will not survive for long. For example, it appears that some interleukins and interleukin receptors are expressed by neuroepithelial cells *in vivo* suggesting that these interleukins may have intrinsic roles within the nervous system. Furthermore, tumor necrosis factor (TNF) has been identified as a potential adipose tissue regulatory factor which is both produced and acts locally. The third volume titled *Systems* deals more directly with the role of these factors in both normal physiology and the disease processes resulting from the deficiency or excess of growth factors/cytokines and their receptors.

This third volume deals with some of the multiple systems that growth factors and cytokines affect. the role of growth factors and cytokines in fetal development, in the immune and hematopoietic systems, as well as in the skeletal and reproductive systems are covered. Various cancers are covered in a number of the chapters. While we have attempted to be as comprehensive and inclusive as possible, there will always be some regrettable omissions.

This is the third and final volume and it gives us the opportunity to thank the authors for their excellent contributions, and the staff at JAI Press for the patience.

Derek LeRoith
Carolyn Bondy
Editors

GROWTH FACTORS AND CYTOKINES IN THE FETUS AND PLACENTA

David J. Hill and Victor K. M. Han

ABSTRACT

Prenatal development depends on a well orchestrated interaction of locally-produced growth factors and cytokines with classical endocrine hormones such as insulin,

Growth Factors and Cytokines in Health and Disease
Volume 3A pages 1-53.

ISBN: 0-7623-0118-X

glucocorticoids, and thyroid hormones. Prior to implantation of the embryo the insulin-like growth factors (IGFs) are widely expressed. A temporal and anatomical expression pattern of various IGF binding proteins (IGFBPs) suggests that these control IGF availability to target organs during early embryo hyperplasia and tissue morphogenesis. Several growth factors, such as fibroblast growth factors (FGFs), and transforming growth factor-β (TGF-β) are key players in morphogenesis. Similarly, the placenta expresses a number of growth factors, including IGF-II, FGFs, transforming growth factor-α (TGF-α), and platelet-derived growth factor (PDGF) which jointly regulate growth and functional maturation in an axis that is linked but separate from that controlling fetal development. The placenta may also contribute growth factors to the maternal circulation. Several Cytokines such as colony stimulating factor-1 (CSF-1), interleukins, and GM-CSF are expressed within placenta. Increased levels of these and other cytokines in amniotic fluid are elevated in pre-term labor associated with infection. The local expression of growth factors in the placento-fetal unit is influenced by environmental factors such as hypoxia or impaired placental function.

I. INTRODUCTION

Classical endocrine hormones of fetal origin, with the exception of insulin, are not major determinants of size at birth. In some instances, as in the maturational effects of thyroid hormones on the central nervous system, transplacental passage of maternal hormones is sufficient to satisfy the basic needs of the conceptus. Other hormones, such as glucocorticoids, have defined roles in late gestation as the liver and lungs mature in preparation for postnatal existance. The fundamental mechanisms of embryonic and fetal growth appears to be orchestrated by the widespread expression, regulation of availability, and interaction of a number of peptide growth factors. Recent studies of gene manipulation have shown peptide growth factors to be obligatory for specific events of embryonic morphogenesis, and for growth of the conceptus in toto. Since the expression of most growth factors is not restricted to individual tissues, but is widespread, prenatal development can be considered mainly as an integrated network of paracrine and autocrine events. This chapter will review recent findings on the contributions of peptide growth factors to embryogenesis, placental growth, and function, and to somatic and visceral growth of the fetus. Other reviews on the roles of particular growth factors in development, and on the integration of the growth factors, classical endocrine systems, and environmental factors have recently been published (Han, 1993; Han and Hill, 1994; Snider, 1994). We°zill concentrate on the major growth factor families listed in Table 1 in the embryo and the fetus, and those in Table 2 in the placenta.

Table 1. Peptide Growth Factors Implicated in Embryonic and Fetal Development

Insulin-like growth factor	IGF-I, IGF-II
	IGF binding proteins: IGFBPs-1-6
Fibroblast growth factor	FGFs 1-9
Platelet-derived growth factor	PDGF-A, PDGF-B
Epidermal growth factor	EGF
	Transforming growth factor-α (TGF-α)
Transforming growth factor-β	TGF-β 1-3
	Activin
	Bone morphogenetic proteins: BMPs 1-8
	Vg1
Nerve growth factor (NGF)	
Neurotrophins: NTs -1-6	
Brain-derived neurotrophic factor (BDNF)	
Ciliary neurotrophic factor (CNTF)	

Table 2. Growth Factors and Receptors Expressed in the Placenta

Growth factor/receptor	*Species*
Epidermal Growth Factor	
EGF	Human
EGF receptor	Human
	Mouse
Transforming growth factor-α	
	Human
	Rat
Insulin-like growth factors	
Peptide	
IGF I, IGF II	Human
	Rat
	Sheep
Receptor	
IGF I receptor (Type 1 IGF receptor)	Human
IGF II/Mannose 6-Phosphate receptor	Rat
IGF Binding Proteins	
IGFBP-1 to -6	Human
	Rat
	Guinea pig
	Sheep
	Human
Transforming growth factor-β	
TGF-β1	Human
TGF-β receptor	Rat

(continued)

Table 2. (Continued)

Growth factor/receptor	*Species*
Platelet derived growth factor	
Peptide	Human
Receptor	
Fibroblast growth factors	
Basic fibroblast growth factor	Human
FGF receptor	Mouse
Interleukins and cytokines	Human
IL-1 to IL-6	Rat
Colony stimulating factors	
Peptide	Human
Receptor	
Tumor necrosis factor	
Peptide	Human

II. EMBRYO AND FETUS

A. Growth Factor Ontogeny

Pre-implantation

Peptide growth factors appear very early in development. In the oocyte and fertilized egg mRNAs for TGFα, FGF2, and PDGF-A chain are present as products of the maternal genome (Rappolee et al., 1988). However, shortly after fertilization these maternally-encoded transcripts are rapidly degraded so that products of the embryonic genome may predominate. In the mouse embryo the fetal genomic products are first apparent at the 2-4 cell stage, and mRNAs encoding TGFα, TGF-β, activin, and IGF-II are immediately detectable following amplification by reverse transcriptase polymerase chain reaction. Two members of the FGF family, FGFs 3 and 4, are also present at the four-cell stage, and are maximally expressed prior to organogenesis. Platelet-derived growth factor-A is detected as an embryonic mRNA transcript in the eight-cell embryo. IGF-I mRNA appears later, around the time of implantation at 8-9 days gestation (Rotwein et al., 1987). Pre-implantation mouse embryos express IGFBPs-2, - 3, -4, and -6, but not IGFBPs-1 or -5 (Hahnel and Schultz, 1994). IGFBP-2 mRNA was detected throughout the epiblast of the egg cylinder as early as day 7, at a time when IGF-II expression is limited to the trophectoderm (Wood et al., 1992). On days 10 and 11, IGFBP-2 mRNA was localized to the rostral brain of the primary neural tube, and in neuroepithelium of the tail during secondary neurulation. Some of the first products of the activated fetal genome are therefore peptide growth factors, suggesting fundamental roles in morphogenesis.

The expression of IGF-II mRNA in human embryos was studied by Brice et al. (1989) using *in situ* hybridization. Blastocysts produced by *in vitro* fertilization did not express IGF-II, but mRNA was detectable in primitive trophoblasts obtained as early as 35 days post-conception. This suggests that IGF-II may be activated concurrent with cell differentiation pathways, and this is supported by studies with embryonal carcinoma cell lines. Little IGF-II is expressed by rat embryonal carcinoma cells until differentiation is induced by exposure to retinoic acid (Nagarajan et al., 1985; Van Zoelen et al., 1989). An increase in IGF-II synthesis is accompanied by a down-regulation of expression of members of the FGF family.

Post-implantation

Considerable evidence suggests that growth factors such as IGF-II, FGF2, and TGF-β are expressed in defined cell populations in almost every tissue of the rat and mouse embryo (Brown et al., 1986; Gonzalez et al., 1990), but that gene transcription may not be a primary level at which biological control is exerted. Rather, the growth factors are bound to specific carrier proteins or extracellular matrix molecules and may require proteolytic processing by target tissues to liberate or activate them, and enable interactions with high affinity signaling receptors. During mid-gestation IGFBP-2 was abundant in surface ectoderm, particularly that of the branchial arches (Wood et al., 1992). This complimented the expression of IGF-II in the adjacent mesenchyme of the branchial arches. IGFBP-2 was seen in a restricted number of mesodermal tissues which did not express IGF-II, such as the mesonephric tubules and the notochord, and was typically seen at sites of mesodermal/ectodermal interaction which directed tissue growth and differentiation. Whether this implies intrinsic morphogenic actions for IGFBP-2, or a key role in regulating IGF availability, is not known. The distribution of IGFBP-5 mRNA in the post-implantation mouse embryo is distinct from that of IGFBP-2 (Green et al., 1994), being abundant in muscle precursor cells, all cells of the anterior pituitary, and axial regions of the neuroepithelium. During late gestation in the mouse, IGFBP-3 mRNA is seen in liver and in vertebrae, while IGFBP-4 mRNA is strongest in kidney, lung, liver, and intestine (Babajko et al., 1993). IGFBP-6 mRNA can only be detected in late gestation in liver, lung, vertebrae, and ribs; while IGFBP-1 mRNA is limited to liver. In summary, while the expression of IGF-II in early gestation and both IGF-I and -II in later gestation, are ubiquitous, the expression of the IGFBP species is precisely anatomically and developmentally regulated. This strongly suggests that the major level of control of the IGF axis in development is the regulation of its bioavailability and actions by IGFBPs.

In the human fetus, in late first and early second trimester, IGF-II and much lower levels of IGF-I mRNAs are widely expressed, but are mostly localized to mesenchymal cells (Han et al., 1987a, 1988). However immunohistochemical localization of the IGF peptides showed that they are predominantly associated with epithelia of the lung, gut, kidney, liver parenchymal cells, and adrenal cortex, and with differentiated muscle (Han et al., 1987b), suggesting disparate sites of synthesis and action. The sites of IGF peptide presence agree with the localization of mRNA or protein for IGFBPs (Hill et al., 1989; Hill and Clemmons, 1992; Delhanty et al., 1993), and suggest that the growth factors are present *in vivo* complexed with their specific binding proteins. The IGFBP mRNAs are most prevalent in regions of active cell replication and differentiation, such as the epidermis of skin, the crypt epithelia of developing gut, and the ureteric bud of the kidney (Delhanty et al., 1993). At a cellular level, the IGF:IGFBP complexes are associated with the plasma membranes and/or extracellular matrix (Hill and Clemmons, 1992), suggesting that the specific binding of IGFBPs to matrix components demonstrated *in vitro* is also widespread *in vivo*.

Both IGF-I and -II are detectable in human amniotic fluid and in extraembryonic coelomic fluid between nine and 12 weeks gestation, as are IGFBPs-1 and -3 (Nonoshita et al., 1994). Levels of IGF-II are 30 times greater than those of IGF-I, and a similar relative abundance of IGF-II persists in fetal blood throughout second and third trimester. In the human infant subject to intrauterine growth retardation IGF-I concentrations are lower than in age-matched control infants, while levels of IGF-II are unaltered (Lassarre et al., 1991). Conversely, in macrosomic infants of diabetic mothers circulating levels of IGF-I are elevated (Delmis et al., 1992). While this suggests that fetal IGF-I expression may be closely related to growth rate in the last trimester of pregnancy, this association may not be the determining biological parameter since circulating levels of IGFBP-1 are substantially elevated in the circulation of the growth retarded infant (Wang et al., 1991). This may limit IGF availability to its high affinity IGF receptors. A specific protease exists in maternal serum which is able to modify IGFBP-3 so as to reduce its binding affinity for IGF-I (Giudice et al., 1990). A similar IGFBP-3-degrading protease exists in extraembryonic coelomic fluid (Nonoshita et al., 1994) but its relevance to IGF bioactivity in the fetal membranes is unknown.

The observations in humans with regard to perturbations in the IGF axis during abnormal pregnancy have been reproduced and extended in animal studies. When fetal growth in the rat is restricted, either by uterine vessel ligation or by maternal fasting, there is a reproducible reduction in IGF-I mRNA levels in fetal liver and other tissues, an increase in IGF-II mRNA, a reduction in IGF-I but an increase in IGF-II in plasma, and an increase in the hepatic expression and circulating levels of IGFBP-1 and -2 (Vileisis and D'Ercole, 1986; Unterman et al., 1990; Straus et al., 1991; Price et al., 1992). Acute hypoxia in the ovine fetus induces a rapid but selective reduction in DNA synthetic rate in a number of tissues including adrenal and lung (Hooper et al., 1991). This is associated with only small changes in the circulating levels of IGF-I or -II, but substantial and prompt increases in circulating levels of IGFBP-1, and its levels of steady state mRNA in liver and kidney (Iwamoto et al., 1992; McLellan et al., 1992). The rapid rise in circulating IGFBP-1, and its increased hepatic expression, could be reproduced in the normal ovine fetus by infusion of catecholamines, which are known to rapidly increase in the hypoxic fetus (Hooper et al., 1994). Collectively these data suggest that tissue growth rate can be rapidly altered by a local or widespread change in IGFBP synthesis, and a limitation of the bioavailability of IGF-I and -II. A more sustained insult to the fetus may also induce a down-regulation of IGF-I synthesis.

Messenger RNA for TGF-β1 is found throughout the mouse embryo and is particularly abundant in bone and megakaryocytes of liver, which are also sites of TGF-β1 peptide synthesis (Pelton et al., 1990). Elsewhere in the embryo, TGF-β1 mRNA is associated predominantly with epithelia, while the peptide is localized to adjacent mesenchymal cells. This growth factor is particularly abundant at sites of mesenchymal/epithelial interaction during morphogenesis, such as in secondary palate formation and hair follicles. In the human fetus, also, TGF-β1, -β2, and -β3 have distinct spatial and temporal patterns of expression (Gatherer et al., 1990) which predominate during morphogenic events. The translocation of TGF-β1 peptide from epithelia to mesenchymal tissues suggests binding to extracellular matrix molecules such as the cell surface proteoglycan, betaglycan, and the extracellular matrix molecule decorin, which is a dermatan sulfate.

Messenger RNA for FGF-2 increases steadily in the embryo during development until day 16, expression being greatest in the tail, face, and developing limbs (Herbert et al., 1990). Conversely, FGF-3 is expressed in parietal endoderm, primitive mesoderm, the pharyngeal pouches, and neuroepithelium of the hind brain between 7.5 and 9.5 days of gestation. Different members of the FGF family appear to have distinct anatomical and ontological patterns of expression which together cover almost every embryonic and fetal tissue. However, outside of the central nervous system FGF peptides are predominantly associated with extracellular matrix, especially[1] the basement membranes underlying epithelia, and in this form may be inaccessible to target tissues without liberation as a result of proteolysis (Gonzalez et al., 1990). While FGF peptides are normally absent from the the postnatal circulation, FGF-2 is present in human fetal serum from at least 18 weeks gestation, and could act as an endocrine factor. FGF-2 also appears in the maternal serum during pregnancy with highest levels being seen at the end of the second trimester (Hill et al., 1995). At term, levels of FGF-2 were positively related to fetal and placental size. Maternal circulating FGF-2 may not be accessible to maternal tissues since it is associated with a high affinity carrier protein representing a solubilized form of the extracellular domain of the FGFR1 receptor.

In the mouse embryo EGF receptors are present on the first differentiated cell type, the trophectoderm (Adamson and Meek, 1984). The major ligand appears to be TGF-α, although this may be available from the maternal decidua in addition to embryonic sources around and following implantation (Han et al., 1987c). Early in the second trimester of human fetal development peptides of the EGF family are present within cells at the base of the gastric and pyloric glands of the stomach, Brunner's glands in the duodenum, the epithelium of the distal convoluted tubules of the kidney, anterior pituitary, skin, and the placental cytotrophoblasts (Kasselberg et al., 1985).

B. Growth Factors and Embryonic Morphogenesis

Mesoderm Induction

One of the best-studied morphogenic events is the induction of mesoderm in the *Xenopus* embryo, and several peptide growth factors have been implicated in the process. Mesoderm induction derives from interactions between animal pole ectoderm cells and vegetal pole endoderm during blastulation, and is thought to be initiated by mesoderm-inducing signals derived from the vegetal pole cells (Kessler and Melton, 1994). In addition to this, vegetal endoderm is able to organize a dorso-ventral pattern on the mesoderm, the dorsal structures giving rise to the notochord and muscle while ventral and lateral mesoderm form mesenchyme, blood, and some muscle. Further delineation of mesoderm continues through gastrulation.

FGF-2 (basic FGF) was able to induce the development of elements of the ventral mesoderm in explanted animal pole ectoderm from *Xenopus* embryos (Kimelman and Kirschner, 1987). Messenger RNA encoding FGF-2 was identified in the embryo at the time of mesoderm induction (Kimelman et al., 1988), while endogenous mesoderm-inducing activity could be neutralized with antiserum against FGF-2 (Slack and Isaacs, 1989). Proof that the endogenous members of the FGF family were involved in the induction of ventral mesoderm was provided by Amaya et al. (1991) who discovered a dominant negative mutant of a high affinity FGF receptor in the *Xenopus* embryo which would bind FGFs, interfered with intracellular signaling, and down-regulated the endogenous receptors. Serious defects in gastrulation and ventral mesoderm formation resulted, including a developmental failure of ventral somites leading to the absence of a tail. The notochord was also absent, which since it is derived from dorsal mesoderm is inconsistent with members of the FGF family being important inducers of ventral mesoderm. However, FGF action is coordinated with other factors which regulate dorsal mesoderm formation, members of the TGF-β family being preeminent. The actions of activin, which is a dorsal mesoderm inducer, are dependent on tissues possessing functional FGF transduction pathways (LaBonne and Whitman, 1994).

Exogenous TGF-β2 was able to induce dorsal mesoderm formation in isolated animal pole ectoderm from the *Xenopus* embryo, while the actions of the endogenous morphogen(s) could be blocked by exposure to TGF-β2 antiserum (Rosa et al., 1988). Despite these results the failure of the *Xenopus* embryo to express endogenous TGF-β at the appropriate time suggested that other members of the TGF-β-related family were biologically relevant. One possibility is the vegetal pole-derived factor, Vg1, which is expressed in cleavage stage embryos. However, while the precursor protein is present the mature, cleaved, biologically active Vg1 has not been identified (Tannahill and Melton, 1989). When Vg1 mRNA was injected into *Xenopus* embryos no mature Vg1 protein resulted and no mesoderm-inducing effects were seen, which suggests that the post-transcriptional processing of Vg1 is developmentally-regulated (Dale et al., 1993). Since treatment of animal pole explants with exogenous, mature Vg1 formed embryoid structures with a complete dorsal mesoderm axis, and therefore head development, it is possible that a transient production of mature Vg1 by dorsal vegetal cells may be sufficient to delineate this axis, which is then developed further by other morphogens. Transfection of embryos with a dominant, inhibitory activin type II receptor inhibited the mesoderm-inducing actions of exogenous Vg1 (Schulte-Merker et al., 1994).

Treatment of isolated animal pole explants with activin induces dorsal mesoderm structures, but unlike Vg1 only a partial dorsal axis is formed. Since the structures formed vary enormously with activin concentration it is possible that an endogenous gradient of activin may contribute to dorsal mesoderm formation. Both activin BA and BB chain mRNAs are expressed in the *Xenopus* embryo, the BB form first appearing in late blastulation, increasing in abundance during gastrulation, and being present in greatest amounts in the tadpole (Thomsen et al., 1990). The BA form of activin mRNA did not appear until late gastrulation. Despite this, analysis of the endogenous dorsal mesoderm-inducing activity in conditioned medium from embryonal vegetal pole cells revealed this to be related to mammalian activin BA. One of the activins is therefore a likely contributor to dorsal mesoderm formation in amphibians, and parallel studies suggest that this is also so in the chick and mouse (Mitrani et al., 1990; Smith et al., 1990). Expression of a dominant negative activin type II receptor in *Xenopus* embryos prevented activin-induced mesoderm production (Hemmati-Brivanlou and Melton, 1992), however, overexpression of follistatin, an activin-binding protein which inhibits its biological actions, did not block mesoderm formation (Schulte-Merker et al, 1994). The bone morphogenetic proteins (BMPs), also structurally related to TGF-β, are maternally expressed in *Xenopus* embryos, and BMP4 induces ventral mesoderm in isolated animal pole tissue. Additionally, BMP4 represses the development of anterior mesoderm structures (Jones et al., 1992). The observation that transfection of embryos with a dominant negative, truncated BMP receptor did not block mesoderm formation but caused excessive dorsalization suggests that competing gradients of activin and BMP4 may contribute to dorso-ventral patterning (Suzuki et al., 1994). Since BMP4 mRNA is not detected in significant amounts until mid-gastrulation, it may be involved relatively late in mesoderm formation (Dale et al., 1992).

Lessons from Gene Deletion Studies

Gene deletion by the process of homologous recombination (gene targeting or knockout) is a powerful way of examining the morphological and anatomical implications of growth factor deficiency. Transforming growth factor-β is an inhibitor of mitogenesis for most epithelial cell types *in vitro*, and for some mesenchymal cells (Roberts and Sporn, 1990). This, coupled with evidence that TGF-β has a fundamental role in embryonic morphogenesis, made it likely that disruption of TGF-β genes would have profound phenotypes. Initial reports showed this not to be the case, since targeted disruption of the TGF-β1 gene resulted in homozygous mice with normal birth size and body function (Shull et al., 1992; Kulkarni et al., 1993). However, three to four weeks after birth animals underwent a widespread and lethal inflammatory response with massive infiltration of lymphocytes and macrophages into the heart, lungs, and other organs. Both class I and II antigens of the major histocompatibility complex were overexpressed. A recent report shows that the lack of embryonic phenotype is likely due to maternal rescue, with a demonstration that radiolabeled latent TGF-β1 could cross the placenta from mother to fetus and distribute rapidly into the extracellular fluids, extracellular matrix, and fetal cells (Letterio et al., 1994). TGF-β1 was also present in maternal cells within milk, and in milk fluid, and could be transferred across the neonatal gut to peripheral tissues within 15 minutes. When homozygous TGF-β1-deficient female mice were protected against autoimmune disease by glucocorticoid treatment in order to allow pregnancy and fetal development, the homozygous offspring, with no access to maternal TGF-β1, showed serious malformations of the heart. The arterioventricular junctions were poorly developed with disordered myocyte proliferation, and abnormal ventricular lumina. Earlier studies with mouse embryos showed that TGF-β1 mRNA was strongly expressed in endocardial cells, cardiac mesenchyme, and around the cardiac cushion tissue during heart formation between days 7 and 9.5 of gestation (Akhurst et al., 1990). TGF-β1 mRNA colocalized with the extracellular matrix protein tenascin, suggesting that some of the morphogenic actions of TGF-β1 may have been indirectly mediated by selective matrix deposition.

Platelet-derived growth factor has also been implicated directly in embryonic morphogenesis. The Ph/+ mutant mouse is viable in the heterozygous condition but in the homozygous state gives rise to grossly malformed fetuses consistent with a failure of neural crest cell migration (Morrison-Graham et al., 1992). Embryos can exhibit an open neural tube, clubbed limbs, a lack of thymus, no dermal layer to the skin, a lack of connective tissue within the organs, and a failure of craniofacial development. The condition has been linked to a deletion of one of the PDGF receptors, PDGF-α, suggesting an important role for PDGF in neural crest cell migration (Orrurtreger et al., 1992). Both PDGF-A and the PDGF-α receptor are known to be expressed within the neural crest area in amphibian embryos.

A fundamental role for IGFs in the regulation of birth size, and particularly in muscle development, has been conclusively demonstrated using "gene knock-out" technology. Homologous recombination has been used to disrupt either the IGF-I, the IGF-II, or the type-1 IGF receptor gene loci in mice. By interbreeding, combination gene "knock-outs" have then been obtained. Deletion of the IGF-I gene yielded homozygotes which had a birthweight about 60% that of normal, of which some died within 6 h of birth (Liu et al., 1993). However, some of the mutant mice survived to adulthood, but females were infertile due to a failure of ovarian follicular development. Using a similar strategy to delete IGF-II, it was found that the IGF-II gene is parentally imprinted and is only transmitted from the male allele in the majority of tissues, exceptions being the choroid plexus and meninges where the gene is active on both alleles (De Chiara et al., 1990). IGF-II-deficient homozygotes had a similar growth deficiency at birth to animals lacking IGF-I, demonstrating that both isomers have a role in prenatal growth. However, IGF-II deficient mice were fertile.

Deletion of the type-1 IGF receptor, which is primarily responsible for the signaling of both mitogenesis and differentiation by both IGF-I and IGF-II, yielded homozygous animals which were only 45% of normal weight at delivery, and died within minutes of birth (Baker et al., 1993). This was due to a failure to breath and probably resulted from a widespread muscle hypoplasia, including that of the respiratory muscles. There was an increase in neuronal cell density in the spinal cord and brainstem of the mutant animals, while the skin was thinner due to a reduction in the stratum spinosum, and bone ossification was delayed by about two fetal days. Double gene knock-out involving both the IGF-I and type-1 receptor genes resulted in a similar phenotype to that found after deletion of the receptor alone, however, co-deletion of IGF-II and the type-1 receptor yielded a subgroup of animals with only 30% of normal birth weight at term and grossly retarded skeletal development. This suggests that an additional receptor to the type-1 form may also contribute to IGF-II signaling. The type-II/mannose-6-phosphate receptor is deleted in a naturally-occurring gene deletion identified by the lack of the imprinted locus "Tme," and results in lethality at the embryonic stage (Barlow et al., 1991). Whether this receptor can contribute to IGF-II signaling *in vivo* is not clear.

The intracellular signaling initiated by ligand binding to either the type-1 IGF receptor or the insulin receptor is thought to be routed largely through phosphorylation of the insulin receptor substrate-1 (IRS-1). Targeted deletion of IRS-1 resulted in viable offspring which nonetheless exhibited up to 60% fetal growth retardation (Araki et al., 1994; Tamemoto et al., 1994); a phenotype very similar to that of the type-1 IGF receptor knock-out animal. The IRS-1-deficient mice were normoglycemic, and did not exhibit a hypoglycemic response following injection of either insulin or IGF-I. This partial metabolic control and postnatal viability suggest that IRS-*1* is not entirely responsible for the transduction of insulin actions, and the recent isolation of a related IRS-2 molecule may provide an alternate pathway (Araki et al., 1994). Mutations of the β-subunit of the insulin receptor in man which disrupt autophosphorylation result in leprechaunism, typified by extreme intrauterine growth retardation and retarded postnatal growth (Jospe et al., 1994). Comparable mutations in the type-1 IGF receptor have not been reported.

The above series of studies demonstrates that neither IGF-I or IGF-II are crucial for key morphological events in early development, but that they act as "true" growth factors, contributing to the expansion of stem cell populations and the progression of cell differentiation. While both IGF-I and -II contribute to fetal growth, the size of the placenta was normal following deletion of the IGF-I and type-1 receptor genes while reduced after deletion of IGF-II (Baker et al., 1993).

This suggests that a high expression of IGF-II in placenta may contribute to its development as an autocrine or paracrine agent.

Given widespread distribution of TGF-α expression in the fetus, it was surprising that targeted disruption of the TGF-α gene resulted in largely normal embryonic and fetal development, although the neonatal mice had curly hair and whiskers and abnormal eyelid maturation (Luetteke et al., 1993). However, it is possible that TGF-α is rendered redundant by the presence of other EGF family members, which may be up-regulated as a compensatory mechanism when TGF-α is lacking. This would be supported by the failure of implantation that occurs following the disruption of the EGF receptor gene (ER-1). Use of targeted EGF antiserum in mice as a means of eliminating EGF peptide bioactivity has been shown to cause a failure of tooth eruption and poor lung morphogenesis (Slavkin et al., 1992; Yasui et al., 1993).

Embryonic Limb Formation

Formation of the limb buds and the subsequent skeletal structure of the limbs has been well studied in the chick and mammalian embryo, and involves the integrated actions of several peptide growth factors. The limb buds first develop as a thickening of the body wall mesenchyme, the surface ectoderm of which is induced by the underlying mesenchyme to form a specialized structure called the apical ectodermal ridge. The mesenchyme beneath the apical ectodermal ridge, the progress zone, is maintained in an undifferentiated, rapidly proliferating state and enables outgrowth of the limb to occur. Limb outgrowth is promptly arrested following removal of the apical ectodermal ridge. As mesenchyme moves distally to the progress zone, so it undergoes a condensation and morphogenic change to become cartilage. Sub-periosteal bone then develops on the surface of the cartilage immediately below the perichondrium to give rise to primary ossification structures. Increase in length of the long bones then continues by epiphyseal chondrogenesis and subsequent ossification. Superimposed upon this sequence of outgrowth and differentiation is the formation of the pentadactyl pattern of the limb in the dorso-ventral plane. One of the most important morphogenetic factors for dorso-ventral patterning is expressed within a specialized area of mesenchyme on the ventral aspect of the progress zone called the polarizing region, and is now recognized as a member of the hedgehog gene family.

Limb outgrowth and patterning is controlled by combinations of Hox gene expression products, diffusible morphogens such as retinoic acid, and several peptide growth factors. Genes of the *Hoxd* complex are temporally expressed in nested sets on the posterior side of the developing limb bud, while genes in the Hoxa complex are expressed on the proximo-distal side. Hox gene overexpression or knock-out studies have shown that these genes dictate the rate and position of cartilage condensation, but also result in some digit duplications, suggesting a partial role in pattern formation (Maden, 1994). This is particularly so for one Hox gene, *Hoxb-8*, which is expressed in the posterior half of the mouse fore-limb bud. Aberrant expression of *Hoxb-8* in the anterior region of the limb caused the appearance of an ectopic polarizing region, the expression of the sonic hedgehog gene (*Shh*) within these cells, and mirror image formation of limb structures (Charite et al., 1994). The hedgehog gene product was first identified in *Drosophila* as a secreted factor controlling body segmentation. Vertebrate homologs are expressed in several embryonic tissues and regulate cell pattern formation in the notochord, the neural tube floor plate, and the posterior limb buds. Secreted hedgehog precursor protein is proteolytically modified to at least two biologically active forms, the first acting on the neural tube and the second on the limb bud. The protease responsible is within the hedgehog precursor protein itself, acting by autoproteolysis. Expression of *Shh* in the polarizing zone of the limb bud induces dorso-ventral pattern formation, and transplantation of cells transfected with *Shh* causes mirror image structures (Peifer, 1994).

FGF-2 was present at both mRNA and protein levels during limb bud formation in the chick and mouse (Herbert et al., 1990; Munaim et al., 1988), peptide levels in the chick limb being greatest on day 3 of gestation (stage 18) when the cell proliferation rate was highest. During this rapid proliferation mesenchymal cells secrete an extracellular matrix rich in hyaluronic acid. Using isolated chick limb bud cells, FGF-2 was shown to potentiate hyaluronic acid release to form pericellular coats (Munaim et al., 1991). A loss of hyaluronic acid synthesis *in vitro* coincided with the timing of condensation of mesoderm into the chondrogenic and myogenic regions of the limb bud, and a decline in FGF-2 abundance, at stages 22-26. Within the mouse embryo, Gonzalez et al. (1990) used immunocytochemistry to localize FGF-2 at 18 days gestation. Positive staining was apparent in chondrocytes of the hyaline cartilage and in the perichondrium. Within ossification centers FGF-2 was absent from hypertrophic cells but present within the extracellular matrix, osteoblasts, and vascular endothelial cells. A different experimental approach was that of Liu and Nicoll (1988) who transplanted fetal rat paws, harvested on day 10 of gestation, under the kidney capsule of adult hosts which were then infused with FGF-2 or anti-FGF-2 antiserum via the renal artery. Infusion of FGF-2 antiserum significantly retarded the growth of the explants and their ossification. Conversely, paw size was increased by administration of FGF-2. Other isomers of FGF may also be involved in limb development, FGF5 mRNA appearing within the limb mesenchyme between embryonic days 12.5 and 14.5 in the mouse (Haub and Goldfarb, 1991). Expression was limited to a patch of cells ventral to the presumptive femur which was undergoing cartilage formation. The above evidence suggests a mitogenic role for FGF-2 in mesenchyme proliferation, and a possible morphogenic role for FGF-5 during cartilage induction. However, the strongest evidence linking FGFs to limb formation does not involve the mesenchyme but the apical ectodermal ridge.

As soon as the apical ectodermal ridge is formed, at day 10 of gestation in the mouse, a high expression of FGF-4 mRNA is seen in the posterior half, and expression persists until day 12 (Niswander and Martin, 1992). Several members of the FGF family can substitute for the apical ectodermal ridge and maintain limb bud outgrowth *in vitro* in both the mouse and chick (Niswander and Martin, 1992, 1993), suggesting that an FGF is involved in the endogenous signaling between the epithelium and the underlying mesenchyme. Recently, Niswander et al. (1993) demonstrated that recombinant FGF-4 could substitute for the ridge *in ovo*, and not only maintain limb bud outgrowth but signal the correct spatial information to achieve normal pattern formation. This would imply that FGF-4 is capable of regulating an appropriate expression of *Shh* from the polarizing zone within the mesenchyme. Contradictory evidence was provided by Fallon et al. (1994) who showed that only FGF-2 was detectable in the chick limb bud, and that exogenous FGF-2 could substitute for the apical ectodermal ridge. Recently, Niswander et al. (1994) showed that retinoic acid applied to the anterior margin of the limb bud could induce the expression of FGF-4 in the apical ectodermal ridge. In turn, FGF4 was able to up-regulate the expression of Shh and establish a polarizing zone within the mesenchyme. Once established, FGF-4 was able to maintain Shh expression, and thus control pattern formation, without a further requirement for retinoic acid. However, *Shh* was also able to maintain the expression of FGF-4 within the apical ectodermal ridge, suggesting that a positive feedback loop between FGF-4 and *Shh* can control both the outgrowth of the limb and pattern formation in a classical epithelial-mesenchymal interaction. Both *Shh* and retinoic acid also activated *hoxd* gene cluster expression. The emerging picture is of an integrated expression of *hox* genes, *Shh*, and FGF-4 which together dictate limb structure, and the possibility that this cascade is initiated by the existence of a retinoic acid gradient in the presumptive limb bud.

It is now possible to predict which receptor types are involved in FGF signaling within both mesenchyme and ectoderm in the developing limbs. In the mouse embryo, the FGFR2 receptor was first expressed on day 9.5 in limb bud mesenchyme, with a concentration gradient increasing in a posterior and proximal direction. At this time the expression of FGFR1 was more diffuse than that of FGFR2, within the limb bud mesenchyme, the somites, and organ rudiments (Peters et al., 1992). By day 11.5 FGFR2 mRNA was localized to mesenchymal aggregates corresponding to the future bones, and in the surface ectoderm of the limb, being strongest in the interdigital web. At day 12.5 gestation FGFR2 mRNA located to chondrification centers, and at 14.5 days to the bodies of the distal bones. This temporal pattern of expression strongly suggests that FGFR2 mediates FGF actions on the chondrogenic pathways, while FGFR1 may mediate FGF actions on the surrounding undifferentiated mesenchyme. The FGFR4 receptor mRNA was found by *in situ* hybridization to map to areas of cartilage condensation, while FGFR3 mRNA was abundant in the resting cartilage during the subsequent process of endochondrial calcification (Stark et al., 1991; Peters et al., 1993).

In the rat embryo, Beck et al. (1987) localized IGF-II mRNA to pre-cartilaginous mesenchymal condensations, perichondrium, and immature chondrocytes, in addition to the periosteum and centers of intramembraneous ossification. Both IGF-I and -II mRNAs were also localized to limb bud mesenchyme in the rat fetus by Streck et al. (1992), who additionally showed that while IGF-II expression was strongest in the presumptive skeleton and muscle at the centre of the limbs, IGF-I mRNA was absent from these areas but strongly expressed in the peripheral mesenchyme beneath the epithelium. Neither IGF isomer was strongly expressed in the rapidly dividing mesenchymal cells of the progress zone. While IGF-II is a potent mitogen for isolated limb bud mesenchyme from the rat *in vitro* (Bhaumick and Bala, 1991), the above findings suggest that the role of IGFs is not primarily as mitogens, but involves the initiation or progression of differentiation pathways for skeletal and muscular elements. In the first trimester human fetus, also, IGF-II mRNA was abundant in perichondrial areas (Han et al., 1987a). We localized IGF peptides by immunocytochemistry in the chick embryo limb buds (Ralphs et al., 1990). At stages 20-24 a uniform presence of IGFs was seen in undifferentiated mesenchyme, but this disappeared in the pre-chondrogenic areas of condensation. As chondrocytes appeared around stage 28, so immunoreactive staining for IGFs returned. By stage 36 endochondrial calcification had begun and intense IGF staining was associated with hypertrophic chondrocytes, as well as osteoblasts in the sub-periosteum of the membraneous bone.

During the early development of the limb bud, IGFBP-2 expression is seen within the anterior-posterior strip of ectoderm which will become the apical ectodermal ridge, and IGFBP-2 continues to be expressed here until outgrowth is complete (Streck et al., 1992). It is possible that the function of IGFBP-2 in the apical ectodermal ridge is to negate an IGF-II-dependent drive toward differentiation in the underlying progress zone mesenchyme, and to maintain a stem cell population.

TGF-β1,-2, and -3 isomers are all expressed within the developing skeleton of the mouse embryo. Heine et al. (1987) observed the distribution of TGF-β1 peptide by immunocytochemistry from 11 to 18 days gestation. Strong staining was seen in all mesenchyme undergoing condensation and cartilage formation, which persisted during ossification within the newly-formed osteoblasts. Analysis of TGF-β1 mRNA distribution by *in situ* hybridization revealed a high expression in perichondrial osteocytes involved with membraneous calcification (Lehnert and Akhurst, 1988). However, differentiated cartilage contained little TGF-β1 mRNA, although this is a site of peptide synthesis. A similar pattern has been described for TGF-β2 and -β3 in developing skeletal tissues (Pelton et al., 1990). In the human embryo of 32-57 days gestation, TGF-β2 and -β3 mRNAs were localized to chondrogenic areas (Gatherer et al., 1990). TGF-β2 mRNA was located within the pre-cartilaginous blastoma of limb bud mesenchyme, and in later development in actively proliferating chondroblasts at the epiphyseal/diaphyseal boundary. Messenger RNA for TGF-β3 was first seen in the developing intervertebral disks and in the perichondrium of cartilage associated with the vertebral column, but not with long bones. An intense site of TGF-β1 abundance was found in areas of membraneous bone formation and, in fetuses of 10-12 weeks gestation, in osteogenic cells at sites of endochondrial calcification in the long bones. No TGF-β mRNA was located in the hypertrophic chondrocytes which immediately precede the area of provisional calcification. A divergent expression of TGF-β isoforms in skeletal primordia suggests distinct biological roles, and that of TGF-β2 would support a role in cartilaginous induction. Evidence for this is provided by the observations that mammalian TGF-β1 and -β2 induced the appearance of phenotypic chondrocytes, associated with increased sulfated mucopolysaccharide and type II collagen synthesis, in chick embryo mesenchyme cultures. Other members of the TGF- β family, namely bone morphometric proteins-2B and -3 (osteogenin), will also induce cartilage formation from embryonic mesoderm in the chick (Carington et al., 1991; Chen et al., 1991) making the identity of the endogenous active ligand unclear. When TGF-β1 or -β2 were injected into the sub-periosteal region of the femurs from newborn rats, local intramembraneous bone and cartilage formation resulted (Joyce et al., 1990). After injections were terminated the new cartilage underwent endochondrial calcification. These results strongly suggest that TGF-β isomers are key players in the formation of cartilage from undifferentiated mesenchyme, and in the subsequent primary ossification process.

Muscle Development

Our understanding of how peptide growth factors contribute to muscle cell growth and differentiation has been helped enormously by the availability of immortalized cell lines. The IGFs interact with other peptide growth factors to precipitate the differentiation of rat L6 cells into postmitotic, contractile myotubes. This will occur spontaneously as the myoblasts grow to high density, but can be precipitated prematurely by incubation with IGF-I, or high concentrations of IGF-II or insulin (Ewton and Florini, 1981). While IGF-I will initially induce cell proliferation, a commitment to terminal differentiation is invoked, which may be mediated by an IGF-dependent activation of specific genes controlling differentiation such, as myogenin and MyoD (Florini et al., 1991a). However, other studies have suggested that IGF-II regulates muscle cell differentiation at intracellular points distal to the activation of myogenin (Rosen et al., 1993). A recent report suggests that the actions of IGF-II on muscle cell differentiation may involve the type-II/mannose-6-phosphate receptor, while those of IGF-I are mediated primarily by the type-1 IGF receptor (Rosenthal et al., 1994).

In contradistinction to the actions of IGFs, incubation with FGF-2 will potentiate the proliferation of myoblasts and prevent commitment to terminal differentiation (Linkhart et al., 1981; Clegg et al., 1987). Further, FGF-2 can suppress transcription of at least two of the myogenic regulatory genes, myogenin and MyoD1 (Vaidya et al., 1989; Brunetti and Goldfine, 1990), suggesting that a widespread presence *in vivo* serves to prevent a premature differentiation of muscle. TGF-β also prevents terminal differentiation, but has little effect on proliferative rate. The onset of differentiation can therefore be finely controlled by relative changes in the abundance of particular growth factors in the microenvironment.

These growth factor signals derive, in part, from the cells themselves *in vitro*. Proliferating myoblasts synthesize FGFs and high affinity FGF receptors, but synthesize little IGF-I or -II. Upon muscle differentiation the synthesis of FGF(s) and its receptor(s) decline, while the expression of IGF-II dramatically increases (Florini et al., 1991b; Rosen et al., 1993). This results in a down-regulation of the type-1 IGF receptor. Insulin-like growth factor-II gene expression by differentiating myoblasts is down-regulated by exogenous IGF-I, IGF-II, or high concentrations of insulin (Magri et al., 1994). The actions of IGFs in muscle cell differentiation are also likely to involve endogenous IGFBPs. The mouse muscle cell line, C2C12, expresses abundant IGFBP-2 mRNA and releases IGFBP-2 peptide in the undifferentiated state, but IGFBP-2 expression declines substantially as differentiation proceeds (Ernst et al., 1992). Conversely, IGFBP-5 expression increases during muscle cell differentiation. Overall, a change in IGF:IGFBP ratio in favor of the former during differentiation may maximize the availability of endogenously produced IGFs.

Studies of the ontogeny of growth factor expression *in vivo* lend general support to the conclusions drawn from tissue culture experiments. The skeletal muscles of the trunk and limbs arise in the somites, which in term derive from segmentation of mesoderm in an anterior to posterior sequence, beginning in the mouse embryo on day 8. Limb muscle precursor cells migrate out from the newly-formed myotome in the trunk and enter the limb buds on days 9 and 10. Several FGF isomers are expressed during muscle development, and may together coordinate both the proliferation of precursors, and the programmed differentiation of the myotome. FGF-2 peptide, detected by immunocytochemistry, is located throughout cardiac muscle, somite myotome and limb bud muscle in the developing chick and rat (Joseph-Silverstein et al., 1989; Herbert et al., 1990). When applied to myogenic cell lines *in vivo*, FGF-2 enhances proliferation and prevents differentiation (Clegg et al., 1987). Further, FGF-2 can suppress transcription of at least two of the myogenic regulatory genes, myogenin and MyoD1 (Vaidya et al., 1989; Brunetti and Goldfine, 1990), suggesting that a widespread presence *in vivo* serves to prevent a premature differentiation of muscle. However, a specific expression of FGF-4 mRNA occurs in embryonic mouse muscle immediately prior to activation of myogenin and MyoD in the developing myotome (Niswander and Martin, 1992). FGF-5 is also expressed in developing muscle, appearing in the myotomes of the trunk on day 10 of mouse gestation following expression of muscle-specific genes such as actin (Haub and Goldfarb, 1991), but never appearing in tail region myotomes. As trunk myotomal cells migrate ventrally and laterally, including those entering the limb buds, so expression of FGF-5 continues. One explanation for this pattern is that FGF-5 selectively suppresses the differentiation of these cell lineages while migrating. Thus, different FGF species may coordinate the amplification of myoblast populations, their migration schedules, and their eventual differentiation.

This is unlikely to depend on any one class of growth factor in isolation. As has already been described, deletion of the IGF signaling pathways in transgenic mice resulted in a severe deficiency in muscle mass, while TGF-β isomers are expressed within developing muscle, and are thought especially to contribute to cardiogenesis (Akhurst et al., 1990).

Skeletal Development

Longitudinal skeletal growth arises from new bone formation as a result of epiphyseal chondrogenesis, and subsequent replacement of cartilage by bone. Adjacent to the epiphysis is a stem cell population of precursor chondrocytes, which gives rise to a zone of closely packed and highly proliferative chondrocytes. Cells within the proliferative zone secrete an extracellular matrix consisting mainly of type-II collagen fibrils and sulfated glycosaminoglycans, the most abundant of which are chondroitin-4 and -6 sulfates, and hyaluronic acid. After a number of cell replications, cells are pushed toward the diaphysis within longitudinally-arranged columns. Proliferative activity gradually ceases while the cells hypertrophy and increase their rate of macromolecular synthesis. As matrix synthesis increases, the cells become postmitotic and terminally differentiated. Cells within the hypertrophic zone have an increased rate of collagen and sulfated mucopolysaccharide synthesis, with an increased ratio of protein to DNA. Increases or decreases in growth rate are predominantly due to alterations in hypertrophic activity rather than changes in chondrocyte proliferation rate (Hunziker, 1988). Terminal differentiation in the lower hypertrophic zone is characterized by a dramatic fall in mucopolysaccharide synthesis but a maintenance of collagen synthesis, the expression of a distinctive type-X collagen, and the appearance of alkaline phosphatase activity. Mineralization then occurs between the columns of chondrocytes, and the chondrocytes are replaced by macrophages and osteocytes. Cartilage matrix is eroded and replaced by hydroxyapatite at a similar rate to new cartilage formation as a result of proliferation and hypertrophy. Calcification is initiated in matrix vesicles in the lower hypertrophic zone coincident with vascular invasion by capillary sprouting, and the delivery of monocytes and chondroclasts which degrade cartilage matrix. The hypertrophic chondrocytes are thought to be deleted by apoptosis. Growth factors may therefore influence epiphyseal chondrogenesis by altering the rate of chondrocyte proliferation, by modulation of the synthetic rate of extracellular matrix molecules, or by changing the rate of terminal differentiation and calcification. There is substantial evidence that FGFs, IGFs, and TGF-β isomers are expressed locally within the growth plates during both prenatal and postnatal epiphyseal chondrogenesis.

The presence of FGF-2 within cartilage has been recognized for over 10 years (Bekoff and Klagsbrun, 1982), and its production by isolated chick growth plate chondrocytes has been described (Rosier et al., 1991). Evidence of its relevance to ordered epiphyseal chondrogenesis in human development was recently provided by Rousseau et al. (1994) who found that patients with achondroplasia invariably exhibited mutations in the FGFR3 receptor. Extensive analysis of IGF-I expression, control, and action within epiphyseal growth plate cartilage in postnatal life has provided conclusive proof for a substantial paracrine component in longitudinal skeletal growth control (Andersson et al., 1986; Nilsson et al., 1986, 1987; Isgaard et al., 1988), and this is also true prenatally. In the human fetus IGF-II is expressed within the proliferative and differentiating chondrocytes of the growth plates within the long bones (Han et al., 1987a, 1987b). Using bovine tissue, IGF receptors were found to be most abundant in the proliferative region but were also found on hypertrophic chondrocytes (Trippel et al., 1986). These observations are in agreement with the observed biological actions of IGFs *in vitro*, which include a predominantly mitogenic action in the proliferative chondrocyte region, and a stimulation of mucopolysaccharide synthesis in the post-mitotic hypertrophic chondrocytes adjacent to newly-formed bone (Hill, 1979). Using *in situ* hybridization, TGF-β1 mRNA was shown to be expressed in abundance in bone cells of the human fetus in first trimester, although relatively little was present within the growth plate (Sandberg et al., 1988). However, TGFβ-immunoreactivity was seen in chondrocytes and the cartilage matrix of fetal and postnatal mice (Ellingsworth et al., 1986; Heine et al., 1987), while bio-detectable TGF-β was released from isolated growth plate chondrocytes from the postnatal chick, this being greater for hypertrophic cells than for cells of the proliferative zone (Rosier et al., 1989; Gelb et al., 1990). TGF-β may be stored within the matrix of bovine articular cartilage as the inactive, latent precursor molecule, which could then be activated by locally produced proteases.

We have studied the regulation of epiphyseal chondrogenesis in the fetus as a model of peptide growth factor interaction during cell proliferation and differentiation, our model being the proximal tibia of the ovine fetus. FGF-2 and its high affinity receptor, FGFR1, were strongly expressed at both mRNA and peptide levels in the proliferative chondrocyte zone of the ovine growth plate, decreased during cell differentiation, and absent from the hypertrophic chondrocytes. No IGF-I mRNA was observed in the fetal growth plate, but IGF-II mRNA and peptide were predominantly associated with the differentiating, but still mitotically active, chondrocytes. Unlike IGF-I in postnatal life, the expression and release of IGF-II in fetal growth plate chondrocytes was not increased in response to growth hormone. An associated expression of IGFBP-2 and IGFBP-3 was found. As chondrocytes began to hypertrophy, so mRNA for IGF-II declined and that encoding TGF-β1 appeared. The terminally differentiated chondrocytes also expressed IGFBP-5, which was not seen in other areas of the growth plate. In summary, as chondrocytes passed from proliferation to differentiation to hypertrophy, they sequentially expressed FGF-2, then IGF-II, and finally TGF-β1. This anatomical distribution allowed testable hypotheses to be formulated with regard to growth factor contribution to epiphyseal chondrogenesis.

The expression of FGF-2 and its receptor in stem and proliferating chondrocyte populations suggests a role as an autocrine mitogen. Using isolated chondrocyte cultures FGF-2 was found to be released and to contribute to DNA synthesis, while a neutralizing antibody against FGF-2 decreased endogenous DNA synthesis in cells by 50% (Hill and Logan, 1992a; Hill et al., 1992a). Exogenous FGF-2 was 100-500 times more potent than IGF-I, IGF-II, or insulin as a mitogen for fetal growth plate chondrocytes (Hill and Logan, 1992b; Hill et al., 1992b). Insulin had an equivalent mitogenic action to IGF-I at low nanomolar concentrations (Hill and De Sousa, 1990), and both were an order of magnitude more effective, on a concentration basis, than was IGF-II, the endogenously produced IGF isomer. Conversely, IGF-II, which was expressed by maturing chondrocytes, was a potent stimulator of glycosaminoglycan and collagen synthesis by chondrocytes, parameters of a differentiated phenotype (Hill et al., 1992a). It is likely that the actions of endogenous IGF-II on chondrocyte growth and maturation are modulated by endogenous IGFBPs. We found that exogenous IGFBP-2 had a biphasic effect on IGF-II-stimulated chondrocyte DNA synthesis, enhancing the actions of IGF-II at concentrations which were approximately equimolar to the added growth factor, but inhibiting IGF-II action when the IGFBP was present in excess. At lower concentrations the IGFBP-2 may bind to cell surface integrins via its consensus RGD binding sequence, and may concentrate IGF-II at the cell surface where it is readily accessible to the high affinity type-1 receptors. At greater concentrations of IGFBP-2 the integrin binding sites may be fully occupied, and soluble IGFBP-2 may then directly compete with the signalling receptor for ligand binding.

Finally, TGF-β1, which was located in terminally differentiated cells, inhibited chondrocyte replication in response to other mitogens yet potentiated extracellular matrix molecule production. TGF-β1 increased sulfated mucopolysaccharide synthesis and enhanced collagenous protein synthesis at the expense of non-collagenous protein (Hill et al., 1992a). Many of the biochemical features of epiphyseal chondrogenesis might therefore be explained by interactions between endogenously produced peptide growth factors. It seems likely that mineralization involves the further interaction with thyroxine, which was found to reverse the mitogenic actions of IGFs on chondrocytes while inducing alkaline phosphatase production, a marker of terminal differentiation (Ohlsson et al., 1992).

Organ Maturation

Growth factors may have key functions in the maturation of organ systems prior to birth. There are reports of growth factor expression in almost all developing organs, but in many the data is anecdotal. Only three organ systems will be considered here in which some understanding of the role of growth factors to tissue function exists, these being the adrenal gland, the lung, and the pituitary. Of the steroidogenic tissues most information exists for the fetal adrenal gland. ACTH is a major trophic factor for the fetal adrenal cortex, in addition to regulating steroidogenesis. Much research on growth factors has focused on whether these might mediate or compliment ACTH action. Han et al. (1987a) showed that the capsule and definitive zone of the human fetal adrenal expressed abundant IGF-II mRNA, and some IGF-I from at least 16-20 weeks gestation. The IGF type-1 receptor has been localized by autoradiography to the fetal zone and medulla of the adrenal gland of human fetuses at 26-33 weeks gestation (Shigematsu et al., 1989). An increase in the levels of IGF-II mRNA in response to ACTH by cultured human fetal zone adrenal cells in second trimester suggests that IGF-II may mediate ACTH effects on adrenal growth (Voutilainen and Miller, 1987a). However, infusion of ACTH into the adult rat caused a decrease in adrenal IGF-II mRNA content (Townsend et al., 1990), while infusion of either ACTH or cortisol into the ovine fetus also reduced IGF-II mRNA and peptide content in the adrenals (Lu et al., 1994). These discrepancies may be explained by the effects of glucocorticoids on IGF-II expression. In the human tissue model fetal zone adrenal cells were used which lack 3β dehydroxysteroid dehydrogenase, and cannot, therefore, synthesize cortisol. Glucocorticoids were also blocked *in vivo* by infusion of metyrapone into the fetal rhesus monkey (Coulter et al., 1993). In this experiment ACTH levels would have increased, and an increase in IGF-II mRNA levels resulted. Thus, ACTH and glucocorticoids are likely to have opposing actions on IGF-II expression in the fetal adrenal. The rise in free cortisol levels which precedes parturition in the fetal sheep may precipitate a reduction in tissue IGF-II mRNA, levels, and circulating IGF-II, since infusion of cortisol to the fetus during the last third of pregnancy was able to reproduce this phenomenon (Li et al., 1993).

Direct evidence that IGFs exert effects on functional maturation of the adrenal have come mainly from studies *in vitro*. Pre-treatment of ovine fetal adrenocortical cells with IGF-I for four days increased the accumulation of cAMP and corticosterone output in response to ACTH stimulation (Naamen et al., 1989). The effects of IGF-I were exerted not only on the ACTH-dependent adenylate cyclase pathway, but at steps beyond this, since cAMP metabolites and IGF-I together also synergized to increase corticosterone release. It has been suggested that the primary action of IGF-I is to increase the uptake of the precursor of steroid synthesis, cholesterol, by adrenal cortical cells. However, work using cultured bovine adrenocortical cells has also shown that IGF-I can increase the activity of 3β hydroxysteroid dehydrogenase (Chatelain et al., 1988). In cultured human adrenal cells, IGF-II mRNA increased in association with increases in mRNA for steroidogenic enzymes such as $P450^{scc}$ and $P450^{C17}$ in response to ACTH stimulation (Voutilainen and Miller, 1987a, 1987b). However, the addition of IGFs to human adrenocortical cell cultures did not effect the mRNA levels of these enzymes. It is therefore unclear whether IGFs influence multiple enzymatic processes in the steroidogenic pathways, or simply maintain cholesterol uptake by the cells.

Using Northern blot hybridization, human fetal adrenal has been shown to express mRNA for FGF-2 (DiBlasio et al., 1987), while studies *in vitro* have shown that FGF-2 is mitogenic for cells of the fetal and adult zones of the human fetal gland (Crickard and Jaffe, 1981). Additionally, the expression of FGF-2 mRNA was increased in response to ACTH (Mesiano et al., 1991), suggesting that FGF-2 may act as an autocrine mediator of the trophic actions of the pituitary hormone. EGF is also a potent mitogen for adrenal cells, but an adrenal source of EGF has not been identified. In contrast to the above growth factors, TGF-β had an inhibitory effect on both basal and ACTH-stimulated growth and steroidogenesis by adrenal cortical cells. The effects on steroidogenesis appear to result from an inhibition of the enzyme responsible for converting cholesterol precursor to cholesterol, although more distal pathways may also be involved since Rainey et al. (1991) showed that TGF-β blocked stimulation of P450 17α hydroxylase mRNA and protein in ovine fetal adrenocortical cells.

Members of the FGF family are potent mitogens for isolated lung pneumonocytes, and both FGF-2 and its receptor have been localized by immunocytochemistry to the airway epithelia of fetal rat lung (Han et al., 1992). A morphogenic role for the FGF family in lung development was shown in studies of functional ablation of the FGFR-2 gene (Williams et al., 1994). In homozygous mice no branching of the central airway occurs, and animals die at birth with an absence of lung development. FGF-7, also called keratinocyte growth factor, is a potent mitogen for isolated type II pneumonocytes from the rat (Ulich et al., 1994). The human fetal lung is rich in mRNA encoding TGF-α in mid-gestation (Strandjord et al., 1993), while the fetal lung is rich in high affinity EGF receptors and exogenous EGF given in the rabbit or lamb induced lung epithelial maturation and surfactant production (Catterton et al., 1979; Sundell et al., 1980). Studies with fetal rat lung explants showed that EGF increased phospholipid biosynthesis, thereby increasing surfactant production, while in the fetal monkey lung EGF caused an acceleration of maturation of type II pneumonocytes (Plopper et al., 1992). The appearance of EGF receptors within fetal rabbit lung is under androgenic control (Klein et al., 1993). Conversely, TGF-β inhibited pneumonocyte development in explants of fetal rabbit lung, the mechanism of which was postulated to involve a reduced synthesis of fibroblast-pneumonocyte factor, a lung fibroblast-derived factor which regulates the maturation of the adjacent epithelial cells (Nielson et al., 1992).

During the embryonic development of the rat, mRNA transcripts for IGF-I and -II are detectable in the lung from at least day 16 gestation (term 22 days; Davenport et al., 1987), with the latter being more abundant. Levels of IGF-II fall in late gestation and postnatally IGF-II is almost entirely replaced by an expression of IGF-I. The sites of expression of IGF-II were examined by *in situ* hybridization in mid-trimester human fetal lung (Han et al., 1987a). The cells containing IGF-II mRNA included the pleura, interlobular septa, and fibroblast cells around the pulmonary vessels. This implies a mesenchymal source of IGF-II. However, immunocytochemistry showed a strong colocalization of IGF peptides with IGFBP mRNA and peptide distribution on or within the pulmonary epithelium of the developing airways (Hill et al., 1989; Delhanty et al., 1993). This would imply a sequestration of IGF-II by IGFBPs and the identification of the lung epithelium as a likely target site of IGF action. IGFBP-2 peptide was localized to the apical membrane of the lung epithelium in fetal rat lung (Klempt et al., 1992). A biological role for IGFs on pulmonary epithelium is supported by the identification of IGF receptors on membrane preparations from fetal porcine lungs (D'Ercole et al., 1976), and the synthesis of functional type-1 IGF receptors by canine tracheal epithelial cells *in vitro.*

The pituitary is a rich source of growth factors and/or their receptors, including the IGFs, EGF, TGFα, FGFs, and TGF-β. While many of these are mitogens for pituitary cells *in vitro*, some have also been demonstrated to have a role in the control of pituitary hormone release. Epidermal growth factor was found to increase ACTH secretion when infused into the fetal sheep or rhesus monkey (Polk et al., 1987; Luger et al., 1988). Immunoreactivity for EGF was seen in both the lactotrophs and corticotroph cells, suggesting a paracrine action. Bondy et al. (1990) showed that IGF-II mRNA was present in the pituitary primordia, and in Rathke's pouch in the rat embryo. Using cultures of human and ovine fetal pituitary, IGF-I or -II were shown to decrease both basal and theophylline-stimulated growth hormone release (Goodyer et al., 1986; Blanchard et al., 1988), and this is likely to result from a direct effect on growth hormone gene transcription. No effects of IGFs on ACTH release from pituitary cultures were found (Goodyer et al., 1986). IGF-1 has been shown to increase cell survival and promote development of hypothalamic cells from the fetal rat (Torres-Aleman, et al., 1990), and its effects were additive to those of FGF-2. Treatment of hypothalamic neuronal cells with FGF-2 increased the abundance of type-1 IGF receptors, IGFBP release, and the release of IGF-I (Pons and Torres-Aleman, 1992). This would suggest that the IGF and FGF axes are interactive in the functional maturation of the hypothalamic neurons.

Central Nervous System

Some of the trophic signals that determine the growth and differentiation of the nervous system, and which permit selective limitation of the neuronal death that occurs during development, have now been identified. Developing neurons may derive trophic support from innervated cells (retrograde influences), from afferent neurons (anterograde influences), from glial cells (paracrine influences), or they may themselves produce supporting factors (autocrine influences) to enhance their survival and differentiation. These trophic factors include the classical "neurotrophins" such as NGF, BDNF, and neurotrophins (NT) -3, -4, -5, and -6. Other growth factors strongly implicated as regulators of neuronal development and function include the FGFs, EGF, the TGFs, and the IGFs. The accumulated information to date reveals an extremely complex pattern of trophic interaction which determines the highly specific connectivity of the nervous system. During embryogenesis, neuroepithelial cells of the neural tube proliferate and give rise to populations of neurons, astrocytes, and oligodendrocytes. Glial and neuronal cell lineages diverge at an early stage of development (for example, days 10-12 for mouse cortex) in response to paracrine/autocrine factors. The pathways of glial cells have been well studied (Raff, 1989). Growth factors are known to regulate the differentiation of astrocytic form from glioblasts. For example, type 1 astrocytes secrete PDGF which keeps O-2A progenitor cells proliferating, thereby preventing their premature differentiation (Noble and Murray, 1984), while ciliary neurotrophic factor (CNTF) promotes the differentiation of bipotential 0-2A cells into type 2 astrocytes (Hughes et al., 1988).

The roles of growth factors in the proliferation and differentiation of neuronal cells are not well defined. FGF-2 is expressed during early embryo CNS development and the expression of the high affinity FGFR1 receptor in the CNS is developmentally regulated, being primarily localized to the ependymal layer of the CNS, which contains mitotic precursor cells (Wanaka et al., 1991). The complexity of the system *in vivo* makes definition of the precise role of growth factors in early neuroblast proliferation and differentiation difficult. Short-term proliferative effects of growth factors including FGF-2, IGFs, and EGF on cultured rat and mouse neuroblasts have been reported (Gensenburger et al., 1987; Murphy et al., 1990; Deloulme et al., 1991) and these factors have been shown to be interactive (Drago et al., 1991). FGF-2 was also reported to stimulate a longer-term proliferation of embryonic hippocampal cells (Ray et al., 1993). EGF was found to have similar proliferative and differentiation effects on cultured striatum neuroblasts (Reynolds et al., 1992). A number of growth factors, including the neurotrophins and FGF-2 are known to enhance neuron survival in cultured neurons (Walicke, 1988; Hatikka and Hefti, 1988; Alderson et al., 1990; Ip et al., 1993).

The pattern of expression of the neurotrophin family and their receptors change during development and each has been implicated in the later development of various subsets of neurons. The prototype neurotrophin, NGF, stands as a good example of the influence of these peptides in the maturing CNS since the principles of their actions hold good for most neurotrophins. Maisonpierre et al. (1990) demonstrated low levels of NGF mRNA in the fetal rat brain, which increase postnatally, approximately 20-fold, until three weeks after birth, when adult levels are attained. This expression correlates with the period of neuronal migration, maturation, and innervation. Highest levels of NGF mRNA and protein were found in the hippocampus, neocortex, and olfactory bulb of the newborn and adult rat brain. Each of these regions serves as a target of basal forebrain cholinergic neurons. It seems that NGF is an essential survival factor during maturation of sympathetic neurons and possibly cholinergic neurons of the basal forebrain. NGF and its mRNA have also been localized to other CNS regions, including the caudate putamen, cerebellum, hypothalamus, and spinal cord, each with a distinct developmental regulation, which suggests that NGF also acts on other cholinergic neuronal populations.

Only recently has it been possible to directly assess the roles of NGF in development, by targeted mutation of its receptors as described earlier. The common low affinity receptor (p75) is widely expressed, both on cells that respond to neurotrophins and others that do not. Mice homozygous for a mutation in the *p75* gene survive with only mild defects to their peripheral sensory nervous system (Davies et al., 1993). It is likely that *p75* plays only an accessory role in mediating neurotrophin function. Both *NGF* and *trkA* gene knock-outs result in dramatic phenotypes in which animals are born viable but die within a month of delivery (Crowley et al., 1994; Smeyne et al., 1994). Neurons are depleted in the DRG and trigeminal nerves and both have almost no sympathetic ganglia by 10 days after birth. The loss of neurons in the DRG involves only a subset of peptidergic neurons which mediate pain and thermoreceptor functions, while neurons which express receptors for other neurotrophins remain viable. The influence of the neurotrophin receptor, *trkB*, on the embryonic and fetal development of mice has been investigated by targeted disruption of the gene (Klein et al., 1993). A mutant receptor carrying a non-functional tyrosine kinase catalytic domain was inserted into the mouse germline. Since *in situ* hybridization to localize *trkB* mRNA had shown a wide expression in multiple structures of the central and peripheral nervous systems, including cerebral cortex, hippocampus, thalamus, brain stem, and spinal cord (Klein et al., 1989), it could be anticipated that gene disruption would have widespread behavioral and motor deficits. Homozygous animals developed to birth, but did not feed and quickly died. Neuroanatomical examination showed a deficiency of neurons in the facial motor nucleus and spinal cord, and in the trigeminal and dorsal root ganglia. The loss of neurons in the facial nucleus probably disables the mastication muscles so that suckling is not possible. Destruction of the *trkC* gene caused a loss of neurons in the superior cervical ganglion, the trigeminal ganglion, and motor neurons in the CNS (Klein et al., 1994). This led to abnormalities of proprioception, and abnormal posture and movements. While the NGF family may play a primary role in directing axonal projections to targets, it seems that the other side of NGF activity during development pertains to the rescue of neurons from the programmed cell death that occurs during development. Competition between specific populations of neurons for target-derived NGF leads to the selection of a subset of neurons for survival. Although the actions of NGF are restricted to a few populations of neurons, the generality of the phenomenon of programmed cell death in the nervous system suggests that most neurons are regulated by growth factors via a target-derived mechanism.

Hence, the principles of retrograde NGF influences during maturation of neurons are broadly applicable across this family of neurotrophins. For example, BDNF demonstrates a very similar pattern of expression in the developing brain to NGF and has retrograde effects like NGF in promoting survival, differentiation, and maintenance of septal cholinergic neurons, suggesting that they may have overlapping effects. Elucidation of the biology of *NT-3* is preliminary, but it seems that this neurotrophin may act like BDNF and NGF as a target tissue-derived survival factor. Targeted mutation of the NT-3 gene in mice had severe neurological dysfunction with a loss of *trkC*-expressing sensory neurons, and abnormalities at the early stages of sensory neuronal development (Tessarello et al., 1994). In contrast to NGF and BDNF, the expression of NT-3 is much higher in the brain during embryonic and fetal development than in the adult. In particular, NT-3 levels are very high in the cerebellum and hippocampus during the peak rate of proliferation of the granule cells that populate these structures. These reciprocal patterns may suggest a major role for NT-3 as a target-derived mitogenic factor during early development, whereas NGF and BDNF may be more important in the regulation of guidance, selection, maintenance, and differentiated function of maturing and mature neurons.

III. PLACENTA

The placenta is the crucial gatekeeper in the transfer of nutrients and substrates between the maternal and fetal compartments. A normally developed and functioning placenta is therefore essential for the normal growth and development of the embryo and fetus. A poorly developed, small, or diseased placenta is often associated with reduced fetal size and puts the fetus at risk to intrauterine death or to poor health leading to a higher perinatal mortality and morbidity. The unfavorable long-term neurologic outcome observed in growth-retarded fetuses which may manifest in childhood as developmental disability is well documented. Recent epidemiologic studies have shown a strong relationship between growth retardation at birth to the incidence of hypertension, coronary vascular disease, stroke, and diabetes in the middle aged adults (Barker et al., 1990). The relationship is stronger if there is a disproportionate size between the fetus and placenta, suggesting that placental size and function in determining fetal size may have far reaching consequences beyond the neonatal and pediatric ages into late adult life.

One of the most interesting as well as confusing facts is the enormous variation in the morphologic and organogenetic pattern in the placentae of various species. Even among primates, which have a similar morphologic placenta (hemochorial), the types and behavior of cells which constitute the organ differ significantly. In this chapter, we will emphasize the role of growth factors and cytokines in the development and function of the hemochorial placenta (human and rodent), with reference to other placental types where applicable (for a detailed description of the development of the hemochorial placenta please see Alpin, 1991). Briefly, the development of hemochorial placenta involves the formation (proliferation and differentiation) of motile and non-motile mononuclear trophoblasts, alternating between faster and slower rates of proliferation and further differentiation into various trophoblast populations. There is also progressive remodeling of the maternal tissues by trophoblasts, the development of the decidua, and remodeling of the maternal arterial walls which allows the placental circulation to become a privileged site independent of maternal vasomodulators. The signals responsible for the complex sequence of developmental events are not known, but growth factors and cytokines play a major role in the development as well as regulation of some physiologic functions of the mature placenta.

The biological activities of growth factors in the placenta are not limited to the regulation of cellular growth, but also are involved in other important functions which include differentiation, survival, programmed cell death, nutrient uptake, cellular metabolism, angiogenesis, and so forth. On the other hand, cytokines, which are initially described as immunomodulatory factors, are now shown to exert local paracrine effects such as cellular proliferation, migration, differentiation, and so forth. Since the placenta is the source of many growth factors and cytokines, their principal biologic action is most likely exerted locally as autocrine or paracrine factors. Certain endocrine growth factors of maternal or fetal origin may influence placental growth or function in an endocrine manner, but most of them are a result of administration in pharmacologic doses and their physiologic significance is still uncertain. However, recent studies using gene targeting have provided evidence for the importance of maternal growth factors in the development of the fetus, suggesting that the placenta may be an important regulator of transfer of growth factors from the maternal compartment to the fetal compartment (Letterio et al., 1994). The multifunctional nature of GFs and the contextuality of their biologic actions suggest that GFs may influence development and function of the placenta at several different stages. Table 2 lists the different GFs that may be involved in the development and regulation of function of the placenta. The stages of placental development in which GFs may be involved are listed in Table 3. A more detailed description of growth factors in the placenta has been published recently (Han, 1993).

Table 3. Stages of Hemochorial Placental Development That May Be Regulated by Growth Factors

1. Formation of outer cell mass and proliferation of primitive trophoblasts.
2. Differentiation into cyto- and syncitio-trophoblasts
3. Migration and invasion of intermediate (extravillous cytotrophoblasts) into maternal endometrium, formation of the trophoblastic shell
4. Modulation of immunity at the feto-maternal interface by interference with the development of immune responses by maternal T and B cells
5. Differentiation of maternal stroma into decidua
6. Proliferation and differentiation of chorionic (extraembryonic) mesoderm
7. Formation of primary stem villi
8. Formation of secondary and tertiary villi
9. Establishment and maintenance of endocrine functions
10. Maintenance of metabolic functions

IV. GROWTH FACTORS

A. Growth Factor Ontogeny

The demonstration of a variety of growth factor transcripts by RT-PCR (Rappolee and Werb, 1991) in mouse and bovine (Watson et al, 1994) preimplantation embryos implies that these growth factors have the potential not only to regulate preimplantation development of the embryo but also the implantation process and development of the placenta. One of the earliest genes expressed in the primitive trophectoderm is IGF-II (Wood et al., 1992). Using *in situ* hybridization, Zhou and Bondy (1992) showed that IGF-II mRNA is abundantly expressed in early implantation tissues in the mouse, particularly in those trophoblasts that are actively invading into the maternal endometrium and interacting with the decidua. We have observed similar expression pattern in human trophoblasts (Han et al,, 1993). Mesodermal core tissue of the chorionic villi and amniotic plate also express abundant IGF-II mRNA (Han et al., 1993). Therefore, explant cultures and fibroblast monolayers from term and preterm human placentae secrete an IGF I-like immunoreactive material as well as IGF II (Fant et al., 1986). The synthesis of IGFs by the placentae is gestationally age dependent. The first and second trimester placentae express more IGF I and IGF II mRNAs compared to term, and placentae from diabetic pregnancies tend to have greater IGF I mRNAs (Wang et al., 1988).

Human and rat placenta have been the source for cloning of IGFBP cDNAs indicating that placenta expresses IGFBP genes. In the mouse, IGFBP-2 is the most abundant IGFBP mRNA expressed, localized predominantly to the spongiotrophoblast cells immediately adjacent to the maternal decidua (Zhou and Bondy, 1992). Recent studies have demonstrated all IGFBP mRNAs (IGFBP-1 to -6) to be expressed in the maternal decidua in both the basal plate region and membranes of humans, whereas only IGFBP-3 mRNA is expressed in the fetal cytotrophoblasts and chorionic mesoderm (Han et al., 1993). Human amniotic fluid is rich in IGFBP-1, and the levels are 10-100 times higher that the maternal serum concentrations (Drop et al., 1984). IGFBP-1 mRNA is the most abundant of all IGFBP mRNAs, and is predominantly expressed in the maternal endometrium and decidua (Pekomen et al., 1988, Murphy et al., 1990) and its mRNA is identified in the fetal membranes and not in the placenta (Brinkman et al., 1989). One of the most exciting findings is the clear demarcation in the cell types expressing IGFs and IGFBPs at the fetomaternal interface: IGF-II mRNA is abundant in the fetal intermediate trophoblasts and IGFBP mRNAs are abundant in the maternal decidua. Since these two types of cells are in close apposition in the basal plate and in the membranes, it is tempting to speculate that this spatially specific expression of the two functionally related peptides is of biological importance. It has been suggested that IGF-II synthesized by the highly invasive trophoblasts is utilized as an autocrine paracrine factor by these cells and the IGFBPs synthesized by the maternal decidua exert an inhibitory influence (Giudice et al., 1994).

The EGF family of peptides are abundantly expressed in the placenta. Our studies showed the predominant expression of TGF-α gene occurs not in the embryo, but in the maternal decidua (Han et al., 1987), and the expression is independent of the conceptus (Bonvissuto et al., 1992) suggesting that the maternal decidua may be the source for the high levels of TGF-α peptide seen in the fetal circulation and may play an important role in placental development at the feto-maternal interface. The intense EGF and TGF-α immunoreactivity is seen in the syncitiotrophoblasts (Hofmann et al., 1991; Lysiak et al., 1993) may originate from the maternal decidua. The placenta is one of the major sites of expression of EGF receptor mRNA and protein (Hock, 1980, Duello et al, 1994) EGF receptors are present predominantly in the syncitiotrophoblasts (in the microvillous surface; Chegini and Rao, 1986), and minimally in the cytotrophoblasts. The expression of EGF receptors in the placenta is gestational age dependent: specific EGF binding sites have been found in placentae as early as four weeks and the numbers increase as gestation progresses to term (Carson et al., 1983). Human fetal membranes from term pregnancies also possess high affinity EGF receptors, although the numbers are lower compared to placenta (Rao et al., 1984). EGF binding is higher in the chorion than in the amnion and the decidua, and appears to be lower in pregnancies delivered by spontaneous vaginal deliveries compared to caesarean sections.

The human placenta contains only basic FGF (bFGF) which has been purified to homogeniety (Gospadarowicz et al., 1985). However, acidic FGF is absent. It is not known if other members of the FGF family exist in the placenta. Cattini et al (1991) have identified bFGF in placental tumor cells and normal placental trophoblasts using immunocytochemistry and demonstrated two patterns of immunostaining-a perinuclear pattern in the normal non-proliferative syncitiotrophoblasts, and an intense homogeneous nuclear and cytoplasmic staining in the actively dividing cells.

PDGF-A chain mRNA and peptide is detectable in pre-implantation mouse embryos from the 4-cell stage (Rappolee et al., 1988). *In vitro*, the expression of PDGF gene and secretion into the media by cultured cells, including those from implantation and embryogenesis (Baird et al., 1988), is constitutive, and the level of gene expression is dependent on the age of the cells when isolated *in vivo*, that is, younger animals secrete more PDGF than other animals. Goustin et al. (1985) demonstrated expression of B-chain mRNA in first trimester human placentas. A large quantity of biologically active PDGF has been purified from human term placenta (Marez et al., 1987). In an ontogenic study of mRNAs encoding PDGF and its receptor, it was shown that both mRNAs are expressed in a coordinated fashion throughout pregnancy, especially in mid-gestation where both PDGF and its receptor are very abundant (Taylor et al., 1988). PDGF mRNA is localized to the highly proliferative and invasive trophoblasts, and not to the differentiated syncitiotrophoblasts. These cytotrophoblasts also express c-myc mRNA. They also demonstrated that the PDGF-B chain mRNA decreases approximately ten-fold between early first trimester and term. More detailed studies showed PDGF-A and

-B chains as well as PDGF receptor β subunit mRNA to be present throughout pregnancy and the abundance to be highest in the second trimester (Taylor and williams, 1988). Placenta c-*fos* mRNA, a gene rapidly induced after interaction of PDGF and its receptor, also showed a similar temporal increase (Cochran et al, 1984). In all gestational ages, PDGF-A chain mRNA is higher than B chain mRNA.

One of the first tissues from which TGF-β was first purified was the placenta (Frolik et al., 1983). Subsequently, human TGF-β cDNA was isolated from the placental cDNA library (Derynk et al., 1987) and the mRNA identified in the placenta (Derynk et al., 1988). Kauma et al. (1991) performed a more detailed study in which they identified TGF-β1 mRNA from the decidua, placenta, and placental membranes of first trimester and term human pregnancies. They showed that TGF-β1 mRNA increased fivefold in decidua compared to proliferative endometrium and 2.5-fold compared to the secretory endometrium. In addition, placenta and membranes contain TGF-β mRNA throughout pregnancy and its levels are higher at term than in the first trimester. A recent immunohistochemical study of human placentae revealed TGF-β1 and TGF-β2 immunoreactivity in all three forms of trophoblasts, (cytotrophoblasts, intermediate trophoblasts and syncytiotrophoblasts), with the most intense immunoreactivity observed in the intermediate trophoblasts from early pregnancies (Selick et al., 1994). In contrast, in murine placenta, TGF-β2 is more abundant than TGF-β1 and is expressed primarily in the maternal decidual and uterine epithelium, not in the placenta and membranes (Cheng et al., 1993). The species differences is unexplained, but methodological differences in the two studies as the reason is possible.

B. Biological Actions of Growth Factors in the Placenta

Human placenta contains abundant IGF-I and IGF-II/man-6 P receptors, whereas rat placenta contain predominantly IGF-II/man-6 P receptors (Daughaday et al., 1980; Nissley and Rechler, 1984) comprising of 1.7% of rat placental proteins (Sklar et al., 1989). Using receptor binding and affinity cross-linking techniques, placental IGF-I receptor has been shown to have a chemical structure similar to insulin receptor, and binds IGF-I with greatest affinity (Chernausek et al., 1981; Stuart et al., 1984). ^{125}I-IGF II could bind to a unique site on the receptor, distinct from the IGF I binding site (Casella et al., 1986). In addition, placenta may contain hybrid receptors which contain one pair of α and β subunits that are products of the IGF I receptor gene and another pair of α and β subunits that are products of the insulin receptor gene (Jonas et al, 1986). The ligand binding properties of these receptors depend on a cooperative interaction between the heterodimers (Tollefsen and Thompson, 1988).

Evidence from gene targeting studies have suggested the existence of a yet undescribed IGF receptor, which may play an important role in the development of the placenta. Heterozygous mutation of the paternal allele of the IGF II gene or a null mutation of the IGF-I gene leads to a 40% decrease in fetal size. A homozygous null mutation of the IGF-I receptor gene causes a 60% reduction in fetal size. The fact that a double knock-out of IGF-II and IGF-I receptor genes leads to a more severe reduction in fetal size than IGF-I receptor null mutation alone suggests that part of the IGF-II biological action on the fetus and placenta may be mediated via a yet undescribed receptor. In addition, the growth restricted phenotype observed in heterozygotes may result as an indirect consequence of impaired placental growth and function.

The biologic actions of IGFs on placental cells are varied. Fant et al. (1986), using fibroblast monolayers cultured from preterm and term placentae, described the capacity of IGF-I to stimulate uptake of non-metabolizable amino acid, aminoisobutyric acid, into the cells, and Bhaumick et al (1987) demonstrated that IGF-I and insulin stimulate hPL production, and IGF-I is 10 times more potent than insulin, suggesting that insulin action may be mediated via the IGF-I receptor. In a series of studies, Nestler demonstrated the effect of IGF-I and insulin on steroidogenesis enzymes in term placental cytotrophoblastic cultures (Nestler and Williams, 1987; Nestler, 1989, 1990). He showed that IGF-I, IGF-II, and insulin decrease aromatase activity by acting on their own receptors, increase the activity of P450 side-chain cleavage enzyme (p450 ^{SCC}) by affecting enzyme synthesis, and increase 3b-hydroxysteroid dehydrogenase (3β-HSD) activity by affecting enzyme synthesis. IGF-I and insulin does not have any effect on steroid sulfatase activity. These studies indicate specific actions of IGFs on steroidogenic enzymes which are different from insulin, and may explain in part the observation of high progesterone levels in diabetic pregnancies. The effects of IGF II can be mimicked by mannose-6-phosphate, and cannot be inhibited by αIR-3, indicating that the biological actions of IGF-II may be mediated via their own receptors. The biological actions of IGFBPs on placental growth or function have not been studied extensively. IGFBP-1 has been shown to inhibit the interaction of IGF-I with its receptor (Pekonen et al., 1988), but the possibility of IGFBP-1 having a potentiating effect on IGF-I action resulting from its capacity to interact with cell surface integrins (Jones et al., 1993) cannot be excluded.

In the placenta, EGF may regulate a wide variety of physiologic functions of various tissues which are not related to growth, for example, stimulation of prolactin production (Schonbrun et al, 1980), and increasing prostaglandin E_2 production[54] and stimulation of inhibin secretion (Qu et al., 1992). EGF binding to its receptor on the placenta is associated with activation of its intracellular tyrosine kinase domain, which in turn stimulates protein phosphorylation. Although EGF does not appear to have any mitogenic actions on the placenta since the majority of the receptors are identified on syncitiotrophoblasts (Huot et al., 1981), a recent study has demonstrated that EGF may stimulate trophoblast proliferation in early (4-5 week gestation) placentae and differentiated trophoblast function in later (6-12 week) placentae (Maruo et al., 1992). Based on studies which demonstrated that syncitiotrophoblasts can transport EGF intact following internalization, it may be speculated that maternal EGF may play a role in embryonic or fetal development. Studies to date have not supported this concept and it is more likely that EGF has an effect on placental development and function since there is a correlation between EGF in the amniotic fluid and placental weight at term (D'Souza et al., 1985).

Basic FGF is usually 10 to 1,000-fold more potent than acidic FGF in most cells, reflecting the higher affinity of basic FGF for the receptor. FGFs are mitogens for many cells. The biologic activity of FGF on a cell which causes the mitogenic cascade includes reversible changes of cell morphology, stimulation of cell transport systems, polyribosome formation, RNA stabilization and synthesis, and protein synthesis and inhibition of protein degradation. Enzymes, such as ornithine decarboxylase which are required for the cell to proceed through the cell division are also induced. One of the most important functions of FGFs is stimulation of angiogenesis demonstrated in a wide variety of *in vivo* models which include implantation and embryogenesis (Gospadarowicz et al., 1985b). This putative role of FGFs may be important in the context of both physiological and pathological regulation. Mouse placenta contains both high and low affinity binding sites for FGF and they are developmentally specific (Hondermarck et al., 1990). At day 12, placenta contains both high affinity (FGF receptor) and low affinity binding sites, whereas at days 14, 16, and 18 of pregnancy, only low affinity binding sites exist. The dramatic decrease of the high affinity receptor number after the twelth day of pregnancy is synchronous with the nine-fold increase in low affinity binding sites. These findings suggest that FGF may have a role in placental development only in early gestation. This finding, together with that of Cattini et al. (1991), indicates that FGF may be involved only during the proliferative phase of placental development and perhaps only in specific cells types of the placenta such as cytotrophoblasts. A recent demonstration of soluble FGF in the maternal circulation during pregnancy, with higher levels observed in women with diabetes bearing large fetuses and placentae, suggest a biologic role of FGF in placental growth and function.

The role of PDGF in the placenta was first identified when its oncogene homologue c-*sis* (B chain of PDGF) was shown to be highly expressed in the placenta throughout pregnancy (Goustin et al., 1985). In the first trimester placenta, the c-*sis* expression is primarily localized to the proliferative and invasive cells of the cytotrophoblastic shell and not in the mesenchymal stroma. The presence of PDGF receptors has also been demonstrated in the term placenta by binding studies (Czyrski and Gawlikowski, 1987) as well as by the capacity of the term cytotrophoblasts to transcribe the PDGF receptor gene (Goustin et al., 1985). However, the major part of placental PDGF receptor is localized to the endothelial cells, suggesting a paracrine interaction between the trophoblasts and the endothelium.

Possibly, one of the more intriguing functions of TGF-β in the placenta is its role as a cytokine in fetomaternal immunity and regulation of invasiveness of fetal trophoblasts into maternal tissues (reviewed by Graham and Lala, 1991). The biologic actions of TGF-β which may be important in this function include (1) suppression of lymphokine activated killer cell activity, (2) inhibition of IL-1 dependent-T and -B cell proliferation, and (3) decreased cytotoxity associated with macrophage activation. Studies in the interactions between maternal decidual cells and trophoblasts *in vitro* have yielded important information on TGF-β as one of the potential molecules synthesized and secreted by maternal decidua that inhibit trophoblast invasiveness (Graham and Lala, 1991). Neutralizing antibodies against TGF-β1 prevented the effect of a soluble factor originating from the decidua on trophoblast invasiveness. It was further demonstrated that TGF-β action may be mediated via the induction of tumor inhibitor of metalloproteinases (TIMP) production by both decidua and trophoblasts and by decreasing the collagenase type IV activity. Trophoblasts themselves are capable of producing TGF-β endogenously as evident by the ability of neutralizing antibody to decrease the invasiveness of trophoblasts. Using immunocytochemistry, these investigators have also localized TGF-β in trophoblastic and decidual tissues from first trimester as well as term pregnancies. TGF-β was identified in the extracellular matrix of the first trimester decidua, syncitiotrophoblasts and cytotrophoblasts. TGF-β also inhibits cytotrophoblastic cells in culture and induces them to form multinucleated cells. Since the latter cells have been shown to be non-invasive, this may be one mechanism by which TGF-β produced at the fetomaternal interface inhibits trophoblastic invasion. Interestingly, TGF-β stimulates proliferation and fails to inhibit invasiveness by choriocarcinoma cells suggesting that malignant cells derived from trophoblasts may have invasive mechanisms different from normal cells. The underlying mechanism is unknown but it may be related to the difference in type and distribution of TGF-β receptors between the two cell types.

Growth Factors in Diseases of Placenta

Newborn infants who are small for gestational age are often found to have decreased serum IGF-I levels compared to the appropriate for gestational age infants, and serum IGF-I levels are positively correlated with birth size (D'Ercole, 1987). In experimental animal models of IUGR, in which uterine artery was ligated, serum IGF I levels and hepatic IGF-I concentrations are decreased (Viliesis and D'Ercole, 1986; McLellan et al., 1992), accompanied by a marked increase in liver IGFBP-1 mRNA abundance. This study implicates that IGFBP-1 is an inhibitor of IGF action and therefore mediates the fetal growth restriction observed in these animals. A recent study in cord bloods from human IUGR infants also demonstrate a similar change in the profile of IGF binding proteins indicating that IGF binding proteins may be involved in the pathophysiology of human fetal IUGR (Crystal & Giudice, 1991). In an experimental model of prolonged fetal hypoxia induced by restricting maternal uterine blood flow, we have demonstrated that IGFBP-1 gene expression is increased in the fetal liver following prolonged hypoxia (McLellan et al., 1992). In this study, the effects were observed mainly in the liver, suggesting that the process of reduced uterine blood flow may have some untoward effect on the placental function which then affect the liver.

Pathological conditions in which placental development is abnormal are associated with alterations in EGF receptors in the placenta. Placentae from streptozotocin treated diabetic rats are morphologically and biochemically immature compared to normal rats. The latter placentae have lower EGF binding compared to controls (30%) because of the failure of the normal developmental increase in EGF receptors, which is mainly due to the appearance of a second class of lower affinity EGF receptors (Sisson et al., 1987). In contrast, placentae from fetuses made growth restricted by unilateral ligation of maternal uterine artery have a higher specific ^{125}I-EGF binding compared to normal controls, even though the placental sizes are similar. These findings suggest that the increased EGF binding in placentae from growth restricted fetal rats may serve to enhance nutrient uptake under conditions of decreased uterine blood flow (Lawrence et al., 1989). A similar finding has been reported in human pregnancies (Hofmann et al., 1988). A more recent study, however, showed that EGF receptor mRNA may be decreased in placentae from both IUGR and diabetic pregnancies compared to normal (Fujita et al., 1991). In smokers, autophoshorylation of the EGF receptor is also impaired (Gabriel et al., 1994). The pathophysiologic significance of these studies is uncertain, but they suggest that EGF is important in placental and fetal growth and development.

V. CYTOKINES

Cytokines, originally known as immunoregulatory proteins, may also influence the development and function of the placenta and may play an important role in parturition. Like growth factors they may also act in autocrine/paracrine mechanism as well as in an endocrine manner to modulate wide range of functions including regulation of proliferation and differentiation, and metabolic activities in the placental cells. Their actions on hormone biosynthesis in the placenta, ovary, and endocrine glands may indirectly influence placental growth and function.

The involvement of the cytokine network in the placenta is recently reviewed by Mitchell et al. (1993). The role of cytokines in the placenta can be defined into two major categories: (1) Early pregnancy-implantation and development of early placenta, and fetomaternal immunity, and (2) later pregnancy-involvement in placental functions in late gestation with particular reference to preterm and term labor.

A. Early Pregnancy

The successful implantation and development of the "fetal allograft" within the maternal uterine environment is still an immunological enigma, and the cytokines of placental and maternal (decidual) origin play a crucial role. It has been shown conclusively that the failure to reject the fetal allograft is not due to the lack of or altered feto-paternal antigens of the placenta nor to the lack of competent immune responsive system of the decidua (Kearns and Lala, 1985). On the contrary, the decidua contains abundant immune effector cells including functional antigen presenting cells, T cells, and natural killer cells, and the immune responsiveness is heightened. These findings suggest that there are powerful regulatory influences that modulate the function of these immune effector cells (Mitchell, 1993), which may include the neutralization of any adaptive immune response against the fetus, and the provision of an environment that ensures growth and differentiation of the fetus. An important feature of the cytokines that may be important in the adaptive response is the functional dichotomy of the cytokines depending on the site and time of expression. A successful implantation and early pregnancy depends on a biological balance between growth factors, cytokines, eicosanoid and steroid milieu, and an imbalance is one of the important reasons for an early fetal loss.

During pregnancy, there is a 1,000-fold increase in the CSF-1 levels in the murine uterus (Wegmann et al., 1989), and approximately sixfold increase in total amniotic CSF-1 in humans (Ringler et al., 1989). The latter appears to be due to a gestational age dependent increase in CSF-1 mRNA expression in the placenta and decidua (Kauma et al., 1991), and appeared to be regulated by estrogen and progesterone (Areci et al., 1989; Pollard et al., 1987). *In situ* hybridization showed CSF-1 mRNA to be localized to the glandular epithelium of the endometrium (Regenstrief and Rossant, 1989). CSF-1 receptor (*c-fms*) is abundantly expressed in the extravillous trophoblast columns emanating from the anchoring villi into the maternal decidua (Areci et al., 1989; Jokhi et al., 1993), but more so in the superficial layers. In the humans, immunoreactive CSF-1 and 4.0 kb CSF mRNA are detectable in the endometrium during the reproductive cycle and placenta and decidua during pregnancy (Kauma et al., 1991). The first trimester decidual tissues contain higher levels of CSF-1 mRNA and peptide than the secretory and proliferative endometrial tissues, whereas *c-fms* mRNA levels are unchanged. Since exogenous CSF-1 stimulates mitogenesis of placental trophoblasts (Athanassakis et al., 1993), and causes differentiation of cytotrophoblasts to differentiate into syncitium and produce hCG and hPL (Garcia-Lloret et al., 1989), it is possible that CSF-1 serves as an important molecule in paracrine interactions between the maternal and fetal cells during placental development. However, high levels of CSF-1 can interfere with implantation and lead to fetal resorption (Tartakovski, 1989), suggesting that a equilibrium of CSF-1 expression in the fetomaternal interface is required for a successful development of the placenta.

GM-CSF is also produced in the decidua in response to the invading trophoblasts (Athanassakis et al., 1987), and is very effective in stimulating the proliferation of early pure ectoplacental cone trophoblasts and less on the later stage placental cells (Armstrong and Chaouat, 1989). *In vitro*, GM-CSF stimulate differentiation of cytotrophoblasts into syncitium and release hCG and hPL (Garcia-Lloret et al., 1989), and promote embryonic implantation into uterine epithelial cells (Wegmann, 1990). G-CSF of maternal source can cross the placenta and induce bone marrow and spleen myelopoiesis in the fetus and neonate (Medlock et al., 1993).

A variety of interleukins, particularly those that are involved in inflammatory and immunomodulatory functions, have been studied intensively (for a recent review please see Mitchell, 1993). IL-1 (Kauma et al., 1990; Baergen et al., 1994), IL-4 (Delassus et al., 1994), IL-6 (Kauma et al., 1993, 1994), TNF-α (Yelavarthi et al., 1991; Hunt et al., 1993), and interferon (Loke and King, 1990) and their receptors (Austgulen et al., 1992; Hampson et al., 1993; Chen et al., 1994) are expressed in a spatial and temporally specific manner. Production of these cytokines occur throughout gestation but each has a highly regulated and specific sequence of expression. The important feature of these cytokines is that they appear to have a dichotomous function. Each may have a physiological role expressed in predetermined sites and in amounts that can influence trophoblastic growth, and may have a pathophysiological role if expressed in abnormal amount or sites (Mitchell, 1994). However, it is also important to note that immunodeficient mice can reproduce without difficulty, suggesting that successful pregnancy is not critically dependent on the activation of maternal intrauterine immune system. Activated macrophages are a major source of cytokines and the temporal relationship in the significant increase in the expression of IL-1, IL-6, and TNF-α during the periimplantation period (Wood et al, 1992) suggests that these cytokines are important in the determination of success or failure of implantation.

B. Late Gestation

A variety of cytokines are detected in the uterus during the second half of pregnancy in the mouse (De et al., 1992), and in humans (Romero and Mazor, 1988). Cytokines are thought of as key mediators in the pathogenesis of preterm or term labor particularly those related to infection. Several studies have shown that, amniotic fluid concentrations of IL-1, IL-6, IL-8, GM-CSF, and TNF are elevated in preterm labor associated with infection (Romero et al., 1989). These cytokines in turn stimulate prostaglandin production by the placenta (amnion and decidua) and and myometrium, which is thought to be responsible for the onset of myometrial contractions in preterm labor. Of the inflammatory cytokines, IL-β is increased in the amniotic fluid of women in term labor, and is thought to play a role in normal term labour. Other cytokines are not detectable at any gestation (TNF) or present in small amounts in late gestation (IL-6, IL-8) in the amniotic fluid, except in preterm or term labor associated with infection, when the concentrations increase significantly (Romero et al., 1991). Amniotic fluid IL-8 levels may also reflect responsiveness to tocolysis.

Cytokines may be synthesized by decidua, chorion, and villi under normal conditions and by the inflammatory cells in infective conditions (chorioamnionitis). A complex interplay between the various cytokines and their receptors follows after the initial cytokine production is initiated (Mitchell et al., 1993). IL-1 and TNF stimulate prostaglandin production by human amnion, chorion, and decidual cells. IL-1 can induce prostaglandin-H synthase-1, a crucial enzyme in prostanoid biosynthesis, or PGHS-2 (Zakar et al., 1995) as well as stimulate substrate (arachidonic acid) availability. They may also stimulate protein kinase-C or cAMP-dependent protein kinase-A, and activate prostaglandin biosynthesis via secondary messengers.

VI. SUMMARY

Development of the fetus and placenta is a highly complex event which involves proliferation and differentiation of several cell types, migration, aggregation, and programmed cell death. In the placenta the following additional specialized devel-

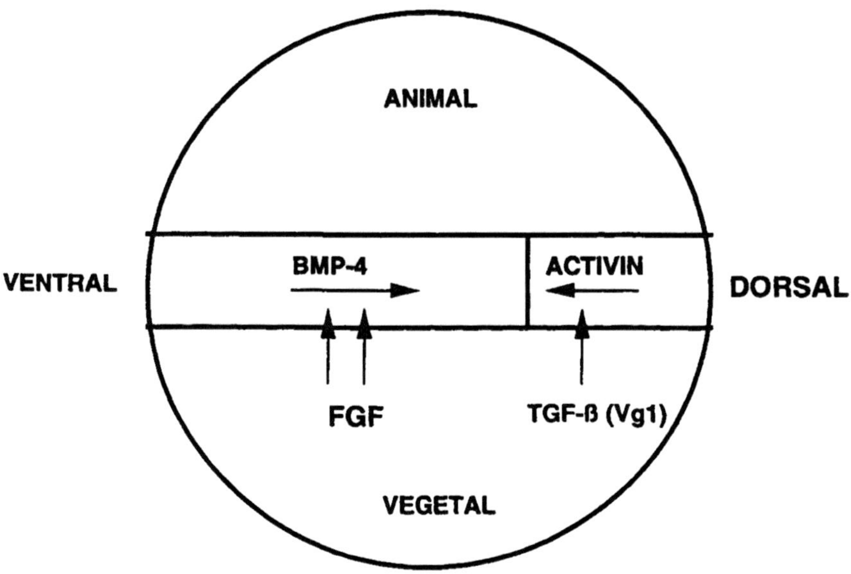

Figure 1. Peptide growth factors implicated in mesoderm formation in the *Xenopus* embryo. Fibroblast growth factors (FGF) and the Vg1 protein (TGF-β family) are expressed in the vegetal pole and induce ventral and dorsal elements of mesoderm respectively from within animal pole ectoderm. Activin and bone morphogenetic protein-4 (BMP-4) may exist as opposing gradients which determine the dorsal-ventral progression of mesodermally-derived structures.

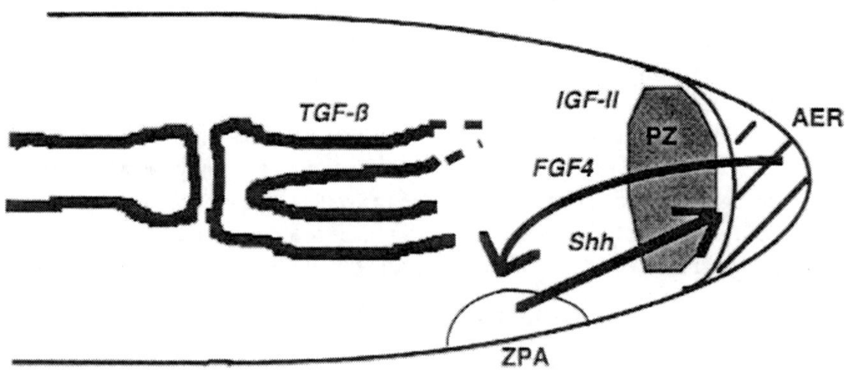

Figure 2. Peptide growth factors implicated in the outgrowth of the chick wing bud. The apical ectodermal ridge (AER) maintains the proliferation of mesenchyme cells in the underlying progress zone (PZ), and thus limb bud outgrowth, by a synthesis of fibroblast growth factor 4 (FGF-4). FGF-4 also maintains the expression of the sonic hedgehog gene (Shh) from the zone of polarizing activity (ZPA). Shh from the ZPA controls dorsal-ventral pattern formation within the limb, and in turn maintains FGF4 expression within the AER. Insulin-like growth factor-II is highly expressed within the proliferating mesenchymal cells of the PZ. As more proximal mesenchyme condenses to express a chondrocytic phenotype, this is associated with expression of transforming growth factor-β isomers (TGF-β), which are known to induce cartilage differentiation.

opmental processes occur: invasion of fetal trophoblasts into maternal endometrium, remodeling of the maternal endometrium and vasculature, interaction between fetal and maternal cells to enable fetal cells to become immunologically privileged within the maternal environment, and differentiation of trophoblasts into special endocrine cells for the maintenance of the pregnancy to term. Growth factors and cytokines, are uniquely qualified not only to regulate the proliferation of placental cells, but also to coordinate and regulate a wide variety of differentiative, endocrine, and immunological functions of the placenta. This is because of their multifunctional biological activities and being synthesized by the developing and mature fetal and placental tissues. The exciting aspect of growth factor biology is its role not only in the normal growth and development of the fetus and placenta but also their potential role in their pathology. The latter may hopefully lead to our better understanding of pathological conditions like intrauterine growth retardation, habitual abortions, pregnancy-induced hypertension, and prematurity, and thus lead to the development of diagnostic tests and therapeutic regimens to treat these conditions. They may also play an important role in the "developmental imprinting" that occurs in the cardiovascular system and endocrine and metabolic systems, in response to fetal growth retardation, which may have far reaching consequences in terms of long term neurodevelopmental morbidity and adult onset diseases.

ACKNOWLEDGMENT

This chapter was supported by a Group grant in Fetal and Neonatal Health and Development from the Medical Research Council of Canada.

REFERENCES

Adamson, E.D., & Meek, J. (1984). The ontogeny of epidermal growth factor receptors during mouse development. Dev. Biol. 103, 62-70.

Akhurst, R.J., Lehnert, S.A., Faissner, A., & Duffie, E. (1990). TGF beta in murinemorphogenetic processes: the early embryo and cardiogenesis. Development 108, 645-656.

Alderson, R.F., Alterman, A.L., Barde, Y-A., & Lindsay, R.M. (1990). Brain-derived neurotrophic factor increases survival and differentiation of rat septal cholinergic neurons in culture. Neuron 5, 297-306.

Amaya, E., Musci, T.J., & Kirschner, M.W. (1991). Expression of a dominant negative mutant of the FGF receptor disrupts mesoderm formation in *Xenopus* embryos. Cell 66, 257-270.

Andersson, I., Billig, H., Fryklund, L., Hansson, H-H., Isaksson, O., Isgaard, J., Nilsson, A., Rozell, B., Skottner, A., & Stemme, S. (1986). Localization of IGF I in adult rats. Immunohistochemical Studies. Acta. Physiol. Scand. 126, 311-312.

Alpin, J. D. (1991). Implantation, trophoblast differentiation and haemochorial placentation: Mechanistic evidence *in vivo* and *in vitro*, J. Cell. Science 99, 681.

Araki, E., Lipes, M.A., Patti, M-E., Bruning, J.C., Haag, B., Johnson, R.S., & Kahn, R. (1994). Alternative pathway of insulin signalling in mice with targeted disruption of the IRS-1 gene. Nature (London) 372, 186-190.

Areci R. J., Shanhan F., Stanley, E. R., & Pollard J.W. (1989). Temporal expression and location of colony stimulating factor-1 (CSF-1) and its receptor in the female reproductive tract are consistent with CSF-1 regulated placental development. Proc. Natl. Acad. Sci. USA 86,8818-8822.

Armstrong, D. T A., & Chaouat, G. (1989). Effects of lymphokines and immune complexes on murine placental growth *in vitro*. Biol Reprod 40,400-406.

Athanassakis, I., Bleackley, R. C., Paetkau, V., Guilbert, L., Barr, P. J., & Wegman, T. G. (1987). The immunostimulatory effects of T cells and T-cell lymphokines on murine fetally derived placental cells. J. Immunol. 138, 37-44.

Athanassakis-Vassiliadis, I., Papamatheakis, J., & Vassiliadis, S. (1993). Specific CSF-1 binding on murine placental trophoblasts and macrophages serves as a link to placental growth. J. Receptor Res. 13, 739-751.

Austgulen, R., Espevik, T., Mecsei, R., & Scott, H. (1992). Expression of receptors for tumor necrosis factor in human placenta at term. Acta Obstet. Gynecol. Scand. 71, 417-424.

Babajko, S., Hardouin, S., Segovia, B., Groyer, A., & Binoux, M. (1993). Expression of insulin-like growth factor binding protein-1 and -2 genes through the perinatal period in the rat. Endocrinology 132, 2586-2592.

Baergen, R., Benirschke, K., & Ulich, T. (1994). Cytokine expression in the placenta: The role of interleukin-1 and interleukin 1 receptor antagonist expression in chorioamnionitis and parturition. Arch. Pathol. Lab. Med. 118, 52-55.

Baird, A., Schubert, D., Ling, N., & Guillemin, R. (1988). Receptor and heparin-binding domains of basic fibroblast growth factor, Proc. Natl. Acad. Sci. USA 85, 2324.

Baker, J., Liu, J-P., Robertson, E.J., & Efstratiadis, A. (1993). Role of insulin-like growth factors in embryonic and postnatal growth. Cell 75, 73-82.

Barker, D. J. P., Bull, A. R., Osmond, C., & Simmonds, S. J. (1990). Fetal and placental size and risk of hypertension in adult life. Bri. Med. J. 310, 259-262.

Barlow, D.P., Stoger, R., Herrmann, B.G., Saito, K., & Schweifer, N. (1991). The mouse insulin-like growth factor type-2 receptor is imprinted and closely linked to the Tme locus. Nature (London) 349, 84-87.

Beck, F., Samani, N.J., Penschow, J.D., Thorley, B., Tregear, G.W., & Coghlan, J.P. (1987). Histochemical localization of IGF-I and -II mRNA in the developing rat embryo. Development 101, 175-184.

Bekoff, M.C., & Klagsbrun, M. (1982). Characterization of growth factors in human cartilage. J. Cell Biochem. 20, 273-245.

Bhaumick, B., & Bala, R.M. (1991). Differential effects of insulin-like growth factors I and II on growth, differentiation and glucoregulation in differentiating chondrocyte cells in culture. Acta Endocrinol. (Copenh.) 125, 201-211.

Bhaumick, B., Dawson, E. P., & Bala, R. M. (1987). The effects of insulin-like growth factor-I and insulin on placental lactogen production by human term placental explants. Biochem. Biophys. Res. Comm. 144, 674.

Blanchard, M.M., Goodyer, C.G., Charrier, J., & Barenton, B. (1988). *In vitro* regulation of growth hormone (GH) release from ovine pituitary cells during fetal and neonatal development: Effects of GH-releasing factor, somatostatin, and insulin-like growth factor-I. Endocrinology 122, 2114-2120.

Bondy, C.A., Werner, H., Roberts, C.T., & LeRoith, D. (1990). Cellular pattern of insulin-like growth factor-I (IGF-I) and type I IGF receptor gene expression in early organogenesis: Comparison with IGF-II gene expression. Mol. Endocrinol. 4, 1386-1398.

Bonvissuto, A. C., Lala, P. K., Kennedy, T. G., Nygard, K., Lee, D. C., & Han V. K. M. (1992). Induction of transforming growth factor-a gene expression in rat decidua is independent of the conceptus. Biol. Reprod. 46, 607-615.

Brice, A.L., Cheetham, J.E., Bolton, V.N., Hill, N.C.W., & Schofield, P.N. (1989). Temporal changes in the expression of the insulin-like growth factor II gene associated with tissue maturation in the human fetus. Development 106, 543-554.

Brown, A.L., Graham, D.E., Nissley, S.P., Hill, D.J., Strain, A.J., & Rechler, M.M. (1986). Developmental regulation of insulin-like growth factor II mRNA in different rat tissues. J. Biol. Chem. 261, 13144-13150.

Brinkman, A., Kortleve, D. J., Groffen, C. A. H., & Drop, S. L. S. (1989). The C-terminal region of the low molecular weight IGF binding protein, IGFBP-1, is essential for IGF-I binding. In: Insulin-like Growth Factor Binding Proteins (Excerpta Medica International Congress Series, 881) (Drop, S. L. S., & Hintz, R. L., Eds.), p. 11. Elsevier, Amsterdam.

Brunetti, A., & Goldfine, I.D. (1990). Role of myogenin in myoblast differentiation and its regulation by fibroblast growth factor. J. Biol. Chem. 265, 5960-5963.

Carington, J.L., Chen, P., Yanagishita, M., & Reddi, A.H. (1991). Osteogenin (Bone morphometric protein-3) stimulates cartilage formation by chick limb bud cells *in vitro*. Dev. Biol. 146, 406-415.

Carson, S. A., Chase, R., Ulep, E., Scrommergna, A., & Benveniste, R. (1983). Ontogenesis and characteristics of epidermal growth factor receptors in human placenta. Am. J. Obstet. Gynecol. 147, 932.

Casella, S. J., Han, V. K. M., D'Ercole, A. J., Svobda, M. E., & Van Wyk, J. J. (1986). Insulin-like growth factor II binding to the type I somatomedin receptor. Evidence for two high affinity binding sites J. Biol. Chem. 261, 9268.

Catterton, W.Z., Escobedo, M.B., Sexson, W.R., Gray, M.E., Sundell, H.W., & Stahlman, M.T. (1979). Effect of epidermal growth factor on lung maturation in fetal rabbits. Ped. Res. 13, 104-108.

Cattini, P. A., Nickel, B., Bock, M., & Kardami, E. (1991). Immunolocalization of basic fibroblast growth factor (bFGF) in growth-inhibited placental cells: A possible role for bFGF in placental cell development. Placenta 12, 341.

Charite, J., de Graaf, W., Shen, S., & Deschamps, J. (1994). Ectopic expression of Hoxb-8 causes duplication of the ZPA in the forelimb and homeotic transformation of axial structures. Cell 78, 589-601.

Chatelain, P., Penhoat, A., Perrard-Sapori, M.H., Jaillard, C., Naville, D., & Saex, J. (1988). Maturation of steroidogenic cells: A target for IGF-I. Acta Paediatr. Scand. Suppl. 347, 104-109.

Chen, P., Carrington, J.L., Hammonds, R.G., & Reddi, A.H. (1991). Stimulation of chondrogenesis in limb bud mesoderm cells by recombinant human bone morphometric protein 2B (BMP-2B) and modulation by transforming growth factor β and B2. Exp. Cell Res. 195, 509-515.

Chen, H-L., Kamath, R., Pace, J. L., Russell, S. W., & Hunt, J. S. (1994). Expression of the interferon-g receptor gene in mouse placentas is related to stage and gestation and is restricted to specific subpopulations of trophoblast cells. Placenta 15, 109-121.

Cheng, H-L, Schnieder, S. L., Kane, C. M., Gollnick, S. O., Grande, C., Thompson, D., Pietrzak, E., & Tomasi, T. B. (1993). TGF-β2 gene and protein expression in maternal and fetal tissues at various stages of murine development. J. Reprod. Immunol. 25, 133-148.

Clegg, C.H., Linkhart, T.A., Olwin, B.B., & Hauschka, S.D. (1987). Growth factor control of skeletal muscle differentiation: Commitment to terminal differentiation occurs in G1 phase and is repressed by fibroblast growth factor. J. Cell Biol. 105,949-956.

Chegini, N., & Rao, C. V. (1986). Quantitative eletron microscopic autoradiographic studies on internalization of 125I-labelled epidermal growth factor in term human placenta. J. Cell. Science 84, 41.

Chernausek, S. K., Jacobs, S., & Van Wyk, J. J. (1981). Structural similarities between human receptors for somatomedin C and insulin: Analysis by affinity labeling, Biochemistry 20, 7345.

Cochran, B. H., Zullo, J., Verma, I. M., & Stiles, C. D. (1984). Expression of the c-*fos* gene and of a fos-related gene is stimulated by platelet-derived growth factor. Science 226, 1080.

Coulter, C.L., Goldsmith, P.C., Mesiano, S., Voytek, C.C., Martin, M.C., & Jaffe, R.B. (1993). *In vivo* regulation of growth and expression of IGF-II and steroidogenic enzymes in the rhesus monkey fetal adrenal. 40th Ann. Meeting Soc. Gynecol. Invest. Toronto, Canada (Abstract S144).

Crickard, J., Ill, C.R., & Jaffe, R.B. (1981). Control of proliferation of human fetal adrenal cells *in vitro*. J. Clin. Endocrinol. Metab. 53, 790-798.

Crowley, C., Spencer, S.D., Nishimura, M.C., Chen, K.S., Pitts-Meek, S., Armanini, M.P., Ling, L.H., McMahon, S.B., Shelton, D.L., Levinson, A.D., & Phillips, H.S. (1994). Mice lacking nerve growth factor display perinatal loss of sensory and sympathetic neurons yet develop basal forebrain cholinergic neurons. Cell 76, 1001-1012.

Crystal, R. A., & Giudice, L. C. (1991). Insulin-like growth factor binding protein (IGFBP) profiles in human fetal cord sera: Ontogeny during gestation and differences in newborns with intrauterine growth retardation (IUGR) and large for gestational age (LGA) newborns. In: Modern Concepts of Insulin-like Growth Factors (Spencer, E. M., Ed.). Elsevier, New York.

Czyrski, J. A., & Gawlikowski, W. (1987) Activity of human, bovine, and porcine platelet-derived growth factor in a radioreceptor assay with human placental membrane protein. FEBS Lett. 219, 331.

Dale, L., Howes, G., Price, B.M., & Smith, J.C. (1992). Bone morphogenetic protein 4: A ventralizing factor in early *Xenopus* development. Development 115, 573-585.

Dale, L., Matthews, A., & Colman, A. (1993). Secretion and mesoderm-inducing activity of the TGF-beta-related domain of *Xenopus* Vg1. EMBO J. 12, 4471-4477.

Daughaday, W. H., Mariz I. K., & Trivedi, B. (1981) A preferential binding site for insulin-like growth factor II in human and rat placental membranes. J. Clin. Endocrinol. Metabol. 53, 282.

Davies, A.M., Lee, K.F., & Jaenisch, R. (1993). p75-deficient trigeminal sensory neurons have an altered response to NGF but not to other neurotrophins. Neuron 11, 565-575.

De, M., Sanford, T. H., & Wood, G. W. (1992). Detection of interleukin-1, interleukin-6, and tumor necrosis factor-a in the uterus during second half of pregnancy in the mouse. Endocrinology 131, 14-20.

De Chiara, T., Efstratiadis, A., & Robertson, E.J. (1990). A growth-deficiency phenotype in heterozygous mice carrying an insulin-like growth factor II gene disrupted by targeting. Nature (London) 345, 78-80.

Delassus, S., Coutinho, G. C., Saucier, C., Darche S., & Kourisky, P. (1994). Differential cytokine expression in maternal blood and placenta during murine gestation. J. Immunol. 152, 2411-2420.

Delhanty, P.J.D., Hill, D.J., Shimasaki, S., & Han, V.K.M. (1993). Insulin-like growth factor binding protein-4, -5 and -6 mRNAs in the human fetus: Localization to sites of growth and differentiation? Growth Regulation 3, 8-11.

Delmis, J., Drazancic, A., Ivanisevic, M., & Suchanek, E. (1992). Glucose, insulin, HGH and IGF-I levels in maternal serum, amniotic fluid and umbilical venous serum: A comparison between late normal pregnancy and pregnancies complicated with diabetes and fetal growth retardation. J. Perinat. Med. 20, 47-56.

Deloulme, J.C., Baudier, J., & Sensenbrenner, M. (1991). Establishment of pure neuronal cultures from the fetal rat spinal cord and proliferation of the neuronal precursor cells in the presence of fibroblast growth factor. J. Neurosci. Res. 29, 499-509.

D'Ercole, A.J., Foushee, D.B., & Underwood, L.E. (1976). Somatomedin C receptor ontogeny and levels in porcine fetal and human cord serum. J. Clin. Endocrinol. Metab. 43, 1069-1077.

D'Ercole, A. J. (1987). Somatomedins/insulin-like growth factors and fetal growth. J. Dev. Physiol. 9, 481.

Derynck, R., Rhee, L., Chen, E. Y., & Van Tilburg, A. (1987) Intron-exon structure of human transforming growth factor-β precursor gene. Nucleic Acids Res. 15, 3188.

Derynck, R., Lindquist, P. B., Lee, A., Wen, D., Tam, J., Graycar, J. L., Rhee, L., Mason, A. J., Miller, D. A., Coffey, R. J., Moses, H. L., & Chen, E. Y. (1988). A new type of transforming growth factor-β, TGF-β3. EMBO J. 7, 3737.

DiBlasio, A.M., Voutilainen, R., Jaffe, R.B., & Miller, W.L. (1987). Hormonal regulation of messenger ribonucleic acids for P450scc (cholesterol side-chain cleavage enzyme) and P450c17 (17 alpha-hydroxylase/17, 20-lyase) in cultured human fetal adrenal cells. J. Clin. Endocrinol. Metab. 65, 170-175.

Drago, J., Murphy, M., Carroll, S.M., Harvey, R.P., & Bartlett, P.F. (1991). Fibroblast growth factor-mediated proliferation of central nervous system precursors depends on endogenous production of insulin-like growth factor. Proc. Natl. Acad. Sci. USA 88, 2199-2203.

Drop, S. L. S., Kortleve, D. J., Guyda, H. J., & Posner, B. I. (1984). Immunoassay of a somatomedin-binding protein from human amniotic fluid: Levels in fetal, neonatal, and adult sera. J. Clin. Endocrinol. Metab. 59, 908.

D'Souza, S. W., Haigh, R., Micklewright, L., Donnai, P., & Keys, A. (1985). Amniotic fluid epidermal growth factor and placental weight at term. Lancet, II, 272.

Duello, T. M., Bertics, P. J., Fulgam, D. L., Van Ess, P. J. (1994) Localization of epidermal growth factor receptors in first- and third-trimester human placentas. J. Histochem. Cytochem. 42, 907-915.

Ellingsworth, L.R., Brennan, J.E., Fok, K., Rosen, D.M., Bentz, H., Piez, K.A., & Seyedin, S.M. (1986). Antibodies to the n-terminal portion of cartilage-inducing factor A and transforming growth factor β. J. Biol. Chem. 261, 12362-12367.

Ernst, C.W., McCusker, R.H., & White, M.E. (1992). Gene expression and secretion of insulin-like growth factor binding proteins during myoblast differentiation. Endocrinology 130, 607-615.

Ewton, D.Z., & Florini, J.R. (1981). Effects of somatomedins and insulin on myoblast differentiation *in vitro*. Dev. Biol. 56, 31-39.

Fallon, J.F., Lopez, A., Ros, M.A., Savage, M.P., Olwin, B.B., & Simandl, B.K. (1994). FGF-2: Apical ectodermal ridge growth signal for chick limb development. Science 264, 104-106.

Fant, M., Munro, H., & Moses, A. C. (1986). An autocrine/paracrine role for insulin-like growth factors in the regulation of human placental growth. J. Clin. Endocrinol. Metab. 63, 499.

Florini, J.R., Ewton, D.Z., & Roof, S.L. (1991a). Insulin-like growth factor I stimulates terminal myogenic differentiation by induction of myogenin gene expression. Mol. Endocrinol. 5, 718-724.

Florini, J.R., Magri, K.A., Ewton, D.Z., James, P.L., Grindstaff, K., & Rotwein, P.S. (1991b). "Spontaneous" differentiation of skeletal myoblasts is dependent upon autocrine secretion of insulin-like growth factor II. J. Biol. Chem. 266, 15197-15923.

Frolik, C. A., Dart, L. L., Meyers, C. A., Smith, D. M., & Sporn, M. B. (1983). Purification and initial characterization of a type beta transforming growth factor from human placenta. *Proc.* Natl. Acad. Sci. USA 80, 3676.

Fujita, Y., Kurachi, H., Morishige, K., Amemiya, K., Terakawa, N., Miyake, A., & Tanizawa, O. (1991). Decrease in epidermal growth factor receptor and its messenger ribonucleic acid levels in intrauterine growth-retarded and diabetes mellitus-complicated pregnancies. J. Clin. Endocrinol. Metab. 72, 1340.

Garcia-Lloret, M., Guilvert, L., & Morrish, D.W. (1989). Functional expression of CSF-1 receptors on normal human trophoblasts. In: Cedend, L., Alsat, E., Charrier, T.-C., Chaduat, G., & Mamassine, A. (Eds.), Placental Communications: Biochemical Morphological and Cellular Aspects. John Libbey Euro text, London.

Gabriel, R., Alsat, E., & Evain-Brion, D. (1994). Alteration of epidermal growth factor receptor in placental membranes of smokers: Relationship with intrauterine growth retardation. Am. J. Obstetr. Gynecol. 170, 1238-1243.

Gatherer, D., TenDijke, P., Baird, D.T., & Akhurst, R.J. (1990). Expression of TGF-β isoforms during first trimester human embryogenesis. Development 110, 445-460.

Gelb, D.E., Rosier, R.N., & Puzas, J.E. (1990). The production of transforming growth factor-β by chick growth plate chondrocytes in short term monolayer culture. Endocrinology 127, 1941-1947.

Gensenburger, C., Labourdette, G., & Sensenbrenner, M. (1987). Brain fibroblast growth factor stimulates the proliferation of rat neuronal precursor cells *in vitro*. FEBS Lett. 217, 1-5.

Giudice, L.C., Farrell, E.M., Pham, H., Lamson, G., & Rosenfeld, R.G. (1990). Insulin-like growth factor binding proteins in maternal serum throughout gestation and in the puerperium: Effects of a pregnancy-associated serum protease activity. J. Clin. Endocrinol. Metab. 71, 806-816.

Gonzalez, A-M., Buscaglia, M., Ong, M., & Baird, A. (1990). Distribution of basic fibroblast growth factor in the 18-day rat fetus: Localization in the basement membranes of diverse tissues. J. Cell. Biol. 110, 753-765.

Goodyer, C.G., Marcovitz, S., Hardy, J., Lefebvre, Y., Guyda, H.J., & Posner, B.I. (1986). Effect of insulin-like growth factors on human foetal, adult normal and tumour pituitary function in tissue culture. Acta Endocrinol. (Copenh.) 112, 49-57.

Gospodarowicz, D., Cheng, J., Lui, G. M., Baird, A., Esch, F., & Bohlen, P. (1985a). Corpus luteum angiogenic factor is related to fibroblast growth factor. Endocrinology 117, 2383.

Gospodarowicz, D., Cheng, J., Lui, G-M., Fujii, D. K., Baird, A., & Bohlen, P. (1985b). Fibroblast growth factor in the human placenta. Biochem. Biophys. Res. Comm. 128, 554.

Goustin, S. S., Betsholtz, C., Pfeifer-Ohlsson, S., Persson, H., Rydnert, J., Bywater, M., Holmgren, G., Heldin, C-H., Westermark, B., & Ohlsson, R. I. (1985). Coexpression of the *sis* and *myc* proto-oncogenes in developing human placenta suggests autocrine control of trophoblast growth. Cell 41, 301.

Graham, C. H., & Lala, P. K. (1991). Mechanism of control of trophoblast invasion *in situ*. J. Cell Physiol. 149, 228.

Green, B.N., Jones, S.B., Streck, R.D., Wood, T.L., Rotwein, P., & Pintar, J.E. (1994). Distinct expression patterns of insulin-like growth factor binding proteins 2 and 5 during fetal and postnatal development. Endocrinology 134, 954-962.

Hahnel, A., & Schultz, G.A. (1994). Insulin-like growth factor binding proteins are transcribed by preimplantation mouse embryos. Endocrinology 134, 1956-1959.

Hampson J., McLaughlin P. J., & Johnson P. M. (1993). Low affinity receptors for tumour necrosis factor-α, interferon-γ and granulocyte macrophage colony-stimulating factor are expressed on human placental syncitiotrophoblast. Immunology 79, 485-490.

Han, R.N.N., Liu, J., Tanswell, A.K., & Post, M. (1992). Expression of basic fibroblast growth factor and receptor: Immunolocalization studies in developing rat fetal lung. Ped. Res. 31, 435-440.

Han, V.K.M. (1993). Growth factors in placental growth and development. In: Molecular Aspects of Placental and Fetal Membrane Autocoids (Rice, G.E., & Brennecke, S.P., Eds.), pp. 395-445. CRC Press, Boca Roton, FL.

Han, V.K.M., & Hill, D.J. (1994). Growth factors in fetal growth. In: Textbook of Fetal Physiology (Thorburn, G.D., & Harding, R., Eds.). pp. 48-69. Oxford Medical Publications, Oxford.

Han, V.K.M., D'Ercole, A.J., & P.K. Lund. (1987a). Cellular localization of somatomedin (insulin-like growth factor) messenger RNA in the human fetus. Science 236, 193-197.

Han, V.K.M., Hill, D.J., Strain, A.J., Towle, A.C., Lauder, J.M., Underwood,L.E. & D'Ercole, A.J. (1987b). Identification of somatomedin/insulin-like growth factor immunoreactive cells in the human fetus. Ped. Res. 22, 245-249.

Han, V.K.M., Hunter, E.S., Pratt, R.M., Zendegui, J.G., & Lee, D.C. (1987c). Expression of transforming growth factor alpha mRNA during development occurs predominantly in the maternal decidua. Mol. Cell. Biol. 7, 2335-2343.

Han, V.K.M., Lund, P.K., Lee, D.C., & D'Ercole, A.J. (1988). Expression of somatomedin/insulin-like growth factor messenger ribonucleic acids in the human fetus: Identification, characterization and tissue distribution. J. Clin. Endocrinol. Metab. 66, 422-429.

Han, V.K.M., Bassett, N., Walton, J., & Challis, J.R.G. (1993). Extravillous trophoblasts are the major sites of expression of insulin-like growth factor-II (IGF-II) gene in the human placenta. Annual Meeting of The Society for Gynecologic Investigation, P114, 239, (Abstract).

Hatikka, J., & Hefti, F. (1988). Comparison of nerve growth factor effects on development of septum, striatum and nucleus basalis cholinergic neurons *in vitro*. J. Neurosci. Res. 21, 352-364.

Haub, O., & Goldfarb, M. (1991). Expression of fibroblast growth factor-5 gene in the mouse embryo. Development 112, 397-406.

Heine, U.I., Munoz, E.F., Flanders, K.C., Ellingsworth, L.R., Lam, H.Y.P., Thompson, N.L., Roberts, A.B., & Sporn, M.B. (1987). Role of transforming growth factor-β in the development of the mouse embryo. J. Cell Biol. 105, 2861-2876.

Hemmati-Brivanlou, A., & Melton, D.A. (1992). A truncated activin receptor inhibits mesoderm induction and formation of axial structures in *Xenopus* embryos. Nature (London) 359, 609-614.

Herbert, J.M., Basilico, C., Goldfarb, M., Haub, O., & Martin, G.R. (1990). Isolation of cDNAs encoding four mouse FGF family members and characterization of their expression patterns during embryogenesis. Dev. Biol. 138, 454-463.

Hill, D.J. (1979). Stimulation of cartilage zones of the calf costo-chondral growth plate *in vitro* by growth hormone dependent rat plasma somatomedin activity. J. Endocrinol. 83, 219-227.

Hill, D.J., & Clemmons, D.R. (1992). Similar immunological distribution of insulin-like growth factor binding proteins -1, -2, and -3 in human fetal tissues. Growth Factors 6, 315-326.

Hill, D.J., & De Sousa, D. (1990). Insulin is a mitogen for isolated epiphyseal growth plate chondrocytes from the fetal lamb. Endocrinology 126, 2661-2670.

Hill, D.J., & Logan, A. (1992a). Cell cycle-dependent localization of immunoreactive basic fibroblast growth factor to cytoplasm and nucleus of isolated ovine fetal growth plate chondrocytes. Growth Factors 7, 215-231.

Hill, D.J., & Logan, A. (1992b). Interactions of peptide growth factors during DNA synthesis in isolated ovine fetal growth plate chondrocytes. Growth Regulation 2, 122-132.

Hill, D.J., Clemmons, D.R., Wilson, S., Han, V.K.M., Strain, A.J., & Milner, R.D.G. (1989). Immunological distribution of one form of insulin-like growth factor (IGF) binding protein and IGF peptides in human fetal tissues. J. Mol. Endocrinol. 2, 31-38.

Hill, D.J., Logan, A., McGarry, M., & DeSousa, D. (1992a). Control of protein and matrix molecule synthesis in isolated ovine growth plate chondrocytes by the interactions of basic fibroblast growth factor, insulin-like growth factor I, insulin and transforming growth factor β. J. Endocrinol. 133, 363-373.

Hill, D.J., Logan, A., Ong, M., DeSousa, D., & Gonzalez, A.M. (1992b). Basic fibroblast growth factor is synthesized and released by isolated ovine fetal growth plate chondrocytes: Potential role as an autocrine mitogen. Growth Factors 6, 277-294.

Hill, D.J., Tevaarwerk, G.J.M., Arany, E., Kilkenny, D., Gregory, M., Langford, K.S., & Miell, J. (1995). Fibroblast growth factor-2 (FGF-2) is present in maternal and cord serum, and in the mother is associated with a binding protein immunologically related to the FGF receptor-1. J. Clin. Endocrinol. Metab. 80, 1822-1831.

Hock, R. A. (1980). Characterization of the receptor for epidermal growth factor urogastrone in human placenta membranes, J. Biol. Chem., 255, 10731.

Hofmann, G. E., Rao, C. V., Brown, M. J., Murray, L. F., Schultz, G. S., & Siddiqi, T. A. (1988). Epidermal growth factor in urine of nonpregnant women and pregnant women throughout pregnancy and at delivery. J. Clin. Endocrinol. Metabol., 66, 119.

Hofmann, G. E., Scott, R. T., Bergh, P. A., Jr., & Deligdisch, L. (1991). Immunohistochemical localization of epidermal growth factor in human endometrium, decidua, and placenta. J. Clin. Endocrinol. Metab. 73, 882.

Hondermarck, H., Courty, J., Ledoux, D., Blanckaert, V., Barritault, D., & Boilly, B. (1990). Evidence of high and low affinity binding sites for basic fibroblast growth factor in mouse placenta, Biochem. Biophys. Res. Comm. 169, 272.

Hooper, S.B., Bocking, A.D., White, S., Challis, J.R.G., & Han, V.K.M. (1991). DNA synthesis is reduced in selected fetal tissues during prolonged hypoxemia. Am. J. Physiol. 261, R508-R514.

Hooper, S.B., Bocking, A.D., White, S.E., Fraher, L.J., McDonald, T.J., & Han, V.K.M. (1994). Catecholamines stimulate the synthesis and release of insulin-like growth factor binding protein-1 (IGFBP-1) by fetal sheep liver *in vivo*. Endocrinology 134, 1104-1112.

Hughes, S.M., Lillien, L.E., Raff, M.C., Rohrer, H., & Sendter, M. (1988). Ciliary neurotrophic factor induced type 2 astrocyte differentiation in culture. Nature (London) 335, 70-73.

Hunt J.S. (1993). Expression and regulation of the tumour necrosis factor-a gene in the female reproductive tract. Reprod. Fertil. Dev. 5, 141-153.

Hunt J., S., Chen, H-L., Hu, X-L., & Pollard, J. W. (1993). Normal distribution of tumor necrosis factor-a messenger ribonucleic acid and protein in the uteri, placentas and embryos of osteopetrotic (op/op) mice lacking colony-stimulating factor-1. Biol Reprod 49, 441-452.

Huot, R. I., Foidart, J-M., Nardone, R. M., & Stromberg, K. (1981) Differential modulation of human chorionic gonadotropin secretion by epidermal growth factor in normal and malignant placental cultures. J. Clin. Endocrinol. Metabol, 53, 59.

Hunziker, E.B. (1988). Growth plate structure and function. Pathol. Immunopathol. Res. 7, 9-13.

Ip, N.Y., Li, Y., Yancopoulos, G.D., & Lindsay, R.M. (1993). Cultured hippocampal neurons show responses to BDNF, NT-3 and NT-4, but not NGF. J. Neurosci. 13, 3394-3405.

Isgaard, J., Moller, C., Isaksson, O., Nilsson, A., Mathews, L.S., & Norstedt, G. (1988). Regulation of insulin-like growth factor messenger ribonucleic acid in rat growth plate by growth hormone. Endocrinology 122, 1515-1520.

Iwamoto, H.S., Murray, M.A., & Chernausek, S.D. (1992). Effects of acute hypoxia on insulin-like growth factors and their binding proteins in fetal sheep. Am. J. Physiol. 263, E1151-E1156.

Jokhi P. P., Chumbley G., King A., Gardner L., & Loke Y. W. (1993). Expression of the colony stimulating factor-1 receptor (c-fms product) by cells at the human uterplacental interface. Lab. Invest. 3, 308-320.

Jonas, H. A., Newman, J. D., & Harrison, L. C. (1986). An atypical insulin receptor with high affinity for insulin-like growth factors copurified with placenta insulin receptors. Proc. Natl. Acad. Sci. USA 83, 4124.

Jones, C.M., Lyons, K.M., Lapan, P.M., Wright, C.V.E., & Hogan, B.J.M. (1992). DVRR-4 (Bone morphological protein-4) as a posterior-ventralizing factor in Xenopus mesoderm induction. Development 115, 639-647.

Jones, J. I., Gockerman, A., Busby, W. H., Wright, G., & Clemmons, D. C. (1993). Insulin-like growth factor binding protein-1 stimulates cell migration and binds to the β1 integrin by means of its ARg-Gly-Asp sequence. Proc. Natl. Acad. Sci. USA 90, 10553-10557.

Joseph-Silverstein, J., Consigli, S.A., Lyser, K.M., & Ver Pault, C. (1989). Basic fibroblast growth factor in the chick embryo: Immunolocalization to striated muscle cells and their precursors. J. Cell. Biol. 108, 2459-2466.

Jospe, N., Zhu, J., Liu, R., Livingston, J.N., & Furlanetto, R.W. (1994). Deletion of 3 basepairs resulting in the loss of lysine-121 in the insulin receptor a-subunit in a patient with leprechaunism: Binding, phosphorylation, and biological activity. J. Clin. Endocrinol. Metab. 79, 1294-1302.

Joyce, M.E., Roberts, A.B., Sporn, M.B., & Bolander, M.E. (1990). Transforming growth factor-β and the initiation of chondrogenesis and osteogenesis in the rat femur. J. Cell Biol. 110, 2195-2207.

Kauma, S. , Matt, D., Strom, S., Eirman, D., & Turner, T. (1990). Interleukin-1b, humna leukocyte antigen HLA-DRa, and transforming growth factor-b expression in endometrium, placenta and placental membranes. Am. J. Obstet. Gynecol. 163, 1430-1437.

Kauma, S. W., Aukerman, S. L., Eierman, D., & Turner, T. (1991). Colony-stimulating factor-1 and c-*fms* expression in human endometrial tisues and placenta during the menstrual cycle and early pregnancy. J. Clin. Endocrinol. Metab. 73, 746.

Kauma, S. W., Herman, K., Wang, Y., & Walsh, S.W. (1993). Differential mRNA expression and production of interleukin-6 in placental trophoblast and villous core compartments. Am. J. Reprod. Immunol. 30, 131-135.

Kauma, S. W., Turner, T. T., & Harty, J. R. (1994). Interleukin-1b stimulates interleukin-6 production in placental villous core mesenchymal cells. Endocrinology 134, 457-460.

Kasselberg, A.G., Orth, D.N., Gray, M.E., & Stahlman, M.T. (1985). Immunocytochemical localization of human epidermal growth factor/urogastrone in several human tissues. J. Histochem. Cytochem. 33, 315-322.

Kearns, M., & Lala, P. J. (1985). Characterization of hematogenous cellular constituents in the murine decidua: A surface marker study. J. Reprod. Immunol. 8, 213-234.

Kessler, D.S., & Melton, D.A. (1994). Vertebrate embryonic induction: Mesoderm and neural patterning. Science 266, 596-604.

Kimelman, D., & Kirschner, M. (1987). Synergistic induction of mesoderm by FGF and TGF-beta and the identification of an mRNA coding for FGF in the early *Xenopus* embryo. Cell 51, 869-871.

Kimelman, D., Abraham, J.A., Haarparanta, T., Palisi, T.M., & Kirschner, M.W. (1988). The presence of fibroblast growth factor in the frog egg: Its role as a natural mesoderm inducer. Science 242, 1053-1056.

Klein, R., Parada, L.F., Coulier, F., & Barbacid, M. (1989). trkB, a novel tyrosine protein kinase receptor expressed during mouse neural development. EMBO J. 8, 3701-3709.

Klein, R., Smeyne, R.J., Wurst, W., Long, L.K., Auerbach, B.A., Joyner, A.L., & Barbacid, M. (1993). Targeted disruption of the trkB neurotrophin receptor gene results in nervous system lesions and neonatal death. Cell 75, 113-122.

Klein, R., Silos-Santiago, I., Smeyne, R.J., Lira, S.A., Brambilla, R., Bryant, S., Zhang, L., Snider, W.D., & Barbacid, M. (1994). Disruption of the neurotrophin-3 receptor gene trkC eliminates 1a muscle afferents and results in abnormal movements. Nature (London) 368, 249-251.

Klempt, M., Hutchins, A-M., Gluckman, P.D., & Skinner, S.J.M. (1992). IGF binding protein-2 gene expression and the location of IGF-I and IGF-II in fetal rat lung. Development 115, 765-772.

Kulkarni, A.B., Huh, C-G., Becker, D., Geiser, A., Lyght, M., Flanders, K.C., Roberts, A.B., Sporn, M.B., Ward, J.M., & Karlsson, S. (1993). Transforming growth factor β1 null mutation in mice causes excessive inflammatory response and early death. Proc. Natl. Acad. Sci. USA 90, 770-774.

LaBonne, C., & Whitman, M. (1994). Mesoderm induction by activin requires FGF-mediated intracellular signals. Development 120, 463-472.

Lassarre, C., Hardouin, S., Daffos, F., Forestier, F., Frankenne, F., & Binoux, M. (1991). Serum insulin-like growth factor binding proteins in the human fetus. Relationships with growth in normal subjects and in subjects with intrauterine growth retardation. Ped. Res. 9, 219-221.

Lawrence, S., Stenzel, W., & Warshaw, J. B. (1989). Increased binding of epidermal growth factor to placental membranes of intrauterine growth restricted fetal rats. Ped. Res. 25, 214.

Lehnert, S.A., & Akhurst, R.J. (1988). Embryonic expression pattern of TGF beta type-1 RNA suggests both paracrine and autocrine mechanisms of action. Development 104, 263-273.

Letterio, J.J., Geiser, A.G., Kulkarni, A.B., Roche, N.S., Sporn, M.B., & Roberts, A.B. (1994). Maternal rescue of transforming growth factor β1 null mice. Science 264, 1936-1938.

Li, J., Saunders, J.C., Gilmour, R.S., Silver, M., & Fowden, A.L. (1993). Insulin-like growth factor-II messenger ribonucleic acid expression in fetal tissues of the sheep during late gestation: Effects of cortisol. Endocrinology 132, 2083-2089.

Linkhart, T.A., Clegg, C.H., & Hauscha, S.D. (1981). Myogenic differentiation in permanent clonal mouse myoblast cell lines: Regulation by macromolecular growth factors in the culture medium. Dev. Biol. 86, 19-30.

Liu, J-P., Baker, J., Perkins, A.S., Robertson, E.J., & Efstratiadis, A. (1993). Mice carrying null mutations of the genes encoding insulin-like growth factor I (Igf-1) and type 1 IGF receptor (Igf1r). Cell 75, 59-72.

Liu, L., & Nicoll, C.S. (1988). Evidence for a role of basic fibroblast growth factor in rat embryonic growth and differentiation. Endocrinology 123, 2027-2031.

Loke Y. W., & King A. (1990). Current topic: Interferon and human placental development. Placenta 11, 291-299.

Lu, F., Han, V.K.M., Milne, W.K., Fraser, M., Carter, A.M., Berdusco, E.T.M., & Challis, J.R.G. (1994). Regulation of insulin-like growth factor-II gene expression in the ovine fetal adrenal gland by adrenocorticotropic hormone and cortisol. Endocrinology 134, 2628-2635.

Luetteke, N.C., Qiu, T.H., Pfeiffer, R.L., Oliver, P., Smithies, O., & Lee, D.C. (1993). TGF alpha deficiency results in hair follicle and eye abnormalities in targeted and waved-1 mice. Cell 73, 263-272.

Luger, A., Calogero, A.E., Kalogeras, K., Gallucci, W.T., Gold, P.N., Loriaux, D.L., & Chrousos, G.P. (1988). Interaction of epidermal growth factor with the hypathalamic-pituitary-adrenal axis: Potential physiologic relevance. J. Clin. Endocrinol. Metab. 66, 334-337.

Lysiak J. J., Han V. K. M., & Lala P. K. (1993). Localization of transforming growth factor-α (TGF-α) in the human placenta and decidua: Role in trophoblast growth. Biol. Reprod. 49, 885-894.

Maisonpierre, P.C., Belluscio, L., Friedman, B., Alderson, R.F., Wiegand, S.J., Furth, M.E., Lindsay, R.M., & Yancopoulos, G.D. (1990). NT-3, BDNF, and NGF in the developing rat nervous system: Parallel as well as reciprocal patterns of expression. Neuron 5, 501-509.

McLellan, K.C., Hooper, S.B., Bocking, A.D., Delhanty, P.J.D., Phillips, I.D., Hill, D.J., & Han, V.K.M. (1992). Prolonged hypoxia induced by the reduction of maternal uterine blood flow alters insulin-like growth factor-binding protein-1 (IGFBP-1) and IGFBP-2 gene expression in the ovine fetus. Endocrinology 131, 1619-1628.

Maden, M. (1994). The limb bud—part two. Nature (London) 371, 560-561.

Magri, K.A., Benedict, M.R., Ewton, D.Z., & Florini, J.R. (1994). Negative feedback regulation of insulin-like growth factor-II gene expression in differentiating myoblasts *in vitro*. Endocrinology 135, 53-62.

Marez, A. N., Guyen, T., Chevallier, B., Clement, G., Dauchel, M. C., & Barritault, D. I. (1987). Platelet derived growth factor is present in human placenta: Purification from an industrially processed fraction. Biochimie 69, 125.

Maruo, T., Matsuo, H., Murata, K., & Mochizuki, M. (1992). Gestational age-dependent dual action of epiderma growth factor on human placenta early in gestation. J. Clin. Endocrinol. Metab. 75, 1362-1367.

Medlock, E. S., Kaplan, D. L., Cecchini, M., Ulich, T. R., Castillo, J. D., & Andersen, J. (1993). Granulocyte colony-stimulating factor crosses the placenta and stimulates fetal rat granulopoiesis. Blood 81, 916-922.

Mesiano, S., Mellon, S.H., Gospodarowicz, D., DiBlasio, A.M., & Jaffe, R.B. (1991). Basic fibroblast growth factor expression is regulated by corticotropin in the human fetal adrenal: A model for adrenal growth regulation. Proc. Natl. Acad. Sci. USA 88, 5428-5432.

Mitchell, M. D., Trautman, M. S., & Dudley, D. J. (1993). Cytokine networking in the placenta. Placenta, 14, 249-275.

Mitrani, E., Ziu, T., Thomsen, G., Shimoni, Y., Melton, D.A., & Bril, A. (1990). Activin can induce the formation of axial structures and is expressed in the hypoblast of the chick. Cell 63, 495-501.

Morrison-Graham, K., Schatteman, G.C., Bork, T., Bowenpope, D.F., & Weston, J.A. (1992). A PDGF receptor mutation in the mouse (patch) perturbs the development of a non-neuronal subset of neural crest-derived cells. Development 115, 133-142.

Munaim, S.I., Klagsbrun, M., & Toole, B.P. (1988). Developmental changes in fibroblast growth factor in the chicken embryo limb bud. Proc. Natl. Acad. Sci. USA 85, 8091-8093.

Munaim, S.I., Klagsbrun, M., & Toole, B.P. (1991). Hyaluronan-dependent pericellular coats of chick embryo limb mesoderm cells: Induction by basic fibroblast growth factor. Dev. Biol. 143, 297-302.

Murphy, L. M., Seneviratne, C., Ballejo, G., Croze, F., & Kennedy, T. G. (1990). Identification and characterization of a rat decidual insulin-like growth factor binding protein complementary DNA. Mol. Endocrinol. 4, 329.

Murphy, M., Drago, J., & Bartlett, P.F. (1990). Fibroblast growth factor stimulates the proliferation and differentiation of neuronal precursor cells *in vitro*. J. Neurosci. Res. 25, 463-475.

Naamen, E., Chatelain, P., Saez, J.M., & Durand, P. (1989). *In vitro* effect of insulin and insulin-like growth factor-I on cell multiplication and adrenocorticotropin responsiveness of fetal adrenal cells. Biol. Reprod. 40, 570-577.

Nagarajan, L., Anderson, W.B., Nissley, S.P., Rechler, M.M., & Jetten, A.M. (1985). Production of insulin-like growth factor II (MSA) by endoderm-like cells derived from embryonal carcinoma cells: Possible mediator of embryonic cell growth. J. Cell. Physiol. 124, 199-206.

Nestler, J. E. (1989). Insulin and insulin-like growth factor-I stimulate the 3b-hydroxysteroid dehydrogenase activity of human placental cytotrophoblasts. Endocrinology 125, 2127.

Nestler, J. E. (1990). Insulin-like growth factor II is a potent inhibitor of the aromatase activity of human placental cytotrophoblasts Endocrinology 127, 2064.

Nestler, J. E., & Williams, T. (1987). Modulation of aromatase and P450 cholesterol side-chain cleavage enzyme activities of human placental cytotrophoblasts by insulin and insulin-like growth factor I. Endocrinology 121, 1845.

Nielson, H.C., Kellogg, C.K., & Doyle, C.A. (1992). Development of fibroblast type-II cell communications in fetal rabbit lung organ culture. Biochem. Biophys. Acta 1175, 95-99.

Nilsson, A., Isgaard, J., Lindahl, A., Dahlstrom, A., Skottner, A., & Isaksson, O. (1986). Regulation by growth hormone of number of chondrocytes containing IGF I in the rat growth plate. Science 233, 571-574.

Nilsson, A., Isgaard, J., Lindahl, A., Peterson, L., & Isaksson, O. (1987). Effects of unilateral arterial infusion of GH and IGF I on tibial longitudinal bone growth in hypophysectomized rats. Calc. Tissue Int. 40, 91-96.

Nissley, S. P., & Rechler, M. M. (1984) Somatomedin/insulin-like growth factor tissue receptors, J. Clin. Endocrinol. Metab., 13, 43.

Niswander, L., & Martin, G.R. (1992). Fgf-4 expression during gastrulation, myogenesis, limb and tooth development in the mouse. Development 114, 755-768.

Niswander, L., & Martin, G.R. (1993). FGF-4 and BMP-2 have opposite effects on limb growth. Nature (London) 361, 68-71.

Niswander, L., Tickle, C., Vogel, A., Booth, I., & Martin, G.R. (1993). FGF-4 replaces the apical ectodermal ridge and directs outgrowth and patterning of the limb. Cell 75, 579-587.

Niswander, L., Jeffrey, S., Martin, G.R., & Tickle, C. (1994). A positive feedback loop coordinates growth and patterning in the vertebrate limb. Nature (London) 371, 609-612.

Noble, M., & Murray, K. (1984). Purified astrocytes promote the *in vitro* division of bipotential glial pogenitor cells. EMBO J. 3, 2243-2247.

Nonoshita, L.D., Wathen, N.C., Dsupin, B.A., Chard, D., & Giudice, L.C. (1994). Insulin-like growth factors (IGFs), IGF-binding proteins (IGFBPs), and proteolyzed IGFBP-3 in embryonic cavities in early human pregnancy: Their potential relevance to maternal-embryonic and fetal interactions. J. Clin. Endocrinol. Metab. 79, 1249-1255.

Ohlsson, C., Nilsson, A., Isaksson, O., Bentham, J., & Lindahl, A. (1992). Effects of triiodothyronine and insulin-like growth factor-I (IGF-I) on alkaline phosphatase activity, [3H]thymidine incorporation and IGF-I receptor mRNA in cultured rat epiphyseal chondrocytes. J. Endocrinol. 135, 115-121.

Orrurtreger, A., Bedford, M.T., Do, M.S., Eisenbach, L., & Lonai, P. (1992). Developmental expression of the alpha receptor for platelet-derived growth factor, which is deleted in the embryonic lethal patch mutation. Development 115, 289-297.

Peifer, M. (1994). The two faces of hedgehog. Science 266, 1492-1493.

Pekonen, F., Suikkari, A. M., Makinen, T., & Rutanen, E-M. (1988). Different insulin-like growth factor binding species in human placenta and decidua, J. Clin. Endocrinol. Metab. 67, 1250.

Pelton, R.W., Dickinson, M.E., Moses, H.L., & Hogan, B.L.M. (1990). *In situ* hybridization analysis of TGFβ3 RNA expression during mouse development: Comparative studies with TGFβ1 and β2. Development 110, 609-620.

Peters, K., Ornitz, D., Werner, S., & Williams, L. (1993). Unique expression pattern of the FGF receptor 3 gene during mouse organogenesis. Dev. Biol. 155, 423-430.

Peters, K.G., Werner, S., Chen, G., & Williams, L.T. (1992). Two FGF receptor genes are differentially expressed in epithelial and mesenchymal tissues during limb formation and organogenesis in the mouse. Development 114, 233-243.

Plopper, C.G., Stgeorge, J.A., Read, L.C., Nishio, S.J., Weir, A.J., Edwards, L., Tarantal, A.F., Pinkerton, K.E., Merritt, T.A., & Whitsett, J.A. (1992). Acceleration of alveolar type-II cell differentiation in fetal rhesus monkey lung by administration of EGF. Am. J. Physiol. 262, L313-L319.

Polk, D.H., Ervin, M.G., Padbury, J.F., Lam, R.N., Reviczky, A.L., & Fisher, D.A. (1987). Epidermal growth factor acts as a corticotropin-releasing factor in chronically catheterized fetal lambs. J. Clin. Invest. 79, 984-988.

Pollard J. W., Bartocci A., Arcecci R., Orlofsky A., Ladner M. B., & Stanley E. R. (1987). Apparent role of macrophage growth factor in placental development. Nature 330, 484-486.

Pons, S., & Torres-Aleman, I. (1992). Basic fibroblast growth factor modulates insulin-like growth factor-I, its receptor, and its binding proteins in hypothalamic cell cultures. Endocrinology 131, 2271-2278.

Price, W.A., Rong, L., Stiles, A.D., & D'Ercole, A.J. (1992). Changes in IGF-I and -II, IGF binding protein, and IGF receptor transcript abundance after uterine artery ligation. Ped. Res. 32, 291-295.

Qu, J., Brulet, C., & Thomas, K. (1992). Effect of epidermal growth factor on inhibin secretion in human placental cell culture. Endocrinology 131, 2173-2181.

Raff, M.C. (1989). Glial cell diversification in the rat optic nerve. Science 243, 1450-1455.

Rainey, W.E., Oka, K., Magness, R.R., & Mason, J.I. (1991). Ovine fetal adrenal synthesis of cortisol: Regulation by adrenocorticotropin, angiotensin II and transforming growth factor-β. Endocrinology 129, 1784-1790.

Ralphs, J., Wylie, L., & Hill, D.J. (1990). Distribution of insulin-like growth factor peptides in the developing chick embryo. Development 109, 51-58.

Rao, C. V., Carman, F. R.,Jr., Chegini, N., & Schultz, G. S. (1984). Binding sites for epidermal growth factor in human fetal membranes. J. Clin. Endocrinol. Metabol. 58, 1034.

Rappolee, D.A., Brenner, C.A., Schultz, R., & Werb, Z. (1988). Developmental expression of PDGF, TGF-a, and TGF-β genes in preimplantation mouse embryos. Science 241, 1823-1825.

Ray, J., Peterson, D.A., Schinstine, M., & Gage, F.H. (1993). Proliferation, differentiation and long-term culture of primary hippocampal neurons. Proc. Natl. Acad. Sci. USA 90, 3602-3606.

Regenstreif, L. J., & Rossant, J. (1989). Expression of the c-*fms* proto-oncogene and of the cytokine, CSF-I, during mouse embryogenesis. Dev. Biol. 133, 284.

Reynolds, B.A., Tetzlaff, W., & Weiss, S. (1992). A multipotent EGF-responsive striatal embryonic progenitor cell produces neurons and astrocytes. J. Neurosci. 12, 4565-4574.

Ringler, G. E., Coutifaris, C., Strauss, J. F., Allen, J. I., & Geier, M. (1989) Accumulation of colony-stimulating factor I in amniotic fluid during human pregnancy. Am. J. Obstet. Gynecol. 160, 655.

Roberts, A.B., & Sporn, M.B. (1990). The transforming growth factor-βs: In: Peptide Growth Factors and Their Receptors (Sporn, M.B., & Roberts, A.B., Eds.), pp. 419-472. Springer, Berlin.

Romero, R., & Mazor, M. (1988). Infection and preterm labor. Obstetrics and Gynecology 51, 227-228.

Romero, R., Sirtori, M., Oyarzun, E., Avila, C., Mazor, M., Callahan, R., Sabo, V., Athanassiades, A.P., & Hobbins, J.C. (1989). Infection and labor: V. Prevalence, microbiology, and clinical significance intra amniotic infection in women with preterm labor and intact membranes. Am. J. Obstet. Gynecol. 161, 817-824.

Rosa, F., Roberts, A.B., Danielpour, D., Dart, L.L., Sporn, M.B., & Dawid, I.B. (1988). Mesoderm induction in amphibians: The role of TGF-β2-like factors. Science 239, 783-785.

Rosen, K.M., Wentworth, B.M., Rosenthal, N., & Villa-Komaroff, L. (1993). Specific, temporally regulated expression of the insulin-like growth factor II gene during muscle cell differentiation. Endocrinology 133, 474-481.

Rosenthal, S.M., Hsiao, D., & Silverman, L.A. (1994). An insulin-like growth factor-II (IGF-II) analog with highly selective affinity for IGF-II receptors stimulates differentiation, but not IGF-I receptor down-regulation in muscle cells. Endocrinology 134, 38-44.

Rosier, R.N., Landesberg, R.L., & Puzas, J.E. (1991). An autocrine feedback loop for regulation of TGF-β and FGF production in growth plate chondrocytes. J. Bone Min. Res. 6 (Suppl.1) S211.

Rosier, R.N., O'Keefe, R.J., Crabb, I.D., & Puzas, J.E. (1989). Transforming growth factor-beta: An autocrine regulator of chondrocytes. Conn. Tissue Res. 20, 295-301.

Rousseau, F., Bonaventure, J., Legeal-Mallet, L., Pelet, A., Rozet, J-M., Maroteaux, P., Le Merrer, M., & Munnich, A. (1994). Mutations in the gene encoding fibroblast growth factor receptor-3 in achondroplasia. Nature (London) 371, 252-254.

Rotwein, P., Pollock, K.M., Watson, M., & Milbrandt, J.D. (1987). Insulin-like growth factor gene expression during rat embryonic development. Endocrinology 121, 2141-2144.

Sandberg, M., Vuorio, T., Hirvonen, H., Alitalo, K.L., & Vuorio, E. (1988). Enhanced expression of TGF-β and c-fos mRNAs in the growth plates of developing human long bones. Development 102, 461-470.

Schonbrunn, A., Krasnoff, M., Westendorf, J. M., & Tashijian, A. H., Jr. (1980). Epidermal growth factor and thytrotropin-releasing hormone act similarily on a clonal pituitary cell strain. Modulation of hormone production and inhibition of cell proliferation. J. Cell. Biol. 85, 786.

Schulte-Merker, S., Smith, J.C., & Dale, L. (1994). Effects of truncated activin and FGF receptors and of follistatin on the inducing activities of BVg1 and activin: Does activin play a role in mesoderm induction? EMBO J. 13, 3533-3541.

Selick, C. E., Horowitz, G. M., Gratch, M., Scott, R. T., Navot, D., & Hofmann, G. E. (1994). Immunohistochemical localization of transforming growth factor-β in human implantation sites. J. Clin. Endocrinol. Metab. 78, 592-596.

Shigematsu, K., Niwa, M., Kurihara, M., Yamashita, K., Kawai, K., & Tsuchiyama, H. (1989). Receptor autoradiographic localization of insulin-like growth factor-I (IGF-I) binding sites in human fetal and adult adrenal glands. Life Sci. 45, 383-389.

Shull, M.M., Ormsby, I., Kier, A.B., Pawlowski, S., Diebold, R.J., Yin, M., Allen, R., Sidman, C., Proetzel, G., Calvin, D., Annunziata, N., & Doetscham, T. (1992). Targeted disruption of the transforming growth factor-94β1 gene results in multifocal inflammatory disease. Nature (London) 359, 693-699.

Sisson, J. F., Stenzel, W. K., & Warshaw, J. B. (1987) Decreased binding of epidermal growth factor in placentas from streptozotocin-diabetic rats, J. Clin. Invest., 80, 242.

Sklar, M. M., Kiess, W., Thomas, C. L., & Nissley, S. P. (1989). Developmental expression of the tissue insulin-like growth factor II/mannose 6 phosphate receptor in the rat. Measurement by quantitative immunoblotting, J. Biol. Chem. 264, 16733.

Slack, J.M.W., & Isaacs, H.V. (1989). Presence of basic fibroblast growth factor in the early *Xenopus* embryo. Development 105, 147-153.

Slavkin, H.C., Hu, C.C., Sakakura, Y., Diekwisch, T., Chai, Y., Mayo, M., Bringas, P., Simmer, J., Mak, G., & Sasano, Y. (1992). Gene expression, signal transduction and tissue specific mineralization during mammalian tooth development. Crit. Rev. Eukaryot. Gene Expr. 2, 315-329.

Smeyne, R.J., Klein, R., Schapp, A., Long, L.K., Bryant, S., Lewin, A., Lira, S.A., & Barbacid, M. (1994). Severe sensory and sympathetic neuropathies in mice carrying a disrupted trk/NGF receptor gene. Nature (London) 368, 246-249.

Smith, J.C., Price, B.M.J., Van Nimmen, K., & Huylebroeck, D. (1990). Identification of a potent *Xenopus* mesoderm-inducing factor as a homologue of activin A. Nature (London) 345, 729-731.

Snider, W.D. (1994). Functions of the neurotrophins during nervous system development: What the knockouts are teaching us. Cell 77, 627-638.

Stark, K.L., McMahon, J.A., & McMahon, A.P. (1991). FGFR-4, a new member of the fibroblast growth factor receptor family, expressed in the definitive endoderm and skeletal muscle lineages of the mouse. Development 113, 641-651.

Strandjord, T.P., Clark, J.G., Hodson, W.A., Schmidt, R.A., & Madtes, D.K. (1993). Expression of transforming growth factor alpha in mid-gestation human fetal lung. Am. J. Respir. Cell Mol. Biol. 8, 266-273.

Straus, D.S., Ooi, G.T., Orlowski, C.C., & Rechler, M.M. (1991). Expression of the genes for insulin-like growth factor-I (IGF-I), IGF-II, and IGF-binding proteins-1 and -2 in fetal rat under conditions of intrauterine growth retardation caused by maternal fasting. Endocrinology 128, 518-525.

Streck, R.D., Wood, T.L., Hsu, M-S., & Pintar, J.E. (1992). Insulin-like growth factor I and II and insulin-like growth factor binding protein-2 RNAs are expressed in adjacent tissues within rat embryonic and fetal limbs. Dev. Biol. 151, 586-596.

Stuart, C. A., Brosnan, P. G., & Furlanetto, R. W. (1984). The somatomedin-C receptor of human placenta. Metabolism 33, 90.

Sundell, H.W., Gray, M.E., Serenius, F.S., Escobedo, M.B., & Stahlman, M.T. (1980). Effects of epidermal growth factor on lung maturation in fetal lambs. Am. J. Pathol. 100, 707-725.

Suzuki, A., Thies, R.S., Yamaji, N., Song, J.J., Wozney, J.M., Murakami, K., & Ueno, N. (1994). A truncated bone morphogenetic protein receptor affects dorsal-ventral patterning in the early *Xenopus* embryo. Proc. Natl. Acad. Sci. USA 91, 10255-10259.

Tamemoto, H., Kadowaki, T., Tobe, K., Yagi, T., Sakura, H., Hayakawa, T., Terauchi, Y., Veki, K., Kaburagi, Y., & Satoh, S. (1994). Insulin resistance and growth retardation in mice lacking insulin receptor substrate-1. Nature (London) 372, 182-186.

Tannahill, D., & Melton, D.A. (1989). Localized synthesis of of the Vg1 protein during early *Xenopus* development. Development 106, 775-785.

Tartakovski, B. (1989). CSF-1 induces resorption of embryos in mice. Immunol. Lett. 23, 65-70.

Taylor, R. N., & Williams, L. T. (1988). Development expression of platelet-derived growth factor and its receptor in the human placenta. Mol. Endocrinol. 2, 627.

Tessarello, L., Vogel, K.S., Palko, M.E., Reid, S.W., & Parada, L.F. (1994). Targeted mutation in the neurotrophin-3 gene results in loss of sensory neurons. Proc. Natl. Acad. Sci. USA 91, 11844-11848.

Thomsen, G., Woolf, T., Whitman, M., Sokol, S., Vaughan, J., Vale, W., & Melton, D.A. (1990). Activins are expressed early in *Xenopus* embryogenesis and can induce axial mesoderm and anterior structures. Cell 63, 485-491.

Tollefsen, S. E., & Thompson, K. (1988) The structural basis for insulin-like growth factor I receptor high affinity binding. J. Biol. Chem., 263, 16267.

Torres-Aleman, I., Naftolin, F., & Robbins, R.J. (1990). Trophic effects of insulin-like growth factor-I on fetal rat hypothalamic cells in culture. Neuroscience 35, 601-608.

Townsend, S.F., Dallman, M.F., & Miller, W.L. (1990). Rat insulin-like growth factor-I and -II mRNAs are unchanged during compensatory adrenal growth but decrease during ACTH-induced adrenal growth. J. Biol. Chem. 265, 22117-22122.

Trippel, S.B., Van Wyk, J.J., & Mankin, H.J. (1986). Localization of somatomedin-C binding to bovine growth-plate chondrocytes *in situ*. J. Bone Joint Surg. 68-A, 897-903.

Ulich, T.R., Yi, E.S., Longmuir, K., Yin, S., Biltz, R., Morris, C.F., Housley, R.M., & Pierce, G.F. (1994). Keratinocyte growth factor is a growth factor for type II pneumonocytes *in vivo*. J. Clin. Invest. 93, 1298-1306.

Unterman, T., Lascon, R., Golway, M.B., Oehler, D., Gounis, A., Simmons, R.A. & Ogata, E.S. (1990). Circulating levels of insulin-like growth factor binding protein-1 (IGFBP-1) and hepatic mRNA are increased in the small for gestational age (SGA) fetal rat. Endocrinology 127, 2035-2037.

Vaidya, T.B., Rhodes, S.J., Taparowsky, E.J., & Konieczny, S.F. (1989). Fibroblast growth factor and transforming growth factor β repress transcription of the myogenic regulatory gene MyoD1. Mol. Cell. Biol. 9, 3576-3579.

Van Zoelen, E.J., Koornneef, I., Holthuis, J.C., Ward van Osterwaard, T.M.J., Feijen, A., De Poorter, T.L., Mummery, C.L., & Van Buul-Offers, S.C. (1989). Production of insulin-like growth factors, platelet-derived growth factor, and transforming growth factors and their role in the density-dependent growth regulation of a differentiated embryonal carcinoma cell line. Endocrinology 124, 2029-2041.

Vileisis, R., & D'Ercole, A.J. (1986). Tissue and serum concentrations of somatomedin-C/insulin-like growth factor I in fetal rats made growth retarded by uterine artery ligation. Ped. Res. 20, 126-130.

Voutilainen, R., & Miller, W.L. (1987a). Coordinate tropic hormone regulation of mRNAs for insulin-like growth factor II and the cholesterol side-chain-cleavage enzyme, P450ssc, in human steroidogenic tissues. Proc. Natl. Acad. Sci. USA 84, 1590-1594.

Voutilainen, R., & Miller, W.L. (1987b). Developmental and hormonal regulation of mRNAs for insulin-like growth factor II and steroidogenic enzymes in human fetal adrenals and gonads. DNA 7, 9-15.

Walicke, P.A. (1988). Basic and acidic fibroblast growth factors have trophic effects on neurons from multiple CNS regions. J. Neurosci. 8, 2618-2627.

Wanaka, A., Milbrandt, J., & Johnson, E.M. (1991). Expression of FGF receptor gene in rat development. Development 111, 455-461.

Wang, C-Y., Daimon, M., Shen, S-J., Engelmann, G.L., & Ilan, J. (1988). Insulin-like growth factor-I messenger ribonucleic acid in the developing human placenta and in term placenta of diabetics.

Mol. Endocrinol. 2, 217.

Wang, H.S., Lim, J., English, J., Irvine, L., & Chard, T. (1991). The concentration of insulin-like growth factor-I and insulin-like growth factor binding protein-1 in human umbilical cord serum at delivery: relationship to birth weight. J. Endocrinol. 129, 459-464.

Watson, A. J., Watson, P. H., Arcellana-Panlilio, M., Warnes, D., Walker, S. K., Schultz, G., Armstrong, S. T., & Seamark, R. F. (1994). A growth factor phenotype map for ovine premiplantation development. Biol. Reprod. 50, 725-733.

Wegmann T. G. (1990). The cytokine basis for cross-talk between the maternal immune and reproductive systems. Curr. Opin. Immunol. 2, 765-769.

Williams, L.T., Peters, K.G., Kikuchi, A., Hu, Q., MacNicol, A., Demo, S., Fields, S., Martin, S., & Muslin, A. (1994). Signalling molecules that mediate the actions of fibroblast growth factors. In: Embo Workshop on Molecular and Cellular Aspects of Fibroblast Growth Factors and Their Receptors. Capri, Italy (Abstract).

Wood, T.L., Streck, R.D., & Pintar, J.E. (1992). Expression of the IGFBP-2 gene in post-implantation rat embryos. Development 114, 59-66.

Yasui, S., Nagai, A., Oohira, A., Iwashita, M., & Konno, K. (1993). Effects of anti-mouse EGF antiserum on prenatal lung development in fetal mice. Ped. Pulmonol. 15, 251-256.

Yelavarthi, K. K., Chen, H-L, Yang, Y., Cowley, B. D., Fishback, J. L., & Hunt, J. S. (1991). Tumor necrosis factor-a mRNA and protein in rat uterine and placental cells. J. Immunol. 146, 3840-3848.

Zakar, T., Hirst, J.J., Mijovic, J.E., & Olson, D.M. (1995). Glucocorticoids stimulate the expression of prostaglandin endoperoxide H synthase-2 in amnion cells endocrinology 136, 1610-1619.

Zhou J., & Bondy, C. (1992). Insulin-like growth factor-II and its binding proteins in placental development. Endocrinology 131, 1230-1240.

THE IMMUNE SYSTEM:

A FUNCTIONAL PERSPECTIVE

Philip J. Morrissey

Growth Factors and Cytokines in Health and Disease
Volume 3A pages 55-125.

ISBN: 0-7623-0118-X

ABSTRACT

The immune system consists of a heterogenous collection of cells that are exquisitely designed to function as a means of protection from colonization by infectious agents. The cells comprising the immune system are described as well as their development and function. The immune system is tightly regulated by a system of signals transmitted by soluble substances (interleukins) as well as modulation of the expression of cell surface proteins. Normally, the immune system is relatively quiescent while performing surveillance. Detection of a pathogen results in the activation of the system, development of effector function designed to eliminate the invader and finally to establish a memory of that particular pathogen so that subsquent encounters result in a response that is both more rapid and of greater magnitude. The immune system has functional safeguards in place to prevent recognition of and reaction to self determinants. Failure in this system results in the development of autoimmune disease. There is a burgeoning effort to manipulate immune reactivity both in a positive and negative manner. It is hoped that immune adjuvants would be helpful in the battle against cancer and infectious disease and that immune depressants would be of use in autoimmune disease and transplantation.

I. INTRODUCTION

The immune system has been the subject of intensive research over the past 25 years. From the time Richard Nixon declared the "War on Cancer" in 1972 when the immune surveillance theory of cancer was popular, to the current era of focus on the pathogenesis of HIV-induced acquired immune deficiency, research to understand and manipulate immune system function has been on the forefront of both the scientific and lay press. In part, this is due to the number of pathological conditions that involve immune dysfunction and the potential to cure and prevent many human diseases if the ability to exogenously regulate the immune system were acquired. Thus, the vast impact of infectious diseases might be minimized or eliminated with appropriate vaccines or the proper immune stimulants. Similarly, the pathology due to an inappropriate and exacerbated immune responses such as occurs in arthritis, systemic lupus erythematosous, or multiple sclerosis to name a few could be ameliorated by *specifically* down-regulating the response to that antigen. Finally, of course, the immune system holds the much debated potential to eradicate neoplasia if the signals to turn on and direct the anti-tumor immune response were better understood. Thus, achieving the capability to modulate immune function would perhaps revolutionize clinical medicine. Indeed, in a popular television series about space explorers in the twenty-fifth century, infectious disease is quite often cured by applying a single dose of a "T cell stimulant." It is hoped that one day our understanding of immune system function will allow such simple manipulations to minimize the source of much suffering for mankind. This review will attempt to synthesize a functional model of the immune system based on current knowledge with an emphasis on the role of cytokines in its response and regulation. This work is not meant to be an all

Table 1. Glossary of Terms

CD	Cluster differentiation-a designation assigned to a cell surface antigen defined by a number of independently derived monoclonal antibodies. These are given numbers, for example, CD1, CD2, and CD3, that identify distinct gene products. There are currently more than 100 CD assignments. Histocompatibility antigens are excluded from this system.
Gene knock-out mice	Mice that have been altered genetically through targeted gene disruption in which a particular gene is mutated to a non-functional form through the process of homologous recombination (see Galli-Taliadoros et al. 1995).
"IL"-as in IL-2	"Interleukin"-the term given to secreted peptide hormones produced by immunocytes that influence the function of the immune system.
"L"-as in CD40L	Ligand.
Monoclonal antibody	An antibody of defined specificity produced by an immortalized, cloned B cell.
"R"—as in IL-2R	Receptor.
Transgenic mice	Mice that have been altered genetically by the introduction of foreign DNA which is stabley integrated into the germ line. The DNA encodes a known gene with a controlling element known as a promoter to direct the expression of the gene (see Palmitter and Brinster, 1986).

encompassing review of immune function either in its scope or in the work cited. In many cases, current reviews have been cited to provide the reader the opportunity to quickly explore and expand areas of interest in greater detail than space allows here. Also, for the non-immunologist, terms that may not be familiar are defined in Table 1.

II. IMMUNE SYSTEM—COMPOSITION AND STRUCTURE

The function of the immune system is to preserve the integrity of the host from colonization by unwanted microbes which find our bodies a hospitable environment in terms of warmth, moisture, nutrient availability, and neutral pH. The immune system consists of a collection of mobile cells and specific clusters of tissue such as lymph nodes and spleen, whose primary function is to detect, isolate, and eliminate foreign substances. The lymph nodes and spleen in effect act as filters, screening both lymph and blood, respectively. Essentially all the tissues of the body (with a few exceptions) have lymphatic channels that drain fluid from interstitial spaces. It is worthwhile noting that the endothelial cells of the terminal lymphatic capillaries, in contrast to the arterial and venous capillaries, are not tightly connected, and thus allow the entry of high molecular weight substances into the lymphatic drainage system. In essence, the lymphatic capillaries are highly permeable. Lymph reenters the circulatory system via the thoracic duct which empties into the vena cava. The lymph nodes are located intermittently along the course of the lymphatic vessels and filtration of the lymph is a primary function of the lymph nodes. Early experiments tested this capacity by perfusing the popliteal lymph node of a dog with a solution containing high numbers of bacteria. One passage through the node reduced bacteria numbers by 99% (Drinker et al., 1934). Thus, the lymph node and spleen function to retain particulate matter in an environment where exposure to lymphocytes is fairly well certain. Rapid assessment of the "foreign" nature (also known as antigenicity) of the retained material occurs through processes detailed below and immune activation can result. For localized foci of infection, immune reactivity occurs primarily in the draining lymph node, and as a consequence of recruitment of lymphocytes and intense proliferation during immune system activation, significant enlargement of the lymph node can occur.

The function of the activated immune system is to accelerate the degradation and elimination of foreign material. The immune system accomplishes this by utilizing an array of potent "effector" functions to kill, incapacitate, or otherwise mark pathogens for elimination (summarized in Table 2). For instance, cells of the immune system have the potent capability of directly killing cells by a number of diverse mechanisms. This is a very useful function if the cell has been determined "infected" by a pathogen. Direct cell mediated cytotoxicity is a function of activated $CD8^+$ T cells, NK cells, and γ/δ T cells. These cells can deliver what has been termed "the kiss of death" to target cells. This process is dependent on cell-cell contact and the release of substances by the killer cell that penetrate and cause the death of the target cell. Macrophages and polymorphonuclear leukocytes also have the capacity to kill microbes by the generation of toxic molecules and degradative enzymes. These cells can either expel the contents of granules into the extracellular space (exocytosis) or attempt to kill an intracellular microbe in a process dependent on phagocytosis of the microbe. Once the microbe is captured and held intracellularly in a phagolysosome, the cells are capable of introducing an array of different toxic substances (reactive oxygen and nitrogen intermediates, hydrolytic enzymes) into the phagolysosome in an attempt to kill the phagocytosed microbe. Another major effector mechanism involves antibodies which, as soluble molecules, function to clear the system of free floating or soluble antigen. It is well known that antigen/antibody complexes are cleared from the circulation significantly quicker than antigen alone. In addition, antibody can bind to antigen on cell surfaces, for instance, viral determinants on normal cells or determinants on the microbes themselves. This antibody coating of cells or microbes essentially marks them for immune elimination in a number of ways. First, macrophages have cell surface receptors for the tail end of the antibody molecule (termed Fc portion) and binding of the Fc portion of antibody by macrophages greatly facilitates the process of phagocytosis. The

Table 2. Effector Function

Macrophage-mediated intracellular killing	In response to bacterial products and cytokines, macrophages up-regulate intracellular killing mechanisms.
Neutrophil exocytosis	In response to inflammatory signals, these cells degranulate and significantly alter the extracellular environment with anti-microbial agents.
Antibody binding	Binding of antibody to antigens displayed on bacteria, viruses, or cell surfaces induces complement binding and induces lysis or phagocytosis.
Complement	These series of serum proteins become activated when antigen-antibody complexes form. Complement activation leads to direct cell death or enhanced phagocytosis by macrophages with receptors for activated complement.
Cell-mediated cytotoxicity	Direct cell contact is required and the killer cell delivers a lethal hit by exocytosis of preformed effector molecules-the cytolysin/perforin proteins and granzyme proteases. Death may also be triggered by activation of the Fas/Fas ligand pathway which induces apoptosis.
Antibody-dependent cell cytoxicity (ADCC)	Effector cells bind Fc portion of antibody molecules bound to antigens expressed on cell surfaces (e.g., viral peptides). This induces the release of effector molecules (see above).

second means of elimination utilizes a series of serum proteins collectively known as complement. The binding of the first component of the complement series to the Fc portion of bound antibody initiates a cascade of protein binding and activation by the other complement components that results in lysis of the cell. Finally, lymphocytes can bind the Fc portion of antibody bound to cells and lyse cells in what is termed ADCC or antibody-dependent cell cytotoxicity. The description of these potent effector mechanisms brings to thought questions as to the mechanisms that control their activation. If the discrimination of infected and normal cells is somehow clouded, these potent effector mechanisms can be directed at normal tissue and dire consequences often occur. One need only to observe a patient with severe rheumatoid arthritis or consider the immediate consequences of a bee sting in a hypersensitized patient to begin to understand the power of the immune system. The remainder of this chapter will describe in further detail the cells and molecules of the immune system and detail our current knowledge of the pathways involved in its regulation.

III. THE IMMUNE SYSTEM AND MHC—MECHANISM FOR SURVEILLANCE

The existence of the major histocompatibility complex (MHC) has been appreciated for many years by transplantation biologists who observed that differences at this genetic locus were the main obstacle to the free exchange of skin grafts in mice. Today transplantation surgeons would affirm that the greater the genetic identity at this locus between donor and recipient the greater the probability of graft acceptance. Much of the early experimental work done to elucidate the genetics of the MHC took advantage of the large number of inbred strains of mice. An inbred strain is one in which the genes at all loci are homozygous and this is accomplished through 20 generations of brother/sister breeding. Due to the relatively short generation time of mice, geneticists were also able to produce an array of interesting congenic (a strain that differs from another at one defined genetic locus) and recombinant inbred lines of mice (a strain often discovered during the generation of a congenic line in which a recombination event has occurred in the loci of interest) that allowed the elucidation of the genetic structure of the MHC. Also, immunologists taking advantage of the availability of these tools began to map genes that influenced immune responses to a defined antigen as well as genes that apparently allowed for cell-cell interaction during immune responses to the MHC. One set of experiments that particularly illuminated the importance of the MHC in this process was performed in a system of host resistance to viral infection in mice (Zinkernagel and Doherty, 1974). In these experiments, MHC strain "A" and MHC strain "B" mice were infected with a virus (V1) as depicted in Figure 1. Here it was seen that cytotoxic T cells that developed in response to virus infection in strain "A" mice could kill virus infected strain "A" target cells, but not virus infected target cells from strain "B" and vice versa. In further experiments, the genes that determined susceptibility to recognition and thus lysis, mapped to the MHC. The important implication of these results was that the cytolytic T cells needed to see the appropriate MHC gene product as well as virus.

Others working with antibodies generated against genetically defined areas of the MHC were able, through the method of immunoprecipitation, to begin to define the structure of MHC molecules. The serologically defined antigens of the MHC are known as H-2 in the mouse and HLA in the human. The deduced genetic organization of these regions that have subsequently been confirmed by sequencing is shown in Figure 2. In the mouse the genes of the H-2 complex are on chromosome 17 and in man the HLA genes are encoded on chromosome 6. From the early immunological and genetic studies, two classes of MHC molecules were defined. The Class I molecules were those that were largely responsible for skin graft

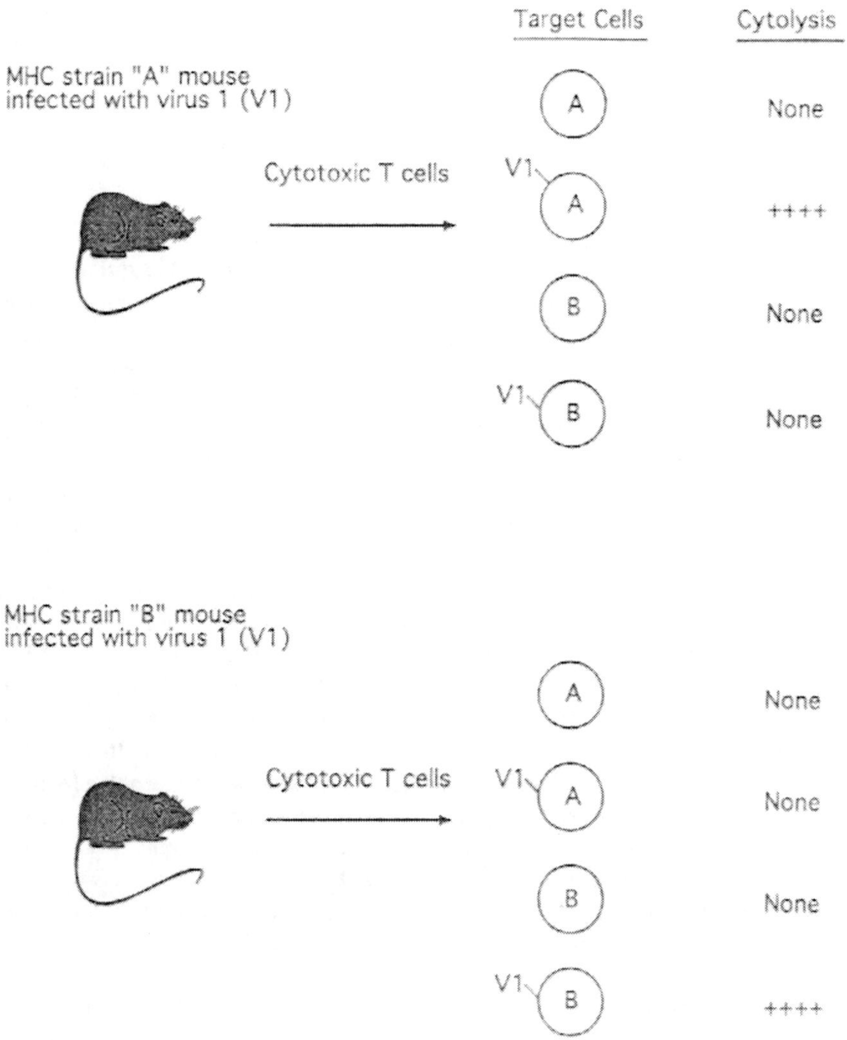

Figure 1. MHC-restricted recognition of virus "V1" by cytotoxic T cells.

rejection between donors differing at these loci as well as susceptibility to lysis by cytotoxic T lymphocytes (Figure 1). The Class II molecules were those that stimulated a strong proliferative response between lymphocytes from donors that differed genetically at this locus. Both MHC Class I and II proteins are structurally members of the Ig supergene family of proteins and are type I integral membrane glycoproteins (Figure 3). Class I MHC is a polymorphic glycoprotein of about 46

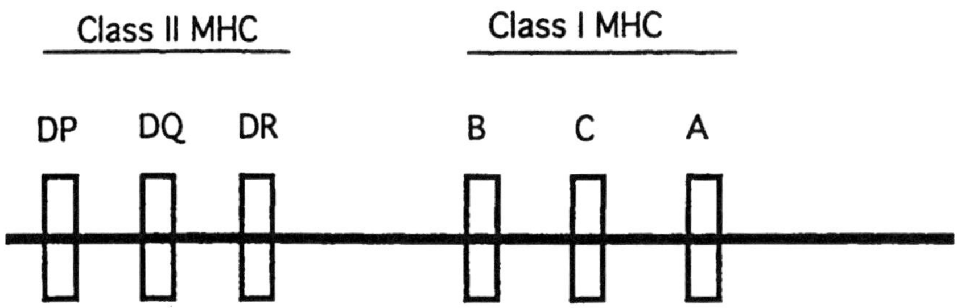

Genetic map of murine MHC.

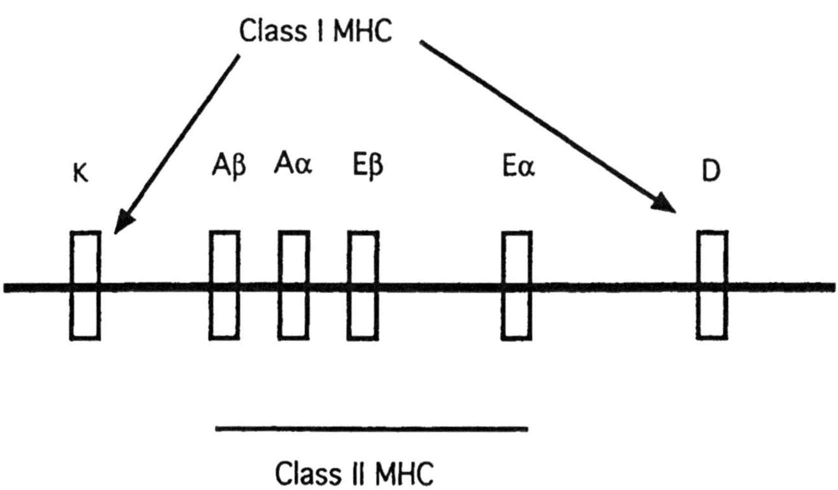

Figure 2. Genetic map of human major histocompatibility complex (MHC) (also known as HLA—human leuckocyte antigens).

kDa noncovalently associated with a 12 kDa glycoprotein known as β_2-microglobulin. Class I MHC is expressed on virtually all cell types. Class II MHC molecules are membrane spanning heterodimers consisting of an α and β polypeptide chain in noncovalent association. Class II MHC is constitutively expressed on

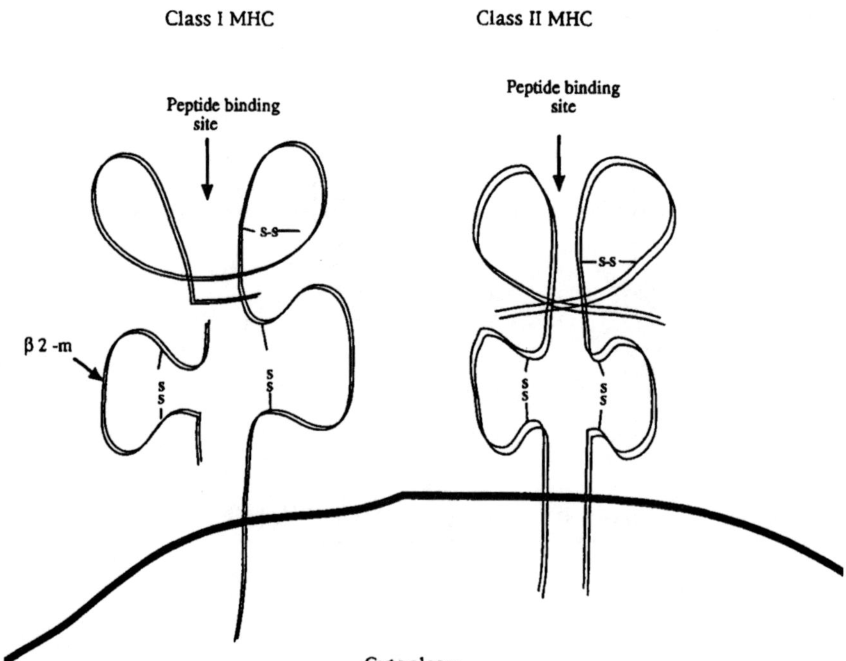

Figure 3. Schematic representation of Class I and Class II MHC molecules.

macrophages, dendritic cells, and B cells. Class II MHC expression can be induced in many other cell types in response to injury, infection, and/or inflammatory cytokines. The genes encoding Class I and II MHC molecules are perhaps the most polymorphic of any gene family yet studied.

Immunologists long appreciated that T cells appeared to recognize antigen in the context of MHC molecules. Also, as the structure of antigens became better understood it was determined that only a small fragment of antigen was recognized by T cells in the context of MHC molecules. Babbit and Unanue demonstrated that an antigen fragment can form a stable complex with purified MHC Class II encoded molecules (Babbitt et al., 1985). They found that the binding was saturable, competeable, and displayed moderate affinity (10^{-6} M). However, the exact nature of the binding was not realized until the three-dimensional crystallographic structure of MHC Class I was determined in the late 1980s (Bjorkman et al., 1987). The structure produced strong evidence for a binding site that could contain a small peptide. This binding site on the Class I MHC molecules consisted of a groove situated on the outermost region of the molecule (Figure 3). It is physically defined by a foundation consisting of a β pleated sheet with walls defined by two parallel α helixes. Interestingly, the polymorphisms found in various MHC Class I alleles are clustered in the areas that determine the putative peptide binding sites. The crystal structure of MHC Class II shows the presence of a similar groove with some unique characteristics that allow binding of longer peptides (Brown et al., 1993). Shortly after the structure of Class I MHC was elucidated it was found that low molecular weight peptides (< 3kDa) could be isolated from purified MHC proteins by decreasing the pH of the solution. The methodology of affinity purifying MHC molecules with specific monoclonal antibodies, isolating the associated peptides with acid extraction and advances in sequencing techniques lead to a rapid explosion of knowledge in this area (Joyce and Nathanson, 1994). Peptides from both Class I and Class II MHC molecules have been isolated and sequenced (Engelhard, 1994; Barber and Parham, 1993). They represent an array of normal cytoplasmic or nuclear peptides (Class I MHC) and extracellular or membrane components for Class II MHC (Falk et al., 1991). Peptides associated with Class I MHC protein are uniformly 9-11 amino acids in length whereas Class II MHC binds peptides that range between 11 and up to 24 amino acids in length. The difference in peptide size is thought to be due to the different geometry of the binding site on the MHC encoded molecules. In addition, studies have led to the designation of residues within a peptide that are critical for MHC binding as well as residues that are critical for T cell recognition of the peptide (Rotzchke and Falk, 1994). It is currently estimated that a given Class I MHC molecule can bind approximately 2-10 x 10^3 different peptides and that given consideration of the level of Class I MHC expression (1-10 x 10^5 molecules per cell), percent occupancy and diversity of the peptide repertoire, a single peptide Class I MHC complex is calculated to be present in the range of 100-4,000 copies/cell.

IV. CELLS OF THE IMMUNE SYSTEM

A. Polymorphonuclear Leukocytes

The term polymorphonuclear (PMN) leukocytes is derived from visual examination of these cells in that they have nuclei that are segmented. This feature is thought to facilitate mobility especially in the ability to exit from blood vessels into connective tissue (extravazation). There are three classes of PMNs based on staining of cytoplasmic granules. These cells have a relatively short life span (days) and are continually produced by differentiation from immature precursors in the bone marrow or spleen. PMNs are generally not discussed in the context of the immune system, however, the singular observation that increased susceptibility to infection correlates strongly with decreased neutrophil numbers in patients undergoing chemotherapy demonstrates that these cells are not an indispensable component of the immune system (Pizzo, 1984). Recent clinical experience with the administration of hematopoietic growth factors that accelerate the recovery of neutrophil numbers in these patients have resulted in a reduced incidence of infection (Reed, 1994). PMNs function to provide a rapid response team acting soon after infection in an attempt to eliminate or minimize the spread of the infectious agent. Their response is rapid and non-specific. Circulating neutrophiles sample the array of adhesion molecules displayed on the surface of vascular endothelial cells. In response to increased expression of existing adhesion molecules, expression of novel adhesion molecules, and a gradient of cytokine concentrations, these cells are capable of transmigration from the blood stream into the connective tissue (Bevilacqua, 1993). Once at the site of infection PMNs can become activated in response to cytokines produced by damaged or infected cells, activated complement components, as well as bacterial derived products (Baggioloni et al., 1993). Upon activation, neutrophiles produce substantial amounts of superoxide (O_{2-}) which is thought to be the mainstay of the anti-microbial arsenal (Badwey and Karnovsky, 1980). Patients whose PMNs are unable to produce oxygen radicals due to a genetic defect have chronic, life-threatening problems with persistent infections (Curnutte, 1992). Interestingly, an exaggerated response by neutrophiles is thought to be responsible for massive tissue damage in response to reperfusion after an ischemic episode. In these situations (e.g., oxygen deprivation due to frost bite or circulatory shock), tissue injury results in increased expression of vascular endothelium adhesion molecules and consequent influx of neutrophiles. Tissue damage is then exaggerated by the activation of these cells, leading to extensive necrosis. Recent experimental strategies have prevented this damage by administering monoclonal antibodies against vascular adhesion molecules (Korthius et al., 1994). This prevents neutrophil extravazation into the damaged tissue by masking vascular endothelial cell adhesion receptors. It is an interesting realization that the destructive tissue injury seen as a consequence of ischemic situations is mediated by an inappropriately exaggerated response by PMNs. In the absence of this response, significant tissue recovery can occur.

B. Macrophages

Macrophages are characterized by an extensive capacity to phagocytose foreign particles as well as dead or damaged cells. They are also capable of killing phagocytosed intracellular microbes (Kaufmann, 1993). Macrophages as well as PMNs are part of the host's early response team that reacts quickly and non-specifically to injury or infection. They are mobile cells that can traverse tissue to the site of inflammation and inflammatory foci contain a significant proportion of macrophages. These cells are well prepared to participate in the early response to infection. They express cell surface receptors for complement components and the Fc portion of antibody molecules as well as Class II MHC. The cell surface expression level of these is increased as a result of activation. They also bind and are activated directly by bacterial components, the best characterized being the response to lipopolysaccharide (LPS) derived from gram negative bacteria. They secrete a wide variety of enzymes and proteins such as elastase, collagenase, and complement proteins. They are also a significant source of cytokines that influence immune function such as IL-1, TNFα, IL-10, and IL-12. Thus, macrophages function, in a sense, as an intermediary, bridging the gap between the use of non-specific phagocytosis as the primary means of host protection by providing important signals to the specific branch of the immune system (i.e., lymphocytes). Mature macrophages isolated from the spleen are well known for their ability to process and present antigen to T cells, resulting in T cell activation, cytokine secretion, and proliferation (Rosenthal and Shevach, 1973; Ziegler and Unanue, 1981).

Macrophages are distributed throughout all tissues of the body (resident macrophages). Some organs such as the liver and spleen have significant macrophage populations. For instance, the sinusoids of the liver are lined with macrophages that have been termed Kupfer cells. These cells are well positioned to insure initial contact with substances absorbed from the gastrointestinal tract. In contrast, the kidney has very few resident macrophages and those are associated with the mesangium of the glomeruli. Thus, there are probably tissue specific macrophages with duties that are tailored to the function of that particular organ. Immunologists have long appreciated the fact that macrophage populations differ in their ability to present antigen. For instance, resident peritoneal macrophages, in the absence of activation, are not as efficient in antigen presentation and some macrophages even suppress lymphocyte function (macrophages from newborn spleen). Macrophages are continually generated from immature precursors in the bone marrow throughout life and are thought to have a moderately long half-life (months in mice). A proportion of circulating WBCs consists of relatively immature macrophages and it is thought that these cells are in transit to tissue.

C. Dendritic Cells

Dendritic cells are specialized cells important in antigen presentation and lymphocyte activation (Steinman, 1991). Like macrophages, they are distributed throughout the body. They are identified morphologically and by a distinct array of cell surface molecules. They are a very rare population of cells that are difficult to purify, grow, and study. Only recently have conditions been described for the routine growth of dendritic cells *in vitro* (Inaba et al., 1992) Purified populations of dendritic cells are, comparatively, the most potent stimulators of T cells. The mechanism of stimulation is not yet clear, but it is known that dendritic cells express high levels of Class II MHC and co-stimulatory molecules. Evidence exists that dendritic cells are motile and can carry antigen from tissue into the draining lymph nodes. Data supporting this derives from studies in which antigen is applied to skin and movement of Langerhans cells (dendritic cells residing in the epidermis) was observed (Macatonia et al., 1987). Dendritic cells can process and present antigen although their phagocytic capacity appears to be minimal. A unique population has been identified in the germinal centers of lymph nodes (follicular dendritic cells). It is thought that these cells play an important role in B cell activation, isotype switching, affinity maturation of Ig, and memory cell formation. These cells seem to have a unique ability to bind and retain antigen in the form of immune complexes.

Thus, although much progress has been made over the past few years concerning this population of intriguing cells, there are still many unanswered questions about the exact role they play in initiating and maintaining immune responses.

D. Lymphocytes

Lymphocytes comprise the arm of the immune system that is responsible for the specific response to foreign substances which are also termed antigens. Individual lymphocytes are specialized in that they recognize, bind, and respond only to a unique structural component (known as an epitope) of an antigen. This is determined by the specificity of membrane receptors (antigen receptors). Importantly, the antigen receptors expressed by an individual lymphocyte all have the same recognition specificity. The antigen receptor is different on individual lymphocytes and the capacity of the immune system to respond to a wide variety of antigens is a consequence of having many lymphocytes expressing antigen receptors with different specificities. Any antigen that enters the system selectively binds to and activates only the lymphocytes with receptors of appropriate specificity. This is known as the "clonal selection" theory of immune function.

Lymphocytes are divided into two broad classes or lineages. The B (bone marrow) lymphocyte are the precursors of the antibody secreting cells. The T (thymus dependent) lymphocytes contain a number of subsets with different functions. These will be discussed separately.

E. B cells

There are at least three distinct B cell populations that have been identified in the mouse. The initial wave of B cell production in the mouse occurs in late fetal and early neonatal development. These B cells (known as B1 cells) have an unusual phenotype (expression of the T cell restricted marker Ly1), unique anatomical site (frequent in the peritoneal cavity, rare in the lymph node and spleen), limited use of immunoglobulin V region genes, extensive capacity for self renewal, a tendency for secretion of autoreactive antibodies, increased frequency in mice predisposed to the development of autoimmune disease, and an apparent increased frequency of neoplastic transformation (Hayakawa and Hardy, 1988; Kantor and Herzenberg, 1993). Most B lymphocyte lymphomas in mice and man display the phenotype of this early B cell population. The production of these B cells seems to be unique to that period in development. These cells do not reappear after being eliminated by lethal whole body irradiation and reconstitution by adult bone marrow. The physiological role that these cells play in the normal or pathological function of the immune system is not clear at this time.

There are two populations of conventional B cells (B2 cells). Both appear ontologically after the initial wave of production of $Ly1^+$ B cells (B1 cells). One of the populations represents the vast majority of splenic or lymph node B cells that produce antibody after primary encounter with antigen, but they do not appear to populate the germinal center or contribute to the memory B cell pool (Klinman, 1994). These B cells are produced continually through adult life by differentiation in the bone marrow from immature precursors. It is estimated that the daily production of B cells from the bone marrow by a young mouse is around 5×10^7 cells. The second population of B cells residing in spleen and lymph node exhibits characteristics of memory B cells (extensive Ig secretion and predominant use of downstream isotypes such as γ heavy chain). These cells have been phenotypically identified among other B cells by their high levels of expression of HSA (Linton et al., 1989). They represent 5-10% of the B cell population in the spleen and lymph node. Not much is known about their development as phenotypic markers to identify these cells at earlier stages of development. Also, it is still possible that this memory cell population is derived in some way from the B cells that form the majority of the sIg+ population in the spleen and lymph node.

F. T cells

One main function of T cells is to regulate the effector function of the immune response. This has been classically known as T cell "help" or T cell "suppression." The T cells in the peripheral immune system (spleen and lymph nodes) can be divided into two distinct populations based on the mutually exclusive cell surface expression of CD4 or CD8. $CD4^+$ T cells recognize antigen in the context of Class II MHC and $CD8^+$ T cells recognize antigen in the context of Class I MHC (Figure 4). It has been shown that the CD4 molecule on the surface of the T cells binds to Class II MHC on the surface of antigen presenting cells (APC). Similarly, CD8 binds to Class I MHC. It is thought that this binding stabilizes or increases the affinity of binding between the TCR and MHC molecules on the APC. The $CD4^+$ population largely contains cells with the capacity to regulate immune function. In fact, $CD4^+$ T cells have been termed the "conductors of the immunological orchestra." Their response to signals provided by other cells (macrophages, dendritic cells) by and large determines the type of effector response generated. Selective elimination of these cells severely compromises antibody production or cytotoxic T cell generation in response to antigen. The $CD4^+$ T cells can also mediate the delayed type hypersensitivity response. $CD8^+$ T cells are the precursors of the classic cytotoxic T lymphocyte (CTL). In response to stimulation with antigen and cytokines (usually provided by the $CD4^+$ helper T cells), these cells develop cytolytic activity.

G. γ/δ^+ T cells

The immune system has a widely dispersed detection system lining the epithelial surfaces of the body where contact with infectious agents is most likely to occur. Unique populations of T cells produced by the thymus early in development home to various epithelial surfaces. These T cells use an alternate set of genes to form

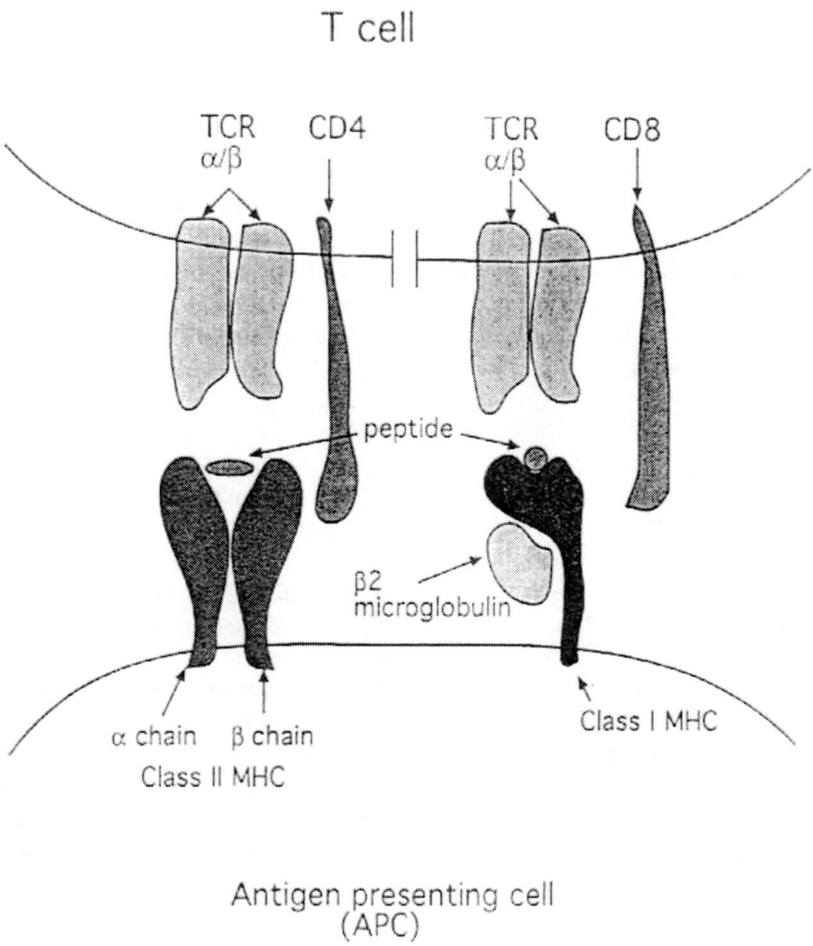

Figure 4. T cell receptor—MHC interactions.

their antigen specific receptor. Most T cells utilize the TCRα and TCRβ gene loci, but these early T cells utilize the TCRγ and δ loci (Allison and Havran, 1991). These loci are unique because of the comparatively limited variable (V) gene repertoire. Uniquely, V gene usage and ultimate tissue localization of the cells occurs in defined sequences during ontogeny (Table 3). These cells are capable of secreting cytokines and lysing target cells in response to activating signals (Haas, 1993). How these cells recognize antigen through their cell surface receptor is not clear although it is becoming apparent that it may be fundamentally different than how

Table 3. V Gene Usage by γδ T Cells

Tissue	*Vγ*	*Vδ*
Skin	V_3	V_1
Reproductive tract	V_4	V_1
Intestine	V_5	$V_{5,4,6,7}$
Lung	V_2, V_4	$V_{6,5,4,7}$

TCRα/β binds to peptide/MHC complexes (Kronenburg, 1994). For instance, Schild et al. (1194) have shown that antigen processing is not required for recognition by TCRγ/δ T cells and that the MHC epitope recognized by the γ/δ T cell is distinct from that recognized by α/β T cells. Recently, the importance of a series of Class I MHC-like molecules has been appreciated. These molecules, known as ClassI-B MHC molecules, are structurally related to but different from Class I MHC. They are also expressed only on selected tissue (Shawar et al., 1994). Recent evidence suggests that the ClassI-B MHC may bind non-peptide antigens, such as bacterial derived lipids (Beckman et al., 1994; Tanaka et al., 1994). This would provide a mechanism for the direct presentation of microbial components in association with MHC molecules that could trigger T cells. In addition, epithelial cells are known to preferentially express MHC ClassI-B molecules. Thus, γ/δ T cells may act as early surveillance units interspersed among the epithelial cells lining the body, detecting and eliminating antigen on infected epithelial cells before it can spread systemically. Quite recently a novel aspect of γ/δ T cell function was revealed. It was shown that intestinal intraepithelial γ/δ, but not α/β T cells can produce keratinocyte-growth factor, a major stimulus of epithelial cell growth. These results suggest that γ/δ T cells can also affect the repair of damaged epithelial tissue or hasten its regeneration.

H. Natural Killer (NK) Cells

NK cells were first identified as their name implies from spontaneous levels of cytotoxicity by freshly explanted spleen cells that were reproducibly noticed on certain tumor cell targets. NK cells are "lymphocyte-like" in morphology and express a number of markers such as CD16, CD56, CD11a, CD11b, and CD2, although none uniquely identify them. Since these molecules are not unique to NK cells, their identification has relied heavily on function. NK cells are negative for CD3 and T cell antigen receptor expression. T cell receptor gene rearrangement is not necessary for NK cell production as they are present in mice genetically incapable of rearranging T cell antigen receptor genes. Also, the production of NK cells is independent of a functioning thymus. Another major function of NK cells is to secrete cytokines and they have been shown to produce TNFα, γ-IFN, and GM-CSF. NK cells respond to IL-2 and IL-12 by proliferating and increasing cytolytic function. To date not much is known about the mechanism or regulation of NK cell production, differentiation, or life span. It is known that they are indeed of lymphoid origin since mice with the targeted disruption of the Ikaros gene, which is a lymphoid-specific transcription factor and the earliest expressed gene detected thus far in cells committed to lymphoid development, lack NK cell function, presumably due to an absence of NK cells (Georgopolous et al., 1994).

The target structure recognized by NK cells has not been conclusively identified. In the experimental system known as "hematopoietic hybrid resistance" in which parental bone marrow is used to reconstitute lethally irradiated F1 mice, radioresistant NK cells are believed to reject the parental bone marrow cells which is contrary to the established laws of transplantation (Cudkowicz and Bennett, 1971). This phenomena is only seen in certain strain combinations that lead to the identification of a gene known as Hh that maps to the MHC. The continual advancements in molecular biology provided tools to further analyze the role of MHC molecules in NK cell mediated lysis. In general, it was found that the susceptibility of cell lines to lysis or bone marrow to rejection by the host was inversely proportional to MHC expression and resistance could be conferred by transfecting and expressing the appropriate MHC gene (Ohlen et al., 1989; Bix et al., 1991). Thus, it appeared as if the recognition of MHC by NK cells downregulated cytotoxic activity and that NK cells react to and lyse cells that are either not expressing MHC class I molecules or expressing them in markedly reduced amounts (Yokoyama, 1993; Raulet, 1992). Recently, genes that are present on subsets of NK cells have been identified and thus far they seem to resemble each other (C-type lectin receptor) and function in the same manner (Trinchieri, 1994). Thus, it is hypothesized that NK cells express a primordial MHC recognition system that binds to a specificity common to a number (but not all) MHC Class I gene products. It is thought that NK cells express recognition structures with different specificities (in contrast to lymphocytes) and that the interaction of receptor with ligand (MHC) results in a negative signal with respect to cytotoxic function.

Through some mechanism though, NK cells must recognize and interact with self MHC and antigen. For instance, it has been shown that NK cells mediate an early response to infectious agents such as *Listeria monocytogenes.* In SCID mice that lack T cells and B cells, the early anti-microbial cytotoxic response by macrophages is stimulated by activated NK cells in a γ-IFN dependent manner (Bancroft et al., 1987, 1989; Wherry et al., 1991). Consistent with the importance of these cells in resistance to microbial infection is the observation that mice lacking NK cell function due to a disrupted Ikaros gene succumb at an early age to fulminant infection (Georgopolous et al., 1994). It is well known that mice lacking T cells (both TCRα/β and TCRγ/δ populations) and B cells can survive quite well (albeit in a pathogen free, but not germ free, environment). Thus, this suggests a unique and important role for NK cells in the resistance of the host to microbial agents.

I. Natural T cells (NT cells)

Recently, another class of T cells has arrived upon the immunological scene (Bix and Locksley, 1995; Bendelac, 1995). These cells, termed natural T cells, were first described years ago as classical T cells that lacked expression of CD4 and CD8. They represent a small percentage of lymphoid cells in the spleen and lymph node, but represent a considerable proportion of T cells in the liver. It is not yet clear if these cells require a functional thymus for maturation. These cells are unique in that they express TCRα/β as well as the putative recognition unit of NK cells. Interestingly, the level of expression of TCRα/β is less than that of normal T cells and a proportion may also express CD4. These cells have attracted significant interest because they secrete significant quantities of cytokines much more rapidly after stimulation than conventional T cells. This initial burst of cytokines may alert and activate non-specific resistance mechanisms as well as influence the eventual development of the specific immune response. Much remains to be learned about this interesting subpopulation of T cells.

V. IMMUNE SYSTEM SPECIFICITY—THE PARADOX RESOLVED

The ability of lymphocytes to recognize and respond to specific antigen relies on the expression of three sets of gene pairs in distinct lymphocyte populations (summarized in Table 4). There are the antibody genes (heavy and light chain genes) expressed in B cells, the T cell receptor α/β genes, and T cell receptor γ/δ genes. The rearrangement and expression of these genes to form recognition units on the cell surface is the essence of the specific immune system.

A. The B cell Antigen Receptor (the Antibody Molecule)

The antibody molecule consists of a heavy and light chain that are disulfide linked (Figure 5). Both chains can be subdivided into C (constant) and V (variable) regions. The antigen binding site is located in the V region at the N-terminal end of the molecule. Contributions from V regions of both heavy and light chains are necessary to confer antigen binding properties. Early sequencing efforts of monoclonal sources of Ig noted that amino acid changes were clustered in the V region

Table 4. Estimated Frequency of Ig and TCR Genes

Gene segment	*Immunoglobulin*		*T cell receptor*	
	heavy	*light*	α	β
V	100-500	10-80	100	30
D	30	—	—	2
J	6	4	50	6

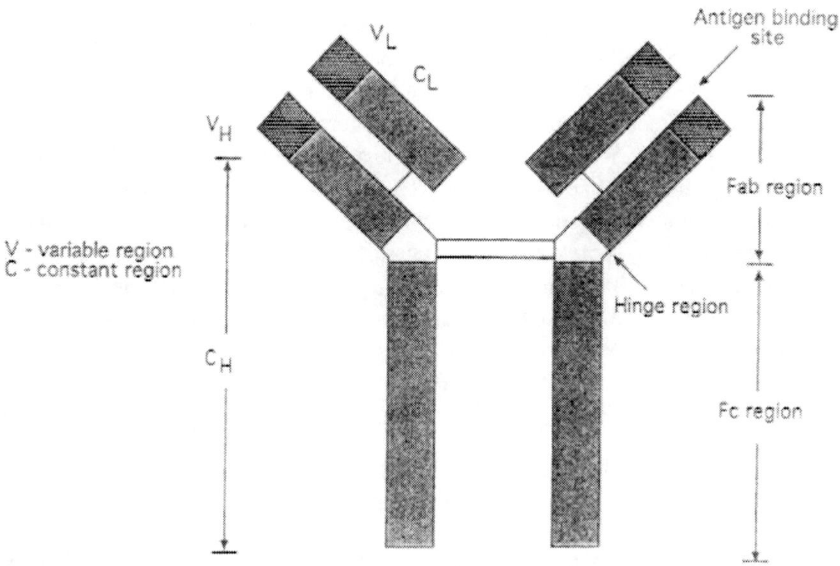

Figure 5. Functional architecture of the IgG molecule.

or antigen binding site of the antibody molecule whereas the C region was significantly more constant and changes could be attributed to defined C region families. Thus, the heavy chain can consist of a number of modular constant regions that confer different functional attributes to the Ig molecules.

For many years the genetic mechanism of the observed specificity and diversity of the antibody response remained a puzzle. There were essentially three different theories put forth to explain this. One postulate was that germ line DNA encoded the information for a large number of complete antibody genes. This would obviously mean that a considerable proportion of the genome would be dedicated to immunoglobulin genes. In contrast there could be a more limited number of genes that underwent somatic mutation as they were expressed in order to generate diversity. Finally it was postulated that antibody diversity was a consequence of gene rearrangement from a more limited number of genes. Each of these theories would predict different numbers of germ line genes for antibody molecules. As a result, many novel experimental strategies were devised in an attempt to accurately count the number of Ig genes. As the experimental prowess in molecular biology increased, it became clear that there was only a single C_H κ gene and multiple V region genes (Seidman et al., 1978a, 1978b). At about the same time, experimental evidence indicated that DNA in myeloma B cells was in a different configuration than in germ line cells (Hozumi and Tonegawa, 1976). One way that this difference could be understood was that if some genes in the B cells had been rearranged. Not long thereafter, the gene families encoding the antibody molecule were cloned and it was found that multiple gene segments code for the antibody molecule and that selected segments rearrange to produce a functional antibody molecule. Thus, multiple genes could contribute to the generation of one protein. The generation of diversity for antibody molecules occurs during B cell maturation via an orderly process of gene rearrangement. Rearrangement in the pre-B cell starts within the IgH (heavy chain) locus of either the paternal or maternal chromosome. The first gene rearrangement detected in B cell precursors is between the D and J regions of the heavy chain (Figure 6). This is followed by addition of one of the numerous V_H genes. If this occurs successfully, the entire VDJ segment and the Cμ gene are transcribed into an Ig heavy chain in the pre-B cell. It has come to be appreciated recently that during the process of VDJ joining different strategies are used to vary the nucleotide sequence and consequently the protein structure. For instance, additional nucleotides may be inserted at both the V_H-D and D-J_H junctions by the enzyme terminal deoxytransferase (TdT). Also, deletion of terminal nucleotides in

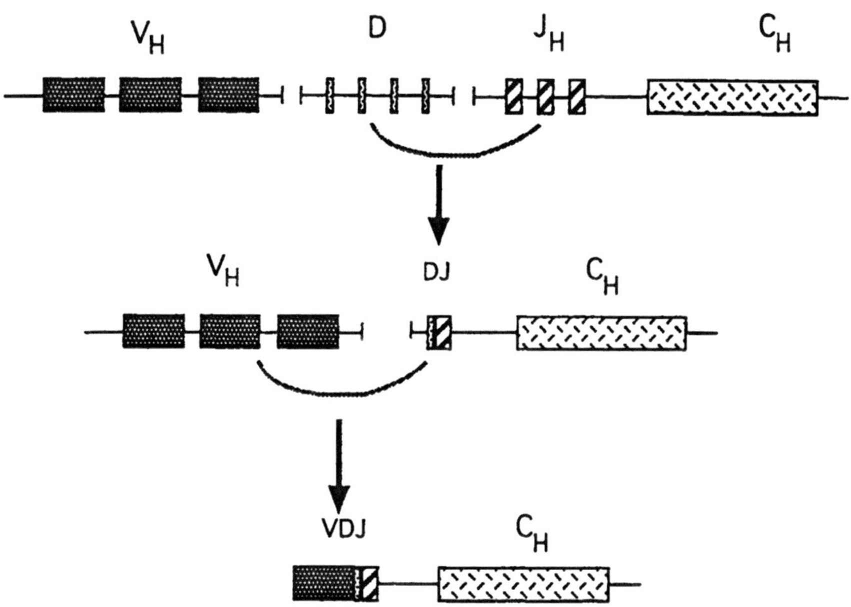

Figure 6. Immunoglobulin gene heavy chain rearrangement.

the rearranging gene segments has been observed as well as D gene segments rearranged in reverse orientation and fused to another D segment. These rather elaborate alterations of genetic material occur most frequently in the gene segments that encode the region of the antibody molecule that is in direct contact with antigen. At the DNA level, this encompasses the 3' end of the V_H gene, all of the D region, and the 5' end of J_H. This ability to insert and trim nucleotides at the part of the gene responsible for antigen binding greatly increases the diversity of antibody binding sites. It has been estimated that up to 10^{14} different binding sites are possible. Also, it means that antibodies with different binding characteristics could arise from identical gene segments that have been modified during recombination.

Light chain gene rearrangement occurs after heavy chain rearrangement and is functionally similar with the exception that there are no D region genes among the Ig light chain genes. Thus, the only rearrangement that occurs is between the V and J regions.

The C region used by the B cell can change with time, dependent on their maturational state and environment (summarized in Table 5). Newly differentiated B cells initially express cell surface Ig utilizing the μ heavy chain C region. Upon emigration from the bone marrow, the B cells coexpress IgD as well as IgM on their cell surface. Switching to other C_H region genes such as γ, ε or α occurs after antigenic stimulation and is dependent upon additional signals, such as co-stimulatory molecules, cytokines, or bacterial products (e.g., LPS; Figure 7). The various C_H regions impart significantly different functional capacities upon the antibody

Table 5. Composition and Function of Immunoglobulin Subclasses

Ig subclass	*Heavy Chain*	*Multimer*	*Function*
IgM	μ	Yes (5)	Low affinity; very effective at agglutination; produced mainly during primary antibody response; efficient at fixing complement.
IgD	δ	No	Present mainly on B cell surface; very low serum concentration; function unknown.
IgG	γ	No	Most abundant serum Ig; efficient at fixing complement; selectively produced during secondary immune response.
IgE	ε	No	Produced in response to parasitic infections; production is focused at external surfaces; binds mast cells and basophils via Fc portion and is responsible for atopic allergic reactions.
IgA	α	Yes (2-3)	Produced in mucosal tissue and is transported through lining epithelial cells into luminal spaces (e.g., intestine).

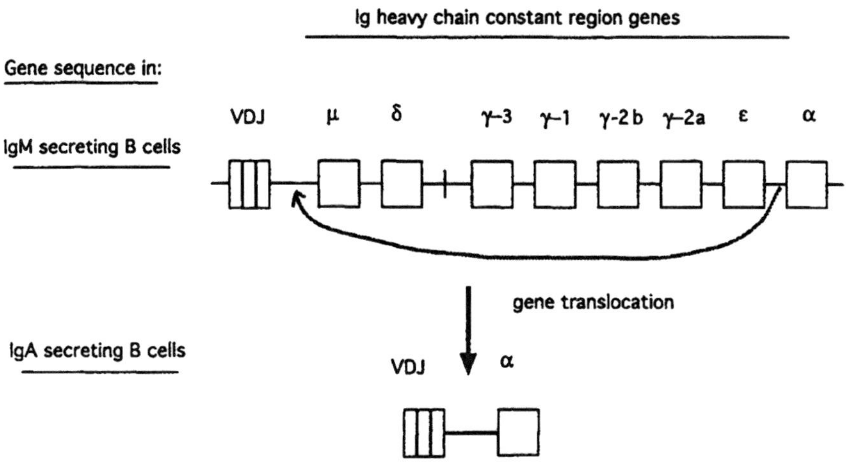

Figure 7. Organization and translocation of IgH chain genes.

molecule. Importantly, during this switching, the V region module used remains the same. For the light chain, the number of constant region genes is less and one is chosen for use during B cell maturation and this does not vary.

Further genetic modification of the V region of the antibody gene is seen in mature B cells after antigenic stimulation; mutations occur in the rearranged VDJ region that are dependent on the activity of enzyme TdT. This process results in the substitution of single nucleotides in the variable, but not constant regions of the Ig gene. It has been estimated that the mutation rate is extremely high (2-4%) for the VH genes. It is thought that this process which occurs in the germinal centers of lymphoid tissue under the influence of T cells and the presence of antigen results in an increase in the affinity of binding between the antibody molecule and antigen.

Thus, C_H region switching combined with the affinity maturation process through genetic alteration of the V region genes provides the immune system with a flexible and diverse effector molecule.

B. T cell Antigen Receptor

The antigen receptor for T cells remained an enigma for approximately 10 more years after the genetic composition of the B cell antigen receptor was elucidated. During this time, the theoretical structure of the receptor was hotly contested. It was well appreciated that antibody molecules could bind antigen in isolation and that T cells bind an antigen fragment in association with or in the context of MHC encoded molecules (see Figure 4). At that time, models existed for a single T cell antigen receptor that bound a combination of antigen and MHC; dual receptors, one with specificity for antigen and the other with specificity for MHC and even a model for a three component T cell receptor was proposed. It is now appreciated that the T cell utilizes a single antigen receptor consisting of two polypeptide chains (α and β) that bind a fragment of antigen bound to MHC encoded proteins. The polypeptide chains are 40-50 kDa in size, disulfide linked, and both have short intracytoplasmic chains. Both chains are required for antigen specificity. The diversity of the T cell antigen receptor is generated during thymic differentiation by programmed recombination of V region, D, J, and constant regions. The TCRα chain consisits is similar to the antibody light chain in that only V and J genes exist and rearrange. There is a D region which contributes to structure of the TCRβ chain, again through rearrangement with V and J regions (Davis and Bjorkman, 1988). Contrary to antibody genes there is only one constant region gene at the TCRβ gene locus and two constant region gene loci for the TCRα gene locus. In contrast to the antibody genes, T cell receptor genes do not undergo C region switching or further somatic mutation at the V region after expression and selection.

VI. SIGNAL TRANSDUCTION

Antigen binding by B and T cells with proper cosignals results in cell proliferation, cytokine gene expression, and, in some cell types, induction of effector function. This complex process utilizes intricate intracellular signaling pathways. It is known that the antigen receptors of both T cells and B cells have short intracytoplasmic domains with no inherent protein tyrosine kinase activity and that observation precludes significant function in intracellular signaling pathways. It has been long established that intracellular protein phosphorylation is associated with lymphocyte activation (Perlmutter et al., 1993). However since structural motifs associated with this pathway were not apparent on the intracellular domains of the antigen receptor, the mechanism of signal transduction upon antigen binding was not readily apparent. For both the T cell and B cell antigen receptor, multimeric protein complexes have been discovered that associate with the antigen receptors on the cell surface and that have appropriate cytoplasmic motifs for signal transduction (Figure 8).

The T cell antigen receptor exists as a complex of four pairs of proteins. In addition to the TCR α/β heterodimer which imparts antigen specificity, there are three dimeric groups of transmembrane proteins that are collectively known as CD3. The CD3 proteins are designated γ, δ, ε, ξ, and η. These transmembrane polypeptides are invariant and have long cytoplasmic tails. Of the proteins, CD3γ, δ, and ε are related evolutionary and exist as noncovalently associated $\gamma\varepsilon$ and $\delta\varepsilon$ heterodimers. The ξ and η components are distinct from the other CD3 proteins and exist as disulfide linked dimers. This family has three related members, CD3ξ, CD3η, and the γ subunit of the high affinity IgE Fc receptor. For thymocytes and the vast majority of peripheral T cells, the CD3 complex consists of $\gamma\varepsilon$, $\delta\varepsilon$, and $\xi\xi$

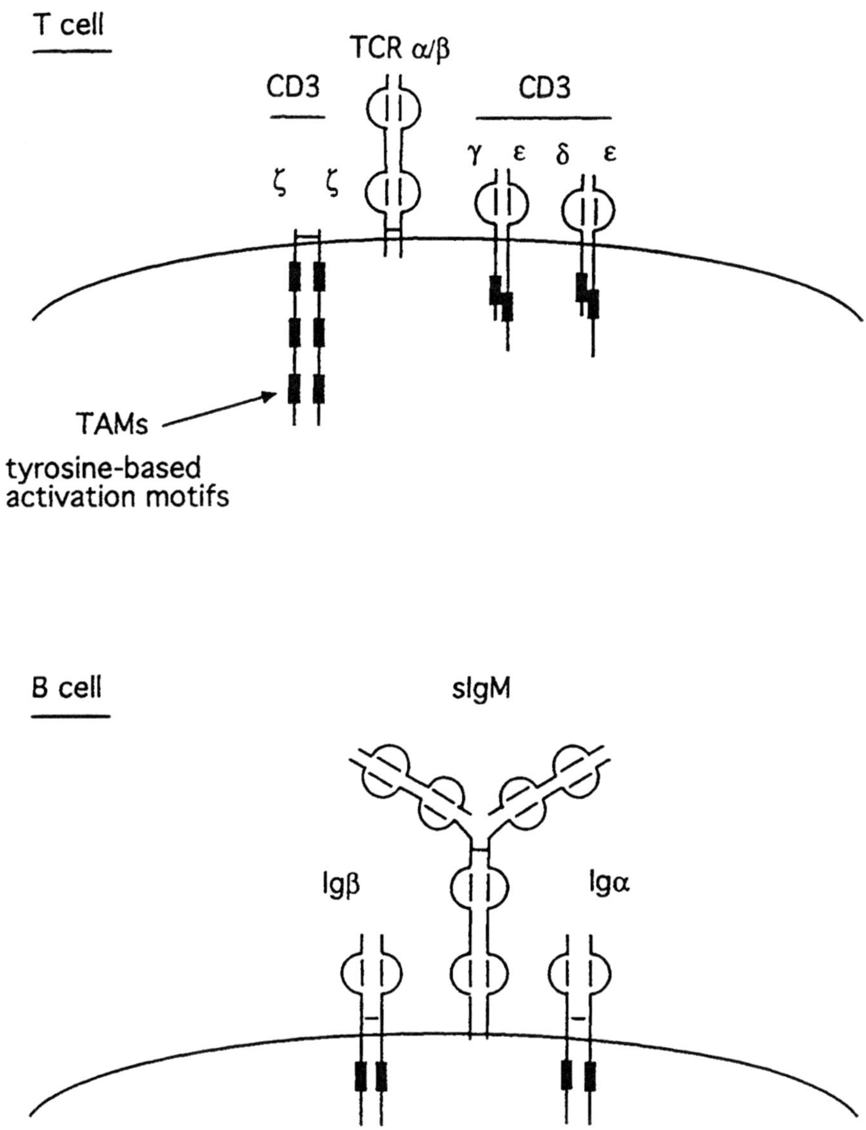

Figure 8. Representation of the T cell and B cell antigen receptor complex.

dimers (Figure 8). These dimers have protein sequence motifs known as TAMs (tyrosine-based activation motif) or ARAM (antigen-related activation motif). These motifs undergo phosphorylation upon antigen binding to the TCR and function to initiate a cascade of signaling events during cell activation. In different situations the composition of the CD3 heterodimers can change, for instance, in intestinal intra-epithelial lymphocytes and lymphocytes from tumor bearing mice, ξ is paired with η or the γ chain of the IgE Fc receptor (Mizoguchi et al., 1992; Guy-Grand et al., 1994). In both cases, the T cells respond poorly to classical activation signals, indicating that alteration of the components of CD3 complex can dampen the strength of the activation signal transduced. T cell activation is a complex and finely tuned process and determining the signals responsible for inducing changes in CD3 components and hence, activation will be vital to our ultimate understanding of T cell function.

B cells express sIg in a stable complex with the invariant disulfide linked homodimers, Igα and Igβ (Figure 8). Igα is the product of the mb-1 locus and Igβ is the product of the B29 locus (Cambier et al., 1994). These proteins have extensive cytoplasmic tails with TAMs that couple these proteins to intracellular phosphotyrosine kinases. It is not yet known how many Igα and Igβ homodimers are complexed with sIg or if the composition of the subunits changes as a reflection of the differentiative state of the B cell. The signal transduction process in the B cell is complicated by the coexpression of sIgM and sIgD. IgD differs structurally from IgM in having a larger and theoretically more flexible hinge region. Thus, it could more easily bind determinants presented at irregular spatial intervals. In various experimental situations, selectively triggering through sIgM has lead to a "negative" signal whereas triggering through sIgD usually tends to activate B cells. It is postulated that the signaling modules sIgM and sIgD associate with may each have a unique component that is capable of altering the signal and there is some recent evidence to support this (Kim et al., 1994). However, it should be remembered that both surface IgM and IgD have the same V region and thus, binding specificity, so signaling uniquely through one Ig type most probably does not occur. Recent evidence with transgenic mice indicates that sIgM negative or sIgD negative B cells develop and function normally (Brink et al., 1992). Also, mice unable to express sIgD due to targeted gene disruption show a very subtle phenotype. B cell numbers and responses are within normal parameters (Nitschke et al., 1993). Although the affinity of the serum antibody was less in the knock-out mice (Roes and Rajewsky, 1993). Interestingly, as B cells mature and become stimulated, levels of sIgM decrease relative to sIgD, consistent with a role for sIgD in stimulating B cell responses. However, all of these observations have yet to reveal a coherent role for both surface IgD and IgM in B cell function.

It is interesting to note that in both B cells and T cells, there is a separation of the ligand binding module from the signal transduction module. This suggests the existence of a mechanism to fine tune responsiveness, but the "nuts and bolts" remain to be identified and precisely understood.

VII. LYMPHOPOIESIS

Both B and T cells are derived from the multipotent hematopoietic stem cell. This stem cell population, which is self-renewing, gives rise to committed precursors that differentiate into cells comprising the functional hematopoietic/immune system. Recent progress in this area involves lineage-specific markers that have been identified and are allowing a better understanding of phenotypic commitment as well as cytokine requirements.

A. B cell Production

The development of B cells primarily involves the process of rearranging Ig genes and expressing cell surface Ig. Included in this differentiation is the process of deleting B cells with receptor specificity to self-determinants as well as an ill-defined selection step. B cell differentiation from immature precursors has been studied in detail by Hardy et al. (1991) who was able to categorize distinct phenotypic stages of development based on cell surface marker expression, Ig heavy chain rearrangement, and stromal cell and cytokine dependence (Figure 9). Immunoglobulin gene rearrangement is necessary for B cell development since disruption of Ig heavy or light chain loci by targeted gene disruption results in a block in B lymphopoiesis at the pre B cell stage (Melchers et al., 1994). Heavy chain rearrangement is detected first in developing B cells. Initially, the Ig heavy chain is not paired with an Ig light chain as is seen in mature B cells or secreted antibody (Figure 5), but it is paired with a non-covalently associated protein known as the surrogate light chain (Melchers et al., 1993). The surrogate light chain consists of two polypeptides that are the products of the V_{preB} and λ5 gene loci. The Ig heavy chain/surrogate light chain complex is expressed on the cell surface of the pre-B cell and is associated with the B cell signal transduction module. Thus, it seems that this complex is capable of transducing signals upon binding of ligand. Suggestive evidence for this arises from an X-linked hereditary immuno deficiency disease in which a mutation in a gene belonging to a family of signal transducing protein tyrosine kinases causes a block in early B cell development (Vetrie et al., 1993). It is not known what the recognition specificity of the IgH/surrogate light chain is. However, expression of the λ5 gene product (and consequently the surrogate light chain) seems to be essential for B cell development as targeted disruption of this gene severely disrupts B cell production (Rolink et al., 1993). Lastly, Ig light chain rearrangement and subsequent expression of surface IgM by the nascent B cell then occurs. Rearrangement of the light chain gene is similar to the process of heavy chain rearrangement, including insertional alterations to increase diversity. Upon successful rearrangement and expression of surface IgM, the very young B cell emigrates from the bone marrow to the spleen. The rearranged Ig gene is not static and genetic changes are made in response to the life experiences of the B cell as described above.

Certain cytokines have been identified that have B cell growth promoting activity (Kincade et al., 1989). IL-7 seems to be required for B cell production since in IL-7 receptor knockout mice, peripheral B cell numbers are dramatically reduced and B lymphopoiesis is halted at an early pre-B cell stage (Peschon et al., 1994).

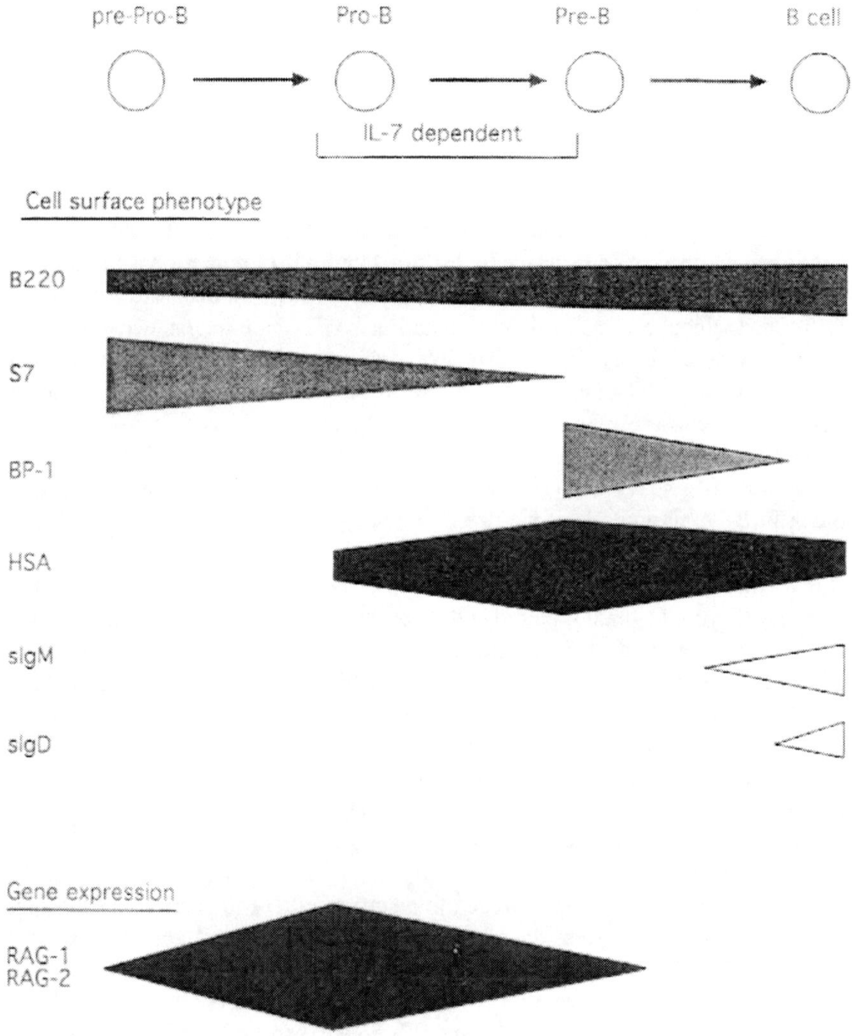

Figure 9. Schematic representation of the phenotypic stages of B cell development.

Also injection of mice with purified recombinant IL-7 for seven days results in significantly increased numbers of B cell precursors in the bone marrow and increased B cell numbers in the spleen with a relatively immature phenotype (Morrissey et al., 1991). After cessation of IL-7 treatment, B cell numbers return to normal levels within two weeks, suggesting that there is a regulatory mechanism to control B cell numbers in the peripheral immune tissue. Other cytokines can enhance B lymphocyte production such as insulin-like growth factor (IGF) and stem cell factor (SCF), but it has not been established whether there is a strict requirement for these factors (Landreth et al., 1992). IL-1, G-CSF, and GM-CSF decrease B lymphopoiesis and it is not known if this is a direct effect or a result of increased myelopoeitic activity in the bone marrow. TGF_{β} and IL-4 have also been shown to be inhibitory to B lymphopoiesis. IL-4 under other conditions can also support the growth of B cell precursors *in vitro*. As mentioned above the number of B cells produced per day by the bone marrow approximately equals the number existing in the spleen. Thus, under normal homeostatic conditions where peripheral B cell numbers are maintained at a constant level, this production must be balanced by cell death. The observation cited above that the increased number of peripheral B cells induced by IL-7 treatment is not long lasting demonstrates that the peripheral B cell population is tightly regulated. It is believed that programmed cell death is important in this process. The molecular mechanisms underlying this are beginning to be understood. For instance, the *bcl-2* protooncogene was discovered as a consequence of chromosomal translocations in follicular B cell lymphomas (Korsmeyer, 1992). The Bcl-2 protein is a 26 kDa integral membrane protein localized intracellularly (mitochondrial and perinuclear membrane) that can prevent programmed cell death in response to a number of stimuli (withdrawal of cytokines, glucocorticoids). Mice that lack a functional *bcl-2* gene (disrupted by targeted mutation) develop normal numbers of T cells and B cells, but by 2-5 weeks of age, there is minimal peripheral lymphoid organ cellularity (Nakayama et al., 1993). Thus, it appears that functional Bcl-2 is necessary for the maintenance of a normal immune system. Similarly, transgenic mice that overexpress Bcl-2 driven by the immunoglobulin promoter have peripheral lymphoid hyperplasia and an increased incidence of autoimmune disease (Strasser et al., 1991). The environmental signals that regulate *Bcl-2* gene expression are unknown. Recently, other genes related to *bcl2* have been described that may function in the regulation of this process (Boise et al., 1993; Oltvai et al., 1993).

B. T cell Production

The production of the vast majority of T cells is dependent on the thymus. Nude mice that lack a functioning thymus are severely, but not entirely, T cell depleted. The thymus is predominantly composed of lymphocytes (>95%) intertwined with a complex mixture of epithelial cells, dendritic cells, and macrophages that guide the nascent T cells along the road to maturation (or death). Thymic T cell production is complex since the resulting mature population of cells that migrate out of the thymus to the peripheral lymphoid organs must be able to recognize antigen *and* self-MHC encoded molecules. Also, the mature population must be tolerant of self determinants. Thus, the process of thymic maturation involves both negative selection (tolerization) and positive selection (ability to recognize self-MHC). It is known that the T cell antigen/MHC recognition repertoire is plastic and that it is molded intrathymically upon encounter with the MHC molecules displayed by the thymic stroma (non-lymphoid cells). Long before the structure of the T cell antigen receptor was elucidated, immunologists were artificially constructing mice with non-functioning immune systems due to different MHC molecules being expressed in the thymus than in the periphery (Figure 10). Thus, when a nude mouse (which lacks a functional thymus) of MHC type "A" is engrafted with a normal thymus from strain "B" MHC, the thymic precursors from the strain "A" nude mouse repopulate the thymus but are selected for recognition of the strain "B" MHC stromal cells of the transplanted thymus. Thus, when they migrate to the periphery and encounter only strain "A" MHC on host-derived antigen presenting cells, they are not able to function. Other experiments using different donor and host combinations of parental and F_1 MHC haplotypes, demonstrated that the MHC molecule that the mature T cell recognized was not the MHC it expressed, but the MHC presented on the radiation resistant cells of the thymus which would be the macrophages and epithelial cells.

The epithelial component of the thymus is seeded early in fetal development by T cell precursors that are derived from the hematopoietic stem cell. The initial wave of T cell maturation involves the production of cells that express products of γ/δ T cell receptor genes. These cells exclusively migrate to epithelial surfaces where their exact functions await to be elucidated. Mice with a disrupted δ T cell receptor

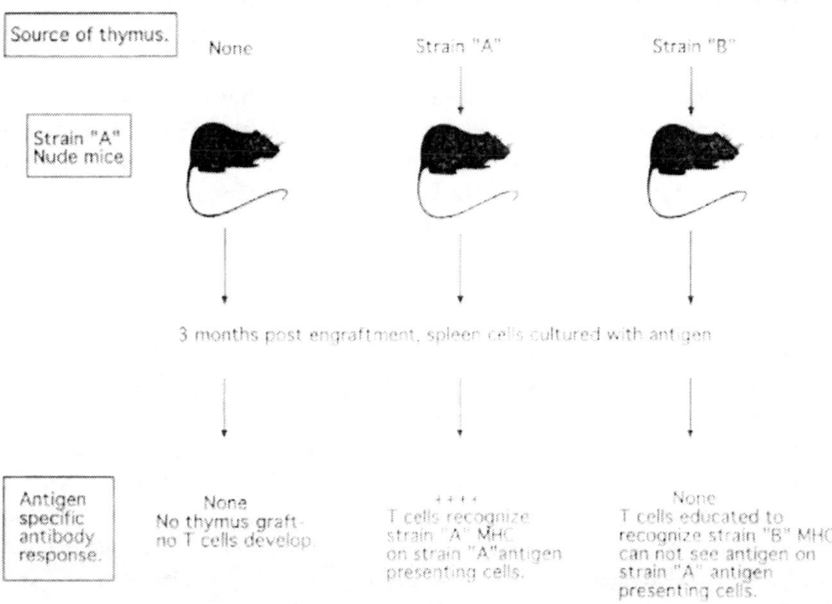

Figure 10. Thymus engraftment of athymic "nude" mice demonstrates the principle of thymic selection of T cells by the MHC encoded antigens of the thymus graft.

gene do not develop these early T cells, but otherwise develop normally. Following the first wave of γ/δ T cell production, the developing thymocytes express products of the α/β T cell receptor gene loci.

Similar to the B cell maturation, stages of development have been defined for thymocytes based on cell surface marker expression and gene rearrangement (Figure 11). Four major populations of thymocytes can be identified based on the expression of CD4 and CD8. The least mature thymocytes are defined by the lack of CD4 and CD8, and comprise less than 5% of the cells. There have been a number of subpopulations of CD4-/CD8-"double negative" cells identified based on the expression of other cell surface markers such as pgp-1, HSA, and IL-2Rα (Nikolic-Zugic, 1991). As the T cell receptor genes rearrange and begin to be expressed, the thymocytes become "double positive," first expressing CD8 and then CD4. Kinetic studies show that there is significant proliferation during this transition state from double negative to $CD4^+/CD8^+$ double positive cells. The stimulus for this proliferation is not known, but IL-7 may be involved since mice lacking the IL-7R have a block in thymocyte maturation at the late double negative stage (Peschon et al., 1994). The double positive population represents 80-90% of thymocytes. The transition from double negative to double positive requires rearrangement and expression TCRβ gene. Rearrangements of the TCRβ gene are detected in the $CD4^-/CD8^-$ cells and mice having a genetically disrupted TCRβ gene have a maturation block at the $CD4^-/CD8^-$ stage and thymic cellularity is drastically reduced (Pfeffer and Mak, 1994). The mechanisms of gene rearrangement in developing T cells are similar to that observed in B cells. For instance, inactivation of the recombinase activating gene (RAG) or in SCID mice with an inability to rearrange genes, both B cell and T cell development are blocked. Initially, TCRβ is expressed on the cell surface with a recently identified subunit that has been termed the pre-TCRα chain (Saint-Ruf et al., 1994). The function of the TCRβ/preTCRα/CD3 complex on these cells is not known, but it is an interesting developmental parallel to the surrogate light chain/IgH complex expressed in pre-B cells. Rearrangement of the TCRα genes and expression on the cell surface with TCRβ then follows. In mice with a genetically disrupted TCRα gene, thymic development proceeds to the double positive stage before arrest and thymocyte numbers are in the normal range. It should be pointed out that TCR expression on developing thymocytes in these mice ranges from very little to the high levels seen on mature T cells. It is thought that as thymocytes mature, TCR expression increases. Thymocytes that successfully complete maturation then lose expression of either CD4 or CD8 depending on whether they have been selected to recognize Class I MHC ($CD8^+/CD4^-$) or Class II MHC ($CD4^+/CD4^-$; see Figure 4). The study of genetically manipulated animals has provided significant insight into the mechanism of thymic differentiation and it has been possible to construct a map of required gene expression during T cell development (Pfeffer and Mak, 1994). As mentioned above thymic maturation of T cells also involves the process of tolerization and positive selection. It was long appreciated that engraftment of

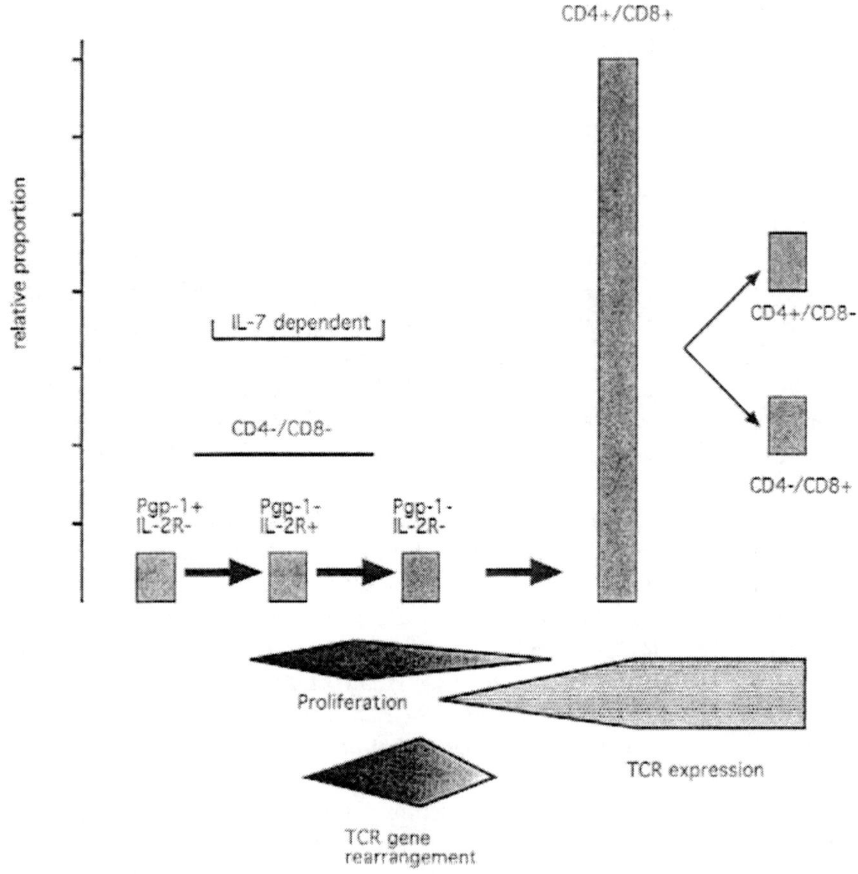

Figure 11. Schematic of thymocyte subpopulations, proliferation, and TCR gene rearrangement.

an allogeneic thymus into a nude mouse resulted in mature T cells that were tolerant of the thymic MHC haplotype. Also, it had been observed that the vast majority of CD4+/CD8+ thymocytes die intrathymically and this was thought to be the result of deletion of potentially autoreactive cells. The actual ability of the thymus to tolerize was first observed by the construction of transgenic animals expressing a T cell antigen receptor of a defined specificity in a manner that preempted the usual generation of TCR intrathymically. The TCR used was cloned from a cytotoxic T cell specific for the male specific H-Y antigen in the context of H-2^b (von Boehmer et al., 1989). In H-2^b female mice (no expression of H-Y), thymic cellularity was normal and all the thymocytes expressed the anti-H-Y TCR. In H-2^b male mice, where H-Y is expressed, thymic cellularity was drastically reduced and maturation was halted at the double negative to double positive transition. Male TCR transgenic mice that are not H-2^b have normal thymocyte cellularity. Other studies involved the use of monoclonal antibodies specific for a certain V region of the TCRβ chain. It was known that T cells expressing this V region specificity could recognize a certain retroviral encoded superantigen. In mice naturally expressing this superantigen, thymocytes expressing the particular V region could be detected in the double positive population, but were deleted in the $CD4^+/CD8^+/TCR^{hi}$ population (Kappler et al., 1987). Tolerance in this case was at a more mature stage of development than seen with the TCR transgenic animals. Thus, intrathymic tolerance can probably be imposed at any stage of thymic development if the interaction signals between the developing T cells and the thymic environment are sufficient.

The process of positive selection of thymocytes essentially results in the guarantee that mature T cells will be able to productively interact with MHC-encoded Class I and Class II proteins. As mentioned above, early studies of thymus engrafted nude mice (Figure 10) demonstrated that it was possible to skew the MHC recognition of T cells by transplanting thymuses of a different MHC haplotype. This was most consistent with a selective process existing during T cell development. More recent results have started to clarify this process. For instance, in gene knock-out mice that lack MHC Class I expression, CD^{8+} T cells fail to mature, and similarly, in MHC Class II negative mice, CD^{4+} T cells fail to mature. Thus, the interaction of the T cell receptor with Class I and Class II MHC is vital for the production of mature T cells. Presentation of peptides by MHC molecules has been discussed and recent data have taken advantage of certain experimental tools to reconstruct a unique model of thymic maturation. In fetal thymus organ cultures established from mice with defective Class I MHC peptide presentation, the Class I MHC can be artificially "loaded" with a peptide of known structure. In addition, the differentiating thymocytes expressed a transgenic TCR of known peptide/MHC specificity. The use of peptide analogs with lower binding affinities to the MHC Class I groove and subsequently to the TCR then allow an investigation as to the effect of binding affinity on positive and negative selection. In both systems, adding the original peptide to the culture system resulted in tolerance (negative selection), but adding the peptides with lower affinity binding resulted in thymocyte maturation (positive selection; Ashton-Rickardt et al., 1994; Hogquist et al., 1994). These results are consistent with a differential avidity selection model of thymocyte maturation in which high affinity binding to a MHC/peptide results in deletion and lower affinity binding results in positive selection and maturation. These results importantly reveal that T cell maturation is dependent on peptide/MHC recognition and not just MHC recognition. However, since *in vivo* any particular peptide is on a very low number of MHC molecules, it is not clear how selection occurs. It has been postulated that selection naturally occurs not to a single peptide but to a mixture, to which the average avidity of binding would allow maturation (Janeway, 1994; Bevan et al., 1994). Nevertheless, these results reveal that the receptors used by the T cells to recognize and respond to foreign peptides are developed and selected by recognition of self-peptide/MHC complexes.

The role of cytokines in thymic maturation is controversial. It is relatively simple to stimulate thymocyte growth with many different cytokines and by molecular analysis both cytokine and cytokine receptor mRNA has been detected in thymic tissue (Montgomery and Dallman, 1991). Interestingly, mice with targeted disruption of cytokine genes, other than the IL-7R as mentioned above, have phenotypically intact thymopoiesis. This includes mice with inactivated IL-2, IL-4, IL-6, IL-1R, and TNF-R genes. It may be that in the absence of a single cytokine, alternative growth pathways are used or T cell maturation may be deficient at a level that is not detectable by current technology.

C. Extrathymic Maturation

The observation that nude mice which do not have a functional thymus, have detectable T cells in the spleen and intestine led to the hypothesis that some T cell maturation could occur without obligate intrathymic maturation. Recent studies have shown that fetal intestine grafts in T cell depleted adult thymectomized mice induce a reconstitution of T cell numbers in lymph node and spleen (Mosley and Klein, 1993). Analysis of normal intestinal intraepithelial lymphocytes reveals a population of $CD4^+/CD8^+$ lymphocytes that would be expected to be seen as a consequence of T cell maturation (Poussier and Julius, 1994). RAG (recombinase activation gene) expression has also been detected in the IEL population. RAG expression is an indication of active gene rearrangement of both Ig as well as TCR genes. Thus, two criteria for T lymphopoiesis are observed in this compartment. Other studies utilizing mice transgenic for a defined T cell receptor specificity have shown that the processes of selection may be different than in the thymus. T cells with receptors for self determinants are not efficiently deleted but are functionally anergic instead. Much remains to be learned about intestinal epithelial T cell differentiation and its role in immune function.

VIII. PROTEIN SYNTHESIS AND DEGRADATION

It is important for the understanding of antigen presentation to review the physiology of protein synthesis and degradation that occur continuously in all cells. The immune system continually surveys for the presence of foreign entities and this is accomplished largely through the sampling of degradation products (i.e., peptides) by T cells (Buus et al., 1988). Protein degradation and peptide association with Class I or Class II MHC is the main pathway of antigen presentation to $CD4^+$ or $CD8^+$ $TCR\alpha/\beta$ T cells. These pathways are depicted in Figures 12 and 13.

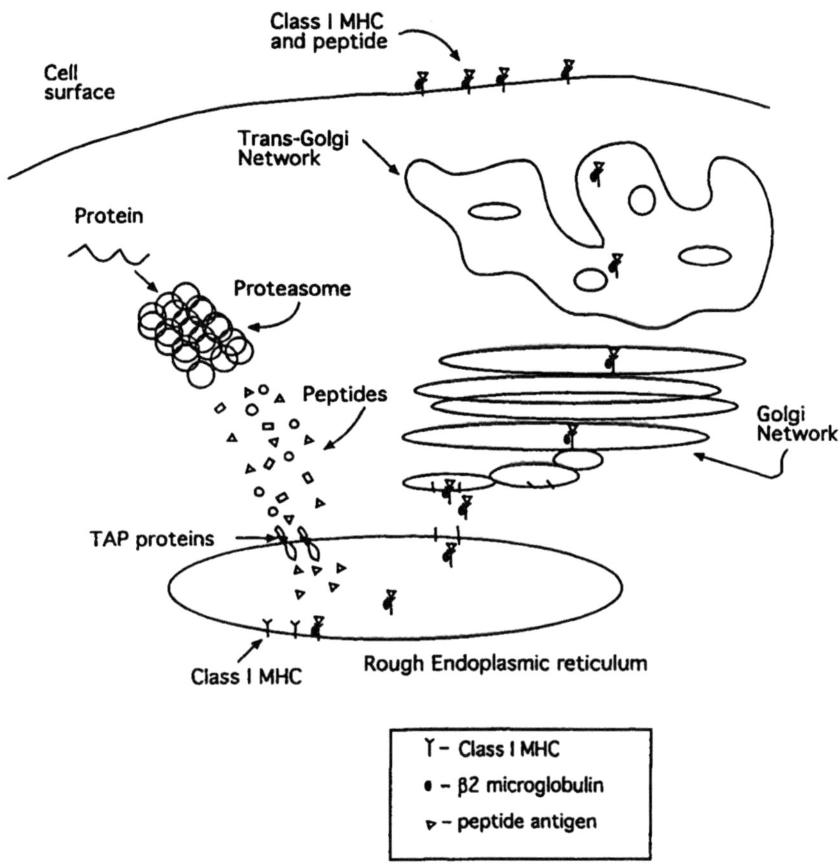

Figure 12. Class 1 MHC synthesis, peptide binding, and transport.

A. Endoplasmic Reticulum (ER)

The ER is a netlike meshwork that extends through the cytoplasm. It consists of a continuous membrane that separates the ER lumen from the cytosol. The membrane is able to selectively mediate the transport of certain molecules into the lumen. The ER is the starting point for the synthesis of all secreted proteins (e.g., cytokines) and cell membrane bound proteins (e.g., MHC, CD3). Thus, the regulatory circuits that exist to control entrance into, synthesis, and egress from the ER have a direct influence on the function of the immune system. The ER captures selectively marked (via a signal peptide) proteins from the cytosol as they are being synthesized. There are two additional proteins involved in this process; a signal

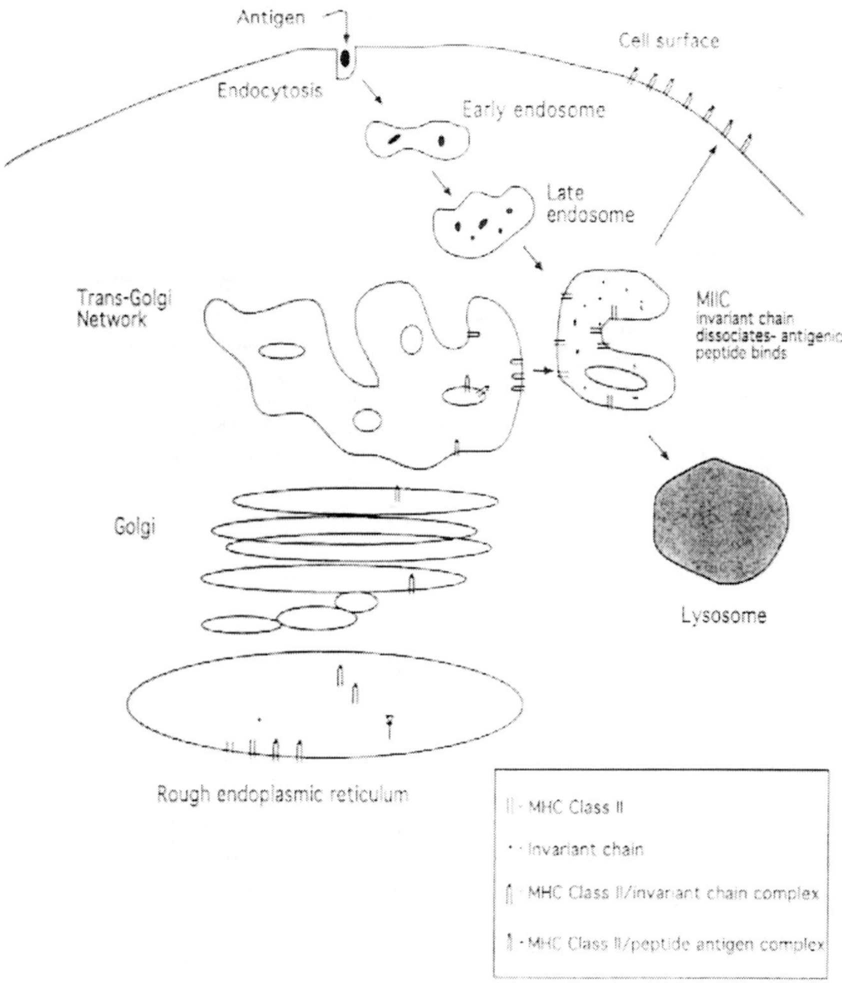

Figure 13. MHC Class II synthesis, transport, and peptide binding.

recognition particle that is present in the cytosol and binds to the signal peptide on the nascent protein and a docking protein that is embedded in the ER membrane. There are generally of two types of proteins that are destined for the ER; transmembrane proteins and soluble proteins destined for secretion. Synthesis of the protein can be completed while attached to the ER membrane or in the lumen. N-linked glycosylation also occurs in the ER.

B. Golgi Apparatus

The Golgi apparatus is a collection of relatively flat membrane enclosed cisternae that resemble a stack of plates. There are many small vesicles seen in close association with the Golgi apparatus that transport proteins between the ER and Golgi and from the Golgi to the ultimate destination. Existing evidence defines three distinct functional activities in the Golgi that include modifications of proteins. N-linked oligosaccharides are often trimmed and additional sugars may be added. O-linked glycosylation also occurs in the Golgi. Finally, the proteins reach the trans Golgi network (TGN) that is responsible for directing the proteins to their final destinations which are either the plasma membrane, secretory vesicles, or lysosomes. Specialized signaling mechanisms too detailed to describe here exist in order to direct the proper proteins to reach their appropriate destinations.

C. Endosomes/Lysosomes

It is an important aspect of cell function to monitor the extracellular environment and to maintain the cell membrane/extracellular interface in proper order. An important mechanism by which this is accomplished is by cell membrane pinching and budding of the vesicle from the plasma membrane into the intracellular space (endocytosis). This functions to sample the extracellular environment as well as to clear cell membrane proteins. The vesicles known as endosomes contain clustered receptors, transmembrane proteins, lipids, and solutes. Endosomes located near the cell membrane are termed early endosomes and there is significant fusion between them at this stage. Subsequently, these vesicles either traffic back to the plasma membrane, to the Golgi complex, or proceed, as late endosomes, to the lysosome compartment near the nuclear membrane (Trowbridge et al., 1993). Early endosomes differ from late endosomes morphologically, biochemically, and functionally in their degradative ability (Gruenberg and Howell, 1989). Lysosomes are acidic, hydrolase-rich vacuoles capable of degrading most biological macromolecules (Knop et al., 1993). They are the final intracellular destination for proteins internalized by endocytosis. Late endosomes are thought to fuse with lysosomes thereby depositing their contents into the degradative environment of the lysosome. There are well over 40 well characterized acid hydrolases known to be contained in lysosomes. Maintenance of the acidic environment within the lysosome is required for proper function of the digestive enzymes.

D. Cytosolic Protein Degradation

It has been long realized that cytoplasmic proteins are subjected to degradation. A major pathway in this process utilizes the conjugation of the protein with ubiquitin (Finley and Chau, 1991). Ubiquitin is a 76 a.a. protein whose sequence is highly conserved. Its amino acid sequence is identical in mammals and ubiquitin in yeast (*S. cerevisiae*) differs in only one amino acid from the mammalian form. Ubiquitin is involved in the regulation of many cellular processes such as DNA repair and cell cycle control. Ubiquitin conjugation of cytosolic proteins marks them for preferential degradation (Hershko and Ciechanover, 1992). The process of ubiquination of the proteins requires at least three different enzymes and ATP. Mutation of the first enzyme in this process is a lethal event in yeast. Various numbers of ubiquitin molecules may be conjugated to a protein either in tandem to form a multi-ubiquitin side chain or on different sites on the protein. However, all conjugation is to the ε-amino groups of lysine residues. Ubiquitin conjugation marks the protein for degradation by a multimeric, non-lysosomal, ATP-dependent proteolytic complex known as the proteasome (Goldberg and Rock, 1992).

Proteosomes are present in the cytoplasm and nucleus of all eukaryotic cells and are estimated to be as much as 1% of cellular protein (Tanaka et al., 1992). This complex is approximately 700 kDa (with a sedimentation coefficient of 20S) and is composed of four stacked rings with a hollow center that forms a barrel-like structure. Each ring is composed of 12-15 different polypeptides with a molecular weight range of 19-36 kDa. The central canal formed by the rings is thought to be the site of proteolysis and this has been termed "digestion in a barrel" although this has not been formally proven. The 20S proteasome is also known as the multicatalytic proteinase because of its multiple proteolytic activities (Orlowski, 1990). Among those described are trypsin-like, chymotrypsin-like, and peptidyl-glutamyl hydrolyzing activity. Inhibitor studies suggest that these activities are a function of different catalytic sites. A number of genes encoding these subunits have been cloned and mutational analysis is now possible to help unravel the mystery of subunit function. It is becoming evident that the subunit composition of the proteasome can differ according to cell type. The role of proteasomes in antigen processing came from the discovery and isolation of two proteins that were co-precipitated with an anti-MHC allosera by Monaco and McDevitt (Monaco, 1992). These proteins known as LMP-2 (low molecular weight protein) and LMP-7 have sequence homology with known proteasome subunits. The genes are polymorphic and map to the MHC. LMP proteins have proteolytic activity and antisera against these two proteins precipitated only a small fraction of the proteosomes. Thus, it is thought that the LMP proteins associate with or are a constituent of proteosomes, replacing other subunits. Expression of LMP-2 and LMP-7 is up-regulated by γ-IFN. Recently, two new genes have been cloned that encode new proteasome subunits. These genes known as MB1 and Delta are closely related to LMP-7 and LMP-2, respectively (Belich et al., 1994). Interestingly, the expression of these genes is reciprocal to that of the LMP genes. This finding may help to explain the observation that LMP negative cell lines are still able to process antigen (Momburg et al., 1992). Recently, mice lacking a functional LMP-7 subunit were produced by targeted gene disruption and these mice have minimal expression of MHC Class I molecules (Fehling et al., 1994). Addition of peptides to lymphoid cells from these mice restores MHC Class I expression and these data are consistent with a vital role of the LMP proteins in the degradation proteins into peptides suitable for binding to Class I MHC molecules.

Several lines of evidence exist that *in vivo* the 20S proteasome forms the core of an even larger structure which is responsible for ubiquitin-dependent degradation and that structure is termed the 26S proteasome. The 26S proteasome contains all the peptides of the 20S particle. The formation of the 26S proteasome requires ATP and at least three additional proteins. Almost a dozen other proteins have been isolated that seem to interact and modify the function of the 20S proteasome. It is thought that the additional proteins play some regulatory function, both positively and negatively in the control of proteolysis. This is clearly a complex system that is involved in many aspects of normal cell function. For instance, study of mutated components of 26S proteasome in yeast show that a temperature sensitive mutation of the *mst2* gene (74% identical in amino acid sequence to subunit 4 of human proteasome) results in the accumulation of ubiquitin conjugated proteins at the restrictive temperature. Also, mutations in two other genes which encode proteins that are subunits of proteasomes result in cell cycle arrest in yeast. Recently, Rock et al. (1994) have produced evidence that rather specific inhibitors of proteasomes reduce the degradation of protein by the proteasomes (Rock et al., 1994). This included long lived proteins as well as abnormal and short-lived proteins. Introduction of ovalbumin into the cytosol of cells treated with these inhibitors resulted in no ovalbumin peptide being presented on cell surface MHC Class I molecules. This evidence supports a major role for the ubiquitin/proteasome system in the non-lysosomal degradation of cellular proteins. Also of interest is the recent report that the ubiquitin/proteasome pathway is required for the processing of the NF-κB1 precursor protein into active NFκB, thus indicating a functional role for this system in processing precursor protein into active forms (Palombella et al., 1994).

E. Peptide Transport and Binding

Evidence on the generation of peptides by the ubiquitin/proteasome pathway still left one puzzle which was how the peptides entered the ER where nascent Class I MHC molecules were being synthesized. Studies on a cell line that was inefficient in the assembling of Class I MHC molecules and had properties consistent with defects in peptide transport lead to the hypothesis that specific peptide transporters existed in the ER membrane that selectively ushered peptides from the cytosol into the lumen of the ER. Two genes have been identified in the MHC that encode proteins similar in size, sequence, and structure to members of a large family of proteins that transport all manner of molecules across cellular membranes (Townsend and Trowsdale, 1993). These proteins, known as TAP-1 and TAP-2, are members of the ABC (ATP-binding cassette) family of proteins (Figure 12). Their expression is localized to the ER and cis-Golgi. Transfection of the TAP genes into cell lines incapable of presenting antigen on Class I MHC corrects the defect. In fact, TAP proteins have been termed "molecular rulers" for Class I MHC-associated peptides (Germain, 1994). Small peptides that would be predicted to bind Class I MHC are preferentially transported. Interestingly, TAP polymorphism (TAP genes map to the MHC) may play a role in determining peptide transport (Powis et al., 1992).

IX. ANTIGEN PROCESSING AND PRESENTATION

The immune system has been able to utilize two aspects of normal cellular house cleaning (protein degradation) to be able to monitor for traces of pathogens. These have been termed the "endogenous" and "exogenous" pathway as a reflection of the degraded starting material (Germain and Margulies, 1993). The endogenous pathway utilizes the ubiquitin/proteasome system of protein degradation to generate peptides from cytosolic proteins which are then transported to the lumen of the ER for binding to Class I MHC (Figure 12). The second mechanism, known as the exogenous pathway, utilizes endosome/lysosome pathway of degradation of extracellular, phagocytosed, or membrane-bound proteins for binding to Class II MHC (Figure 13).

A. Class I MHC Synthesis

As mentioned above, synthesis of Class I MHC and β_2 microglobulin is completed in the ER (Jackson and Peterson, 1993). Evidence suggests that Class I heavy chain forms a stable complex with β_2 microglobulin within minutes of synthesis. It is believed that Class I MHC is associated with a molecular chaperone known as calnexin or p88 in the ER until complete synthesis, folding, and binding has occurred. The heavy chain associates with β_2 microglobulin and peptide while still in the ER and these complexes dissociate from calnexin and are rapidly transported through the Golgi complex to the cell surface. In cells lacking expression of β_2 microglobulin, most of the Class I MHC heavy chain remains in the ER associated with calnexin until degraded. Some, however, can escape and be transported to the cell surface. In RMA-S cells which lack the TAP peptide transporters and cannot accumulate peptides in the ER, the association of calnexin with β_2-heavy chain was prolonged although there was still a very slow rate of transport to the Golgi. Efficient dissociation of calnexin from Class I MHC heavy chains requires the binding of both β_2 microglobulin and peptide. Thus, the chaperone function of calnexin is to retain inappropriately or incompletely folded or assembled Class I from exiting the ER.

B. Class II MHC Synthesis

Similar to Class I MHC, Class II MHC α and β chain synthesis occurs in the ER (Cresswell, 1994). Class II MHC has an associated polypeptide known as the invariant chain that binds to newly formed Class II in the ER and prevents peptide binding. Newly synthesized invariant chain forms a trimer and evidence suggests that the Class II MHC α and β chains bind to invariant chain simultaneously to their dimerization forming a nonamer complex (3 sets of α and β chains in association with the invariant chain trimer). Invariant chain prevents occupancy of the Class II peptide binding site. In studies on invariant chain negative cell lines or mice, it was found that Class II MHC molecules had accumulated in the ER and cell surface MHC Class II levels were drastically reduced, resulting in an inability to present antigen to T cells. Invariant chain negative mice have drastically reduced numbers of $CD4^+$ T cells (Bikoff et al., 1993). Thus, the association of invariant chain to MHC Class II dimers in the ER is vital for processing and transport to the cell surface. The multimeric nonamer appears to be rapidly transported from the ER to the Golgi as additional post translational glycosylations are made. The fate of the MHC Class II/invariant chain complex is cloudy after passage through the Golgi complex. It is thought that the MHC Class II/invariant chain complex is targeted for incorporation into endosomes in the trans-Golgi network, but there is evidence that some of the complex directly proceeds to the cell surface and is subsequently endocytosed. It may be that the sorting mechanism in the TGN is not 100% efficient. Once in the endocytic pathway, invariant chain appears to be proteolyzed and dissociates from Class II MHC. After cleavage and complete dissociation of the invariant chain, the MHC Class II α/β dimer binds a peptide and appears on the cell surface. The specific site in which endogenous peptide-MHC Class II complexes are formed has remained elusive. Several studies have implicated early endosomes, late endocytic compartments, and lysosomes. Recently, a unique compartment has been identified that is distinct from the late endosomes and lysosomes where the bulk of MHC Class II molecules accumulate (Peters et al., 1991; Rudensky et al., 1994). These dense endocytic vesicles, termed MIIC (MHC Class II compartment), are positioned between late endosomes and dense lysosomes. Exogenous protein is most efficiently delivered here via immunoglobulin-mediated binding and not by fluid phase uptake. Once the peptide/MHC Class II complex is formed, it returns to the cell surface. Its exact route is again a mystery. It is not known if peptide/MHC Class II complexes are shuttled back to the TGN for constitutive transport to the cell surface or if another mechanism exists.

The finding of efficient receptor-mediated delivery of antigen to the MIIC vesicles highlights the importance of receptor-mediated uptake in contrast to fluid phase uptake by the antigen presenting cell. Since the concentration of fluid phase extracellular antigen is probably low, mechanisms to increase entry into the endocytic system would be advantageous. This would be mediated by antigen binding to cell surface Ig expressed by B cells or the use of Fc receptor mediated binding of antigen/antibody complexes by macrophages and B cells. This would act to increase the concentration of antigen entering the endocytic pathway and eventually result in more peptide/MHC complexes for lymphocyte stimulation.

Also, it must be realized that these mechanisms for antigen presentation are not mutually exclusive. There is evidence that cytosolic proteins which would be predicted to be presented by Class I MHC, can be presented by Class II MHC (Nuchtern et al., 1990; Weiss and Bogen, 1991). Recently, evidence has also been produced that proteins derived from phagocytosed bacteria can enter the cytosol and be presented on Class I MHC molecules (Kovacsovics-Bankowski and Rock, 1995). Finally, evidence exists that different types of antigen presenting cells can produce different peptides derived from the same antigen (Vidard et al., 1992), and that dependent on the environment, cells can alter antigen processing and presentation pathways (Michalek et al., 1989). Thus, there is significant plasticity in the processing and presentation system to insure that pathogen-derived peptides get plenty of "exposure" to lymphocytes via MHC molecules.

X. CYTOKINES AND THE IMMUNE RESPONSE

Cytokines are soluble secreted proteins that act to modulate immune function. Although originally termed "interleukins" as defined by lymphocyte-specific interactions, it has become apparent recently that many different cell types can influence immune reactivity (i.e., lymphocyte function) by producing cytokines. Volume 2 in this series contains detailed reviews of many of the cytokines that influence immune responses. The reader is referred to this volume for information on interleukins 1-10, interferons, tumor necrosis factor, and the colony stimulating factors. This section will detail information on the recently discovered (cloned) cytokines.

A. Interleukin-11

IL-11 was identified as an activity in the supernatant of a bone marrow stromal cell line, PU-34, that stimulated the growth of the IL-6 dependent cell line, T1165, in the presence of excess anti-IL-6 neutralizing antibodies (Du and Williams, 1994). It was subsequently "expression cloned" from a cDNA library from the PU-34 cell line. The resulting cDNA encodes a predicted protein of 199 amino acids (23 kDa). There is no sequence homology with IL-6. The receptor for IL-11 has not been cloned as of this writing, but from different studies it is clear that IL-11 is among several growth factors (IL-6, leukemia inhibitory factor (LIF), oncostatin M, and ciliary neurotrophic factor) that use gp130 as a component of the receptor signaling mechanism. It has been demonstrated that IL-6 does not compete with IL-11 for binding indicating that another unique receptor subunit confers specificity for IL-11 binding. IL-11 is produced by stromal cells, fibroblasts, and other connective tissue cells (especially after induction with IL-1). Biologically, IL-11 shows multifunctional activity influencing hematopoietic stem cell proliferation and differentiation. IL-11 induces minimal colony formation *in vitro* with normal bone marrow cells (Musashi et al., 1991). The colonies that did form were mostly mixed granulocyte and macrophage types and comparatively IL-11 was much less active than IL-3. Using bone marrow cells from mice treated with the cytoreductive agent 5-FU, IL-11 by itself did not stimulate much of an *in vitro* colony-forming response. However, there was significant synergy with IL-3. The effect was thought to be due to a decrease in the quiescent cell cycle period of stem cells. Administration of IL-11 to normal mice results in increases in absolute numbers of femoral marrow and splenic myeloid progenitor cells and an increase in cycling cells. This was associated with increased neutrophil counts in peripheral blood. Administration of IL-11 to mice compromised by cyclophosphamide or whole body irradiation resulted in a more rapid return of both neutrophil and platelet numbers to normal values (Musashi et al., 1991). *In vitro*, IL-11 also synergizes with IL-3 and SCF in stimulating the growth of many hematopoietic precursor cells. IL-11 also has unique properties in other systems other than hematopoiesis. IL-11 administration to mice with lethal intestinal damage as a consequence of 5-FU and irradiation resulted in a survival advantage that was related to a more rapid recovery of the intestinal epithelial compartment (Du et al., 1994). IL-11 also directly inhibited adipogenesis in a murine cell line and inhibited adipocyte differentiation as well (Kawashima et al., 1991). In addition, IL-11 has been shown to be active in promoting neuronal differentiation of an immortalized neural stem cell and progenitor cell line. IL-11 shares some activities with IL-6, such as induction of acute phase proteins and stimulation of B cell growth. However, IL-11 does not stimulate myeloma cell proliferation or act as a cofactor for T cell proliferation.

B. Interleukin-12

IL-12 was originally described as an activity that stimulated NK (NKSF) or cytotoxic T cell function (Brunda, 1994). The genes encoding this factor were cloned in 1991 (Gubler, et al., 1991). IL-12 is a heterodimer consisting of p35 and p40 subunits. It is produced mainly by macrophages with B cells being a secondary source. The cell receptor for IL-12 has also been cloned. It is a 662 amino acid, type I transmembrane protein that is a member of the hemopoietin receptor gene superfamily. It has homology with gp130 which is a common receptor subunit for IL-6 and IL-11. The receptor is expressed on T cells and NK cells. IL-12 has been shown to play a pivotal role in the early stages of the immune response. It is produced by macrophages that have encountered, ingested, or otherwise been stimulated by antigen. Its prime function appears to be the potent stimulation of NK cells that results in γ-IFN production. The consequences of this are twofold; first, elaboration of γ-IFN results in the up-regulation of antigen processing and priming for intracellular cytotoxic function by macrophages, and second, γ-IFN affects the subsequent development of cytokine profiles by the activated T helper cells. Thus, IL-2 production is increased and IL-4 and IL-10 production is decreased. This results in the development of potent cell-mediated immunity (Scott, 1993). IL-12 administration to experimental animals has increased anti-tumor immunity and resistance to infectious disease. Thus, IL-12 shows promising pre-clinical potential as an immunomodulator. Significant hepatotoxicity has been seen in mice treated long term with IL-12, and its toxicity in the clinic is currently being evaluated.

C. Interleukin-13

IL-13 was first cloned from a library of induction-specific mRNA derived from a murine Th2 cell line (Zurawski and de Vries, 1994). Murine IL-13 is a 131 amino acid protein. Human IL-13 has also been cloned and is similar in length. Interestingly, in both human and mouse, IL-13 maps to a region that encodes IL-3, IL-4, GM-CSF, and IL-5. The IL-13 gene shares a common intron-exon structure with GM-CSF, IL-4, and IL-5. IL-13 is produced by activated $CD8^+$ and $CD4^+$ T cell clones and in the human, IL-13 was produced by Th1, Th2, and Th0 T cell clones. Unlike IL-4, IL-13 does not activate T cells. IL-13 down-regulates the cytotoxic and inflammatory function of macrophages in both the human and mouse. IL-4 has a similar effect on inflammatory functions of macrophages. IL-13 up-regulates antigen presentation by human macrophages, but no effect of IL-13 was seen in a similar study performed in mice. IL-13 enhances the proliferative response of human B cells that have been stimulated with anti-IgM or anti-CD40 antibodies. IL-13, like IL-4, can induce Ig switching to IgE in human B cells. IL-13 does not appear to act on mouse B cells. Thus, IL-13 is a cytokine with considerable, but not total, functional overlap with IL-4.

D. Interleukin-14

IL-14 was recently described as a high molecular weight B cell growth factor (Ambrus et al., 1993). The protein was purified from culture supernatant of PHA-stimulated Namalava cells. A cDNA library was screened with an antibody raised against the purified protein. The gene identified encodes a protein of 483 amino acids (53 kDa). It is distinct from IL-4 and IL-5, two other cytokines that promote B cell growth. Recombinant-derived IL-14 induced a fivefold enhancement of B cell proliferation to the bacterial stimulant, *Staph aureus Cowan*. It also inhibited Ig production. These activities are consistent with a cytokine that augments B cell proliferation.

E. Interleukin-15

IL-15 was originally identified as a background activity in CTLL (IL-2 dependent cell line) assays with culture supernatants of simian kidney epithelial cell line, CV-1/EBNA (Grabstein et al., 1994). The gene encoding IL-15 has been cloned from these cells and encodes a 162 amino acid protein of 14-15 kDa. There is minimal sequence homology between IL-15 and IL-2 although computer modeling supports a four helix bundle-like structure that is typical of the helical cytokine family that includes IL-2. Interestingly, IL-15 seems to perform many of the functions of IL-2. IL-15 can induce proliferation of T cell lines and secondary T cell blasts. It increases cytotoxic T cell activity as well. Uniquely, IL-15 is not produced by T cells as is IL-2. Initial reports indicate that it is produced by many other cell types that are not considered members of the immune system (epithelial cells, myocytes). IL-15 and IL-2 share receptor components (IL-2Rβ and γ chain) and there is functional evidence of a novel receptor since there are cell lines that proliferate to IL-15 and not to IL-2 (Giri et al., 1994). It is of interest to speculate why these molecules with similar profiles of bioactivity were conserved. The evidence thus far suggests that it allows a wide variety of cell types other than lymphocytes to influence and modulate immune reactivity.

XI. T CELL ACTIVATION

The *sine qua non* of the immune response is the activation of $CD4^+$ helper T cells by Class II MHC expressing antigen presenting cells (dendritic cells, macrophages, B cells). It is well known that depletion of T helper cells in experimental animals by various means severely compromises the development of effector function. The decisive event in the process of T helper activation is the binding of the T cell antigen receptor to MHC-Class II/peptide complexes on the surface of an antigen presenting cell. The nature of this interaction has been the subject of intense study in order to understand the number and nature of the events that lead to T cell activation. Experimental evidence indicates that the affinity of the T cell receptor for MHC is quite low ($K_D=10^{-4}$-10^{-5}). For comparison, the affinity of antibody for its ligand is about three orders of magnitude higher. In addition, the antigenic peptide is competing with a vast excess of self-proteins for empty MHC molecules in the APC. Thus, the T cell must be very sensitive to signaling through the antigen receptor. Indeed, evidence suggests that only 30-100 peptide/MHC complexes on APC are required for T cell triggering (Demotz et al., 1990; Harding and Unanue, 1990). It should be noted that activated antigen presenting cells express other cell surface molecules (see below) and produce cytokines that bind to T cells. If the integrated strength of these signals mediated through receptor binding (cell surface expressed co-stimulatory signals, soluble cytokines, TCR receptor) exceed some threshold, then the T cell activation program is initiated (Karjalainen, 1994). The threshold for activation can also vary depending on the life encounters of the lymphocyte (i.e., naive, memory, effector). It has been well accepted that lymphocyte activation is a multistep process. For instance, cross-linking the TCR by monoclonal antibodies results in minimal proliferation or cytokine secretion and instead makes the cell refractory to appropriate activation signals (Schwartz, 1990). This concept originally, formulated by Bretscher and Cohn (1970) states that two signals are required for T cell activation and that T cell antigen receptor ligation in the absence of costimulatory signals results in paralysis. This would be one mechanism to ensure that autoreactive processes are more difficult to initiate. Over the past few years a large number of molecules have been identified that have costimulatory properties in T cell activation assays (Jenkins and Johnson, 1993). It is perhaps a reflection of the multiple regulatory pathways and innate redundancy that exist in the immune system. A number of molecules are ranked below in this author's attempt to assess their relative importance.

A. CD28

CD28 is a cell surface glycoprotein composed of two identical disulfide-linked 44 kDa subunits (Linsley and Ledbetter, 1993). Most mature T cells (both $CD4^+$ and $CD8^+$) constitutively express CD28 and its expression is increased after activation. Another cell surface protein known as CTLA4 is closely related to CD28 (30% amino acid identity) and is expressed mainly on activated, but not resting, T cells. The ligands for CD28 and CTLA4 are known as B7-1 and B7-2. All four proteins are members of the immunoglobulin gene superfamily. Both CD28 and CTLA4 can bind to either B7-1 or B7-2. CTLA4 can bind to B7-1 with a greater (20X) affinity than CD28. B7-1 and B7-2 are expressed on activated accessory cells, such as B cells, macrophages, and dendritic cells. Antibodies to CD28 enhance proliferation, IL-2R expression, and cytokine production of T cells to anti-CD3 stimulation (Schwartz, 1992). Interestingly, the production of IL-2 by anti-CD28 stimulation is not inhibited by cyclosporin A. This indicates that the signaling mechanism involved in CD28 costimulation occurs via a separate intracellular pathway and not an amplification of the CD3 signaling module pathway. The importance of the CD28 pathway is illuminated by studies on CD28 knock-out mice that show relatively deficient T cell responses (Shahinian et al., 1993). The CD28/CTLA4 system may be more complex than initially realized, however, since recent evidence has demonstrated that ligation of CTLA4 results in down-regulation of immune responsiveness. These results come from studies using antibody fragments against CTLA4 that block signaling to show enhancement of an *in vitro* immune response (Walunas et al., 1994). Also, assessment of transgenic mice that constitutively express B7-1 on B cells are markedly depressed in their ability to make an antibody response to a T-dependent, but not T-independent antigen (Sethna et al., 1994). The decreased response to the T-dependent antigen was corrected by coadministration of anti-B7-1 antibodies, suggesting the blocking of a negative signal. Clearly much remains to be understood about this important regulatory system in the immune response.

B. CD40

CD40 is an integral membrane protein expressed constitutively on the surface of B cells, dendritic cells, and some epithelial cells (Banchereau et al., 1994). It is a 40-50 kDa glycoprotein of 277 amino acids and is a member of the TNF receptor gene superfamily. CD40 is a molecule involved in B cell activation. Cross-linking of CD40 with immobilized antibodies results in B cell proliferation. Addition of IL-4 to these cultures results in long-term B cell growth. The ligand for CD40 has been cloned from a murine T cell thymoma cell line (Armitage et al., 1992). Murine CD40L is a type II membrane glycoprotein (extracellular carboxy-terminus) of 260 amino acids (39 kDa) and is a member of the TNF gene superfamily. CD40L is not expressed on resting T cells, but is induced after activation. Cross-linking of CD40L on T cells augments proliferation in response to anti-CD3 (Armitage et al., 1993). It is thought that CD40-CD40L interaction is important in T cell-mediated activation of B cells. Injection of antibodies to CD40L block both the primary and secondary responses to antigen (Foy et al., 1994). Also, the induction of collagen-induced arthritis was blocked by antibodies to CD40L (Durie et al., 1993). Recently, it was discovered that peripheral blood lymphocytes from patients with hyperIgM syndrome fail to bind a CD40 construct (Aruffo et al., 1993). Further analysis revealed that these patients had abnormal CD40L gene structure that resulted in a lack of functional CD40L protein. These patients have excessively high levels of serum IgM, and have drastically reduced levels of all other isotypes of serum Ig. There are no germinal centers in the lymphoid tissue of these patients and they suffer from recurrent infections. This phenotype was recently reproduced in mice by inactivating the gene for CD40L by targeted disruption (Xu et al., 1994). These mice have normal numbers of T cells and B cells, but are only able to produce IgM in response to antigen administration. These findings highlight the importance of CD40/CD40L interactions in Ig isotype switching and memory B cell formation.

C. CD2

CD2 is a member of the immunoglobulin supergene family and is a nonpolymorphic single chain molecule of 351 amino acids with a large cytoplasmic domain of 116 amino acids (Bierer and Hahn, 1993). CD2 can act as an adhesion molecule binding to CD58, CD59, and CD48. The role of CD2 in T cell activation is best understood from studies on human T cells since there are a variety of monoclonal antibodies against different epitopes of the human CD2 molecule. Using a particular combination of monoclonal antibodies, T cells can be activated (proliferation, IL-2 secretion) without stimulation through TCR/CD3. There is evidence that some components of the CD3 complex may associate with CD2 and many $CD3^-$ T cell lines fail to be activated by the anti-CD2 monoclonal antibodies. Interestingly, mice lacking CD2 due to targeted gene disruption can mount normal T helper and CTL responses and no functional defects have yet to be identified in these mice (Killeen et al., 1992). Thus, a redundancy in CD2 function may exist or the defect is subtle.

D. Ligands and Receptors in the TNF Gene Family and Ancillary Molecules

In addition to CD40 and CD40L, other members of the TNF gene family have been identified that have costimulatory properties (Armitage, 1994). Certainly the biological characteristics of TNFα and LTα have been well studied. However, the role CD30, CD27, 4-1BB, OX40, and their respective ligands play in the physiological function of the immune system remains to be elucidated. Without exception these molecules are all induced by cell activation, suggesting that their function is to modulate function in some manner post activation. Gene knock-out mice will be most useful in the assessment of the functional importance of these molecules.

A number of other molecules such as CD4, CD8, HSA, Thy1, and LFA-1 have also been shown to have costimulatory activity for T cells (Collins et al., 1994). Other than CD4 and CD8 which act to stabilize TCR/MHC interactions, it is not known what physiological role these molecules play or in what circumstances their presence or function may be important.

XII. TH1/TH2 T HELPER SUBSETS

A consequence of T cell activation is the production of cytokines that stimulate or augment lymphocyte proliferation as well as induce the development of effector function. It was long realized that in response to immune stimulation, the effector function that developed was either a strong antibody response and weak cell mediated response or vice versa and it was not clear what factors influenced the development of the predominant effector function. The puzzle began to unravel with the observations by Mosman et al. (1986) that murine T cell clones preferentially produced either IL-2 and γ-IFN or IL-4 and IL-5. The clones that produced IL-2 and γ-IFN were termed Th1 and the clones that produced IL-4 and IL-5 were termed Th2 (Figure 14). The fascinating aspect of this finding was that IL-2 and γ-IFN would drive the immune response toward development of cell-mediated immunity while elaboration of IL-4 and IL-5 would skew the response to antibody production. These seminal findings stimulated a veritable avalanche of reports that purported to support or disprove the original observations. However, it is generally accepted these days that naive T cells have a Th0 phenotype in which cytokines of either subset can be produced, most importantly IL-2 (Umlauf et al., 1993). Subsequently, T cells have the potential to be locked into a Th1 or Th2 pattern dependent on the stimulatory microenvironment (i.e., costimulatory molecules and cytokines) in which they encounter antigen (Seder and Paul, 1994). In this microenviroment, cytokines can be produced by NK cells, NT cells, or macrophages that would steer the nascent immune response toward a Th1 (as a consequence of IL-12 and γ-IFN production) or Th2 (as a consequence of IL-4 and IL-10 production). Much credibility to this dichotomy of T helper function was derived from the study of resistance of inbred strains of mice to the intracellular pathogen, *Leishmania major* (Sher and Coffman, 1992). In resistant C57BL/6 mice, γ-IFN production

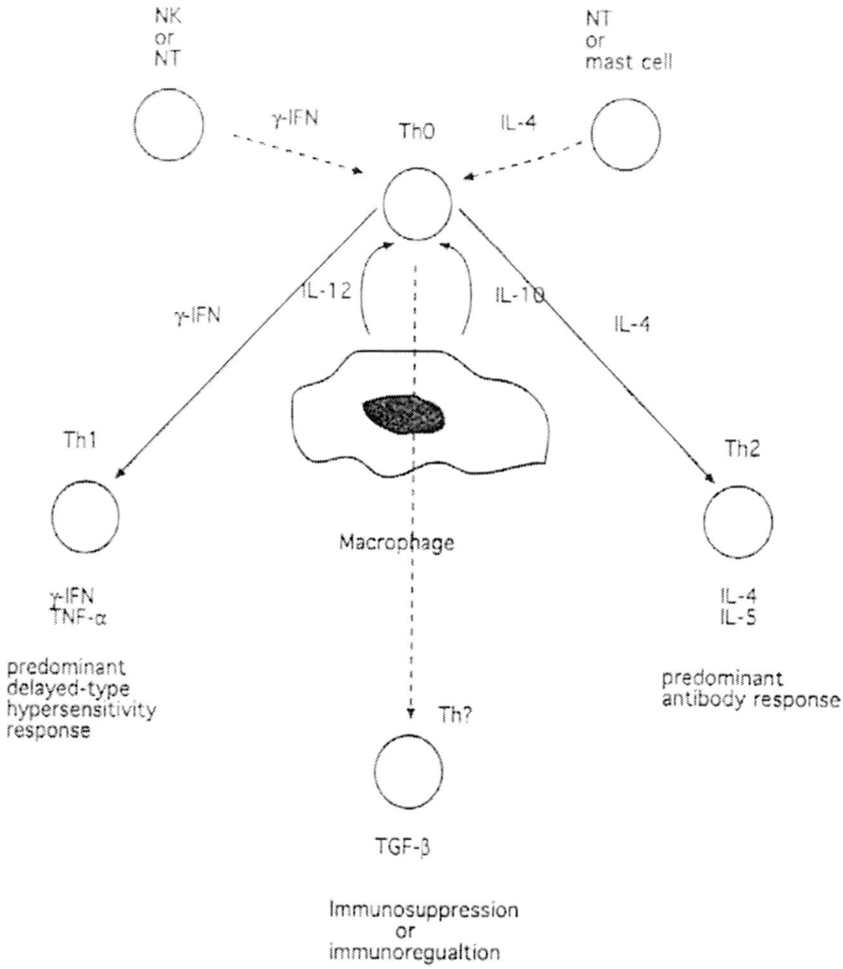

Figure 14. Development of functional T helper cell populations.

predominates after infection, strong cellular immunity is generated and the infection is resolved. In susceptible BALB/c mice, infection with *L. major* results in the production of IL-4 and a progressive infection is observed. Thus, although the immune system in BALB/c mice detected and responded to the pathogen, the antibody response as a consequence of the predominant Th2 response was ineffective in limiting the spread of the intracellular pathogen. In contrast, the T helper cell response of C57BL/6 mice that resulted in a strong cell-mediated immunity was able to eliminate the pathogen. Evidence that cross-regulation of immune responsiveness by cytokines existed was garnered from attempts to affect the response of both strains of mice. C57BL/6 mice injected with neutralizing antibodies to γ-IFN became susceptible and BALB/c mice injected with IL-12 became resistant. In addition, IL-4 inhibits the proliferation of Th1 cells and γ-IFN inhibits the proliferation of Th2 cells. Recently, the Th1/Th2 dichotomy has been extended to the system level (Clerici and Shearer, 1994). Since many different cell types can produce cytokines during an immune response (e.g., macrophages, B cells, basophiles), the development of a predominant CMI or antibody response is not strictly dependent on what cytokines T cells are producing, but to what pattern of cytokines is being produced by the system *in toto*. This information is critical to the understanding of basic immune function and will be vital to attempts to design immunotherapeutics. The progression of a number of human infectious diseases such as leshmaniasis, tuberculosis, toxoplasma, listeriosis, and perhaps even AIDS is influenced by Th1/Th2 response profile of the immune system (Modlin and Nutman, 1993). Still many questions remain, central of which are the conditions that drive the immune response toward a polarized response of the Th1 or Th2 type. The decisive event may well be the initial interaction of the immunogen or pathogen with the macrophage. As our understanding of host/pathogen relationships has grown and we now understand that pathogens have evolved means to evade severe immune rejection, the answer may lie in the initial influence of the pathogen on the activation of the macrophage. Perhaps certain microbes know what we conjecture, the means to down-regulate IL-12 production by the macrophage and up-regulate IL-10 production and diminish a strong cell-mediated response.

Recently, another functional T cell population has been described that is unique in its production of TGFβ. This cytokine is capable of down-regulating immune responses especially of the Th1 type (Wahl et al., 1989). $CD4^+$ T cell populations with antigen specific suppressive capabilities mediated via production of TGFβ have been described (Weiner et al., 1994). The microenvironmental influences that induce T cells to produce TGFβ are unkown, but seem to be present in gut-associated lymphoid tissue.

XIII. EFFECTOR FUNCTION AND IMMUNOLOGICAL MEMORY

After activation of T helper cells, effector function is induced that acts to clear the antigen. The primary response to antigen develops slowly, on the order of a week or so. During this period there is significant lymphocyte proliferation that signifies the period of clonal expansion. Typically, the primary immune response leads to the rapid clearance of antigen. Subsequently, effector function is down-regulated and evidence exists that the effector cells are eliminated. Much of the evidence for this is derived from studies on superantigens which stimulate a significant proportion of lymphocytes that utilize a unique V region gene in the T cell antigen receptor. In this case, the cells can be identified by the use of a monoclonal antibody specific for that V region gene product. This approach has shown that after an initial proliferative phase, the cells bearing the superantigen are eliminated and the responding T cells then disappear (Webb et al., 1990). Death of these cells is seen in the gut and spleen. Similar observations have also been made with mice expressing transgenic T cell receptor genes of known specificity. The elimination of effector cells is thought to reflect a protective mechanism. It is probably wise that these "loaded guns" are not left scattered throughout the body since they may inadvertently be triggered with dire consequences.

One of the major characteristics of the immune system is the ability to react more quickly and intensely to a second exposure to antigen (Gray, 1993). Generally, memory cells have the functional characteristics of being easier to activate, producing greater amounts of cytokine and generating an effector response of greater magnitude. The memory response by definition must rely on the preservation of some cells that responded to the original antigenic challenge. Based on the previously described elimination of cells after the primary response, the cellular basis of the secondary response is a mystery. To date, the precise phenotypic identification of the memory cell population especially of T cells is not known. Evidence suggests that B memory cells are of a separate lineage from the cells that produce a primary antibody response (Klinman, 1994). The formation of a memory B cell population is dependent on signals from T cells and some evidence suggests that CD40/CD40L interactions may be important. Memory B cells develop in the germinal centers of lymph nodes (MacLennan, 1994) There are also important questions to be answered about memory persistence and the role antigen plays in this process (Gray, 1994). Although much progress has been made in the past few years in terms of understanding the components of memory immune responses, it is obvious that many more questions need to be answered.

XIV. TOLERANCE

One characteristic of the immune system is non-responsiveness to self-determinants. Initially only deletion and suppression were studied as mechanisms to minimize or prevent autoimmunity. It is now apparent that there are a number of additional processes designed into the system such as induction of anergy, immunoregulation, and activation-induced cell death that act to prevent the presence, activation, and expansion of cells with autoimmune potential. There has recently been significant advances in understanding the mechanisms by which this is achieved and these will be detailed below.

A. Deletion

A basic mechanism of minimizing the number of lymphocytes with autoreactive potential is to delete them during development and it has been well established that antigen receptor binding during maturation results in tolerance (deletion) and not activation. Interestingly, in some cases, small numbers of mature cells expressing the transgenic receptor could be observed in the periphery. This is consistent with the detection of T cells with "forbidden receptor" specificities in the periphery of normal animals shortly after birth (Bonomo et al., 1995). It is postulated that deletional mechanisms are not 100% effective especially during times of peak lymphocyte production.

B. T cell Anergy

There are theoretical objections to deletion being the only mechanism to insure functional tolerance. For instance, it was speculated that not all self-determinants could be present in the microenvironment during maturation. For example, there was no a priori reason to believe that liver specific or pancreas specific proteins would be present in the thymus or the bone marrow microenvironment to tolerize developing lymphocytes. Also, it was well appreciated that B cells underwent somatic mutation of V region genes after activation giving rise to the possibility that mature cells with autoreactive receptor specificities would continually arise. Thus, the need for additional mechanisms to tolerize lymphocytes with specificity to self-determinants was evident. A possible mechanism for this was revealed as a consequence of *in vitro* studies on antigen reactive T cell clones. These T cells could be grown indefinitely by regular stimulation that required both antigen and a functional antigen presenting cell. It was observed that when these clones were stimulated with MHC Class II molecules and antigen in an artificial lipid planar membrane, they became unresponsive to subsequent attempts at stimulation with antigen and normal antigen presenting cells (Schwartz, 1990). This was reproduced with antibodies to the CD3 complex of the T cell receptor as well. Importantly, this indicated that antigen occupancy of the T cell receptor in the absence of costimulatory molecules was not a neutral event, but had a long-lasting impact on subsequent attempts at stimulation. T cells that have been made "anergic" characteristically neither proliferate nor make significant amounts of IL-2 in response to antigen on competent antigen presenting cells. The induced anergy tends to be long lasting *in vitro*. Recent evidence indicates that CD28/B7 interaction is the vital costimulatory signal that prevents the induction of anergy (June et al., 1990). Thus, costimulation with anti-CD28 prevent the induction of anergy in T cell clones stimulated by anti-CD3 antibodies (Harding et al., 1992). Some of the best evidence that similar mechanisms exist *in vivo* come from studies utilizing transgenic mice that express foreign Class I or Class II MHC molecules on various cell types through the use of tissue-specific promoters. For instance, in one strain of transgenic mice, the gene for a foreign MHC molecule is introduced under the control of the rat insulin promoter and, thus, is only expressed by the insulin-producing β cells of the pancreas (Burkly et al., 1989; Morahan et al., 1989). Interestingly, these mice develop diabetes with age, not as a consequence of an immune response but due to the transgene expression which adversely affects islet cell function. It can be shown through the use of a monoclonal antibody against the V region of the T cell receptor that recognizes the particular MHC gene product that the T cells were present in the spleen. Thus, they were not deleted intrathymically. However, T cells from the spleen, but not the thymus, of these mice were incapable of reacting to the MHC transgene *in vitro*. Addition of IL-2 to the cultures of spleen T cells resulted in the appearance of responsiveness to the transgene suggesting that excess cytokine can overcome anergy. In older mice, T cell responsiveness develops as the pancreatic β cells are lost as a consequence of transgene expression. Thus, continual presence of antigen is required for maintenance of the peripheral tolerance. The induction of functional anergy is related to the interaction of the T cells with antigen on the pancreatic cells in the absence of costimulatory signals. This resulted in a functional state similar to that induced in antigen-specific T cell clones *in vitro*. Recently, the importance of the CD28/B7 pathway in tolerance was assessed by experiments performed in transgenic mice that express the T cell costimulatory molecule B7 (ligand for CD28) under the control of the rat insulin promoter and mating them with transgenic mice that express a foreign Class II MHC protein (I-E^k) with pancreatic-specific expression (Guerder et al., 1994). Neither transgenic line developed diabetes, however, the double transgenic mice developed autoimmune diabetes characterized by T cell-mediated destruction of the pancreatic β cells. Thus, expression of the costimulatory molecule, B7, on the same cells as the foreign MHC was sufficient to trigger the T cells specific for the I-E transgene and initiate tissue destruction. Using different combinations of transgenic mice, these authors were able to show that induction of thymic tolerance by thymic expression of the foreign MHC (I-E^k) did not prevent the development of diabetes. This observation is important because it conclusively demonstrates that although tolerance is induced intrathymically by deleting thymocytes that recognize I-E^k with high affinity, the positive selection that occurs as well in thymic development results in T cells that can recognize and potentially react to tissue-specific peptides presented by the I-E^k transgene in the pancreas. It is the lack of costimulation that results in anergy and not autoreactivity. This induced anergy is potent because grafting pancreatic tissue from the B7/I-E^k double transgenic mice to the transgenic mice that express only I-E^k did not result in graft rejection. Thus, T cells anergized by encountering I-E^k without costimulation did not become responsive when encountering I-E^k and the co-stimulatory molecule B7 on engrafted tissue. These results have clearly demonstrated some important principles concerning the control of potential autoreactivity. Further issues will involve the role of cytokines and other costimulatory molecules in reversing or maintaining anergy. It will be of interest to determine whether the principles established for induction of anergy by MHC transgene expression in the pancreas hold for transgenes expressed in other tissue, such as liver, brain, or the articular collagen of the joints.

C. B cell Tolerance

Ig expressed on B cells can bind soluble as well as cell-bound antigens and, thus, the rules for B cell tolerance induction could be different than those for T cells.

Since B cell activation is almost entirely dependent on T cell help it was postulated that autoreactive B cells could exist *in vivo*, but in the absence of T cell help they would be inconsequential. Tolerance would only be important in the T cell population. Again significant insights into B cell tolerance were garnered from studies on transgenic mice. For instance, mice were constructed that were transgenic for a rearranged Ig gene that recognized hen egg lysozyme (HEL). Another transgenic line was made that led constitutively express HEL in either a membrane-bound or secreted form (Goodnow, 1992). In the double transgenic line that expressed cell surface-bound HEL, no mature lysozyme-binding B cells could be detected in the lymph node or spleen. Thus, the B cells can be deleted during maturation from immature precursors. In the double-transgenic line that expressed low levels of soluble HEL, there was little change in B cell numbers or function. It was calculated that the concentration of soluble HEL would only occupy 5% of the surface Ig on B cells. These B cells were capable of secreting anti-lysozyme antibodies upon adoptive transfer. This condition of immune non-responsiveness was termed "clonal ignorance." In other double-transgenic lines where the amount of secreted HEL was sufficient to occupy up to 45% of the sIg, surprisingly, B cells were not reduced in numbers, but were unable to secrete antibody upon stimulation. In contrast to the deletion of the B cells when exposed to cell-bound antigen, clonal deletion was not observed to high concentrations of soluble antigen. Significant changes were observed in the expression of sIg in these mice, however, in that membrane-bound IgM was reduced by more than one order of magnitude, but surface expression of IgD was unaffected. The induced functional anergy significantly affected the B cells in that they were unable to secrete antibody when stimulated with the T cell independent antigen, LPS. However, they were able to proliferate. Thus, the induction of tolerance and anergy in the B cell population involves a complex consideration of antigen concentration and physical state of the antigen (cell-bound or soluble). The extent that these mechanisms apply to other self-determinants remains to be determined.

D. Activation-induced Cell Death

Recent research results have revealed another mechanism for controlling autoreactivity that involves the deletion of activated cells. This process might act to prevent the expansion of lymphocytes that became activated inappropriately and would be downstream from the induction of anergy which regulates activation. For many years, researchers were intrigued with mice homozygous for the *lpr* gene or the *gld* gene (Cohen and Eisenberg, 1991). Homozygosity for either one of these recessive genes led to the development of a rampant, aggressive autoimmune state that resembled human systemic lupus erythematososus. These mice developed high levels of anti-DNA autoantibodies, vasculitis, and glomerulonephritis. There were also significant abnormalities in T cells with an expanded population that lacked expression of CD4 and CD8. An important advance in understanding these syndromes came from the cloning of both the *lpr* and *gld* genes and the discovery that they were a receptor-ligand pair. The *lpr* gene contained a retroviraltransposon insertion that resulted in premature termination of transcription (Watanabe-Fukunaga et al., 1992a, 1992b; Adachi et al., 1993). Formal proof that the mutant gene was responsible for the autoimmune syndrome was demonstrated by constructing transgenic *lpr/lpr* mice that express the wild type *Fas* gene. These mice do not develop autoimmunity. The *gld* gene that encodes the ligand for the *Fas* gene product differed from the wild type gene by a single point mutation (Takahashi et al., 1994). The Fas/FasL proteins are members of the *TNF/TNFR* superfamily and their function is to induce programmed cell death via apoptosis. Fas is expressed constitutively on many tissues, but the ligand generally requires cell activation for expression, especially in lymphocytes. At this time, there are still many questions to be answered concerning this novel system for controlling potentially autoreactive cells, however, it is important to remember that inactivation of either gene results in severe autoimmune disease. Interestingly, it has been shown that in naive T cells, a monoclonal antibody to *Fas* has costimulatory ability and enhances the proliferative response to anti-CD3 (Alderson et al., 1993). However, in long-term T cell lines, this same monoclonal antibody induced programmed cell death when used as a costimulator. Thus, some internal mechanism must be induced in cells that have undergone a certain number of divisions that initiates a self-destruct program. A cell that continues to receive and respond to proliferative signals may realize that it is either abnormal or has escaped mechanisms of tolerance induction and it is the wisdom of physiology to rid the system of potential cellular non-conformists.

XV. MUCOSAL IMMUNOLOGY—LIFE ON THE FRONT LINE OF THE IMMUNE SYSTEM

The epithelial surfaces of the body form the interface between the external and internal environments of the body. Thus, areas such as the skin, gastrointestinal tract, lungs, and the genito-urinary system are main sites for invasion of the host by infectious agents. In addition, these specific areas of the body must fulfill unique functional roles. For instance, the gastrointestinal tract must primarily absorb food and also fulfill a barrier function against bacteria, gastric acid, and pancreatic secretions among other things. As would be expected, there are significant lymphoid tissue associated with the lungs, skin, and genito-urinary and gastrointestinal tract (Brandtzaeg, 1989). A major effector function of mucosal-associated immune tissue is the production of IgA which is secreted into the external environment (Underdown and Schiff, 1986). The mucosal immune system seems interconnected in this regard in that the specific IgA response to oral administration of antigen will result in antigen-specific IgA production in the lungs. Unique among the mucosal-associated lymphoid tissue is that associated with the gastrointestinal tract (GALT).

It is an unusually complex mixture of cells with a complex job. The GALT must be non-reactive to the extreme variety of food antigens that are absorbed into the body, but yet react to pathogens. The composition of the GALT is described below. There is not a detailed description of the function of these cells because the manner in which the system works is largely unknown. It is estimated that the total cellularity of the GALT makes it the largest lymphoid organ in the body.

A. Peyer's Patches and Lymphatic Follicles

Peyer's patches are rather large, discrete, raised nodules of lymphoid tissue located in the small intestine mucosa. They usually contain a number of lymphatic follicles. Peyer's patches are abundant in the ileum, but do occur in the jejunum and duodenum as well. The follicular area of the Peyer's patch is composed of a germinal center, corona, and dome. It is comprised of largely $CD4^+$ T cells and B lymphocytes. Solitary lymphatic follicles are scattered throughout the intestinal tract. The follicle is separated from the intestinal lumen by a single layer of epithelial cells that are morphologically different from the typical columnar intestinal epithelial cell (Wolf and Bye, 1984). Distributed among the epithelial cells are specialized M cells that protrude through the epithelial layer. These cells take up and transport a wide variety of substances including microorganisms. M cells also express Class II MHC. It is thought that this unique design represents a mechanism by which intestinal luminal contents can be sampled and monitored by the immune system (Owen, 1994).

B. IEL

The intraepithelial lymphocytes (IEL) are found dispersed among the intestinal epithelial cells primarily those lining the villi of the small intestine. They are nestled between the epithelial cells abutting the villous basement membrane and shielded from the intestinal lumen by the apical tight junction of the epithelial cells. It is estimated that one of every 10 cells in the epithelium is a lymphocyte. The IEL are a complex mixture of at least six different phenotypic subsets mostly of the T cell lineage (Lefrancois, 1991). There are virtually no B cells in this unique lymphocyte population. The majority of IEL express the TCRγ/δ and CD8 although TCRα/β, $CD8^+$ T cells are also present. The proportion of the TCRα/β, $CD8^+$ T cell population present in the IEL is thought to depend on stimulation from the environment (e.g., intestinal bacteria). Their numbers are minimal in germ-free animals. Two populations of TCR α/β, $CD8^+$ cells can be observed based on expression of CD8α/α homodimers or CD8α/β heterodimers. Evidence exists that the TCRα/β, CD8α/α cells represent the population of T cells that mature independent of intrathymic residence (Poussier and Julius, 1994). However, this has not been conclusively proven. There are also $CD4^+/CD8^+$ double positive cells among the IEL. The proportion of these cells increases with age. They are different from double positive thymocytes in that they are $CD3^{bright}$. Population kinetics of IELs reveal diversity in that the data support the existence of both a long-lived population as well as a rapidly proliferating population. Interesting data on the population dynamics of IEL were obtained from experiments utilizing parabiosed (shared common circulation) mice that differed in an allelic marker expressed on lymphocytes. It was found that spleen and lymph node and even Peyer's patches T cells rapidly equilibrated between the animals indicating that they freely circulated. The IEL T cells largely remained of host phenotype, suggesting that this compartment was unique and did not recirculate outside the GALT. The exact functions of IEL are not known. Freshly isolated cells tend to be cytotoxic and antibody ligation of CD3 induces significant cytotoxicity. The target antigen is not known, although activated IEL can lyse Class II MHC expressing epithelial cells. Interestingly, the cytotoxicity is minimal in germ-free mice. Recent evidence suggests that IEL may have a role in regulating oral tolerance induction and epithelial cell function (Fujihashi et al., 1992; Boismenu and Havran, 1994) Thus, the many subpopulations of IEL may exist to regulate the induction of tolerance, selectively eliminate infected epithelial cells, or influence epithelial cell function.

C. Lamina Propria

The lamina propria is the tissue contained within the center of the intestinal villi. In addition to connective tissue elements, it contains a loosely associated population of lymphocytes. The composition of the lymphocytes is similar to that seen in normal lymph node with a preponderance of $CD4^+$ T cells and B cells and a minimal representation of $CD8^+$ T cells. Studies using immunohistochemistry have shown that a high percentage of the lamina propria B cells synthesize IgA. This IgA is destined for secretion into the intestinal lumen. The antibody is taken up by the epithelial cells where it is bound to another protein known as the secretory component and secreted into the intestinal lumen. It has been estimated that the daily production of IgA is more than twice that of IgG and IgM combined.

D. Oral Tolerance

Interestingly, it has been long known that feeding an antigen to an experimental animal results in systemic tolerance to that antigen (Chase, 1946). The primary mechanisms by which systemic reactivity is inhibited is by the induction of anergy or active suppression (Weiner et al., 1994). Evidence indicates that low dose feeding induces suppression and high dose feeding induces anergy. The suppression seems to be mediated by T cell elaboration of cytokines such as TGF_{β}, IL-4, and IL-10. This array of cytokines would potently inhibit macrophage antigen presenting function and cytokine elaboration. This may lead to induction of anergy due to the lack of an appropriate second signal for activation. Interestingly, the induction of oral tolerance has been shown experimentally to down-regulate induced autoimmune disease (Chen et al., 1994). For instance, feeding of myelin basic protein to

SJL mice decreases the severity of induced encephalomyelitis. The use of this strategy is currently being investigated as a therapy to alleviate human autoimmune disease.

XVI. IMMUNOTHERAPY

With the advent of recombinant DNA technology that led to the cloning of many novel molecules involved in immune function, the ability to manipulate gene expression, and the intense focus on the immunopathology of AIDS, the last decade has witnessed an explosion of knowledge in immunology. The possibility of affecting the progression of disease states by manipulating immune function is being actively pursued. Below is a partial list of some of the disease states in which immunotherapy applications are actively being investigated.

A. Immunotherapy of Cancer

The potential use of the immune system to eradicate neoplastic cells has long been a dream of immunologists. However, efficient use of immunotherapy has thus far remained elusive. There are a number of different strategies in use in the battle against cancer, ranging from attempts to bolster the immune response toward neoplastic cells with cytokines, to the use of toxin-conjugated monoclonal antibodies to bind to and eradicate neoplastic cells (Rosenberg, 1988; Riethmuller et al., 1993). Early attempts at immune system stimulation involved the administration of a potent cell wall extract from bacteria to patients with cancer. This strategy did not prove entirely useful except for bladder cancer (Jackson and James, 1994). As genes for cytokines were cloned, they became potential therapeutics for cancer (Chapman and Houghton, 1993). For instance, IL-2 which was initially described as "T cell growth factor," generated considerable excitement as a booster of immune function. However, significant toxicities were observed upon clinical testing and this precluded widespread utilization (Lotze et al., 1985). As new cytokines are discovered, there is again renewed hope that administration will augment tumor-specific immunity without significant toxicities. One area that has been the focus of intense clinical research involves the isolation of peripheral blood lymphocytes from the cancer patient and growing them *in vitro* with IL-2 for a period of time. IL-2 stimulates the growth of both T cells and NK cells and also activates their cytolytic machinery. The expanded cells are then reintroduced into the patient and IL-2 therapy initiated. Some of the early results using this methodology produced startling remissions of well established tumors. However, as the data accumulated the overall response rate was not as high as hoped and significant toxicities associated with IL-2 administration were seen. Other efforts included using tumor infiltrating lymphocytes (TILs) isolated from the resected tumor as the source of cells to grow and reintroduce. These therapies are most effective for renal cell carcinoma and melanoma with response rates in the range of 20%. More recent efforts have been in the area of genetic modification of TILs with cytokine genes. Thus, TILs isolated from patients would be transfected with a cytokine gene, expanded and reintroduced into the patient. It is hoped that the tumor cells would home to the residual tumor and produce sufficient quantities of cytokine to allow effector function in that particular microenvironment. This would avoid systemic administration of cytokine and the consequent toxicities. This therapy is still experimental with many regulatory and ethical issues to be resolved. In contrast to genetically manipulating anti-tumor lymphocytes, another strategy has been to manipulate the tumor in order to increase its immunogenicity (Pardoll, 1993). As discussed above, presentation of antigen in the absence of costimulatory molecules or sufficient cytokine results in the induction of anergy. Recently, numerous studies have evaluated the effects of introducing cytokine genes into tumor cells as a means of overcoming anergy. The transfected cells are reintroduced into animals in the hope of overcoming immune anergy. The cytokines tried to date include IL-1, IL-2, IL-4, IL-5, IL-6, IL-7, IL-12, GM-CSF, γ-IFN, and TNFα. In many cases, inhibition of tumor growth has been observed and early clinical trials are in progress. Other investigators have increased immunogenicity of experimental tumors by transfecting genes for Class I MHC, Class II MHC, or the T cell costimulatory molecule B7.

Recent advances in the technology of purifying and sequencing MHC-associated peptides has led to the exciting possibility of identifying the tumor-associated antigen recognized by T cells (Finn, 1993; Slingluf et al., 1994). This has been accomplished for melanomas and a few other different tumor types. It is hoped that these findings will allow novel immunotherapeutic strategies to be defined that would encompass possible vaccine development, immune augmentation, as well as development of assays to monitor-specific immune responses to the tumor.

There is a continuing effort to establish and perfect the use of toxin-conjugated monoclonal antibodies to cell surface antigens expressed by tumor cells (Riethmuller et al., 1993). Initially, there was hope that monoclonal antibody technology would provide the "magic bullet" for cancer therapy, that is, identification of determinant expressed uniquely by the tumor cells. However, this has not been the case as most, if not all, antigens are shared with normal cells to some extent. Recently, the strategy of using lineage-specific markers has proved potentially useful. For instance, the use of toxin-conjugated antibody to a determinant expressed on both B cell lymphoma cells and normal B cells destroys both the tumor as well as normal B cells (Vitetta and Uhr, 1994). However, the patient can survive for a time without B cells, and as described above, B cells would eventually regenerate from precursors in the bone marrow. The use of a combination of toxin-conjugated antibodies to different determinants expressed by tumor cells after cytoreductive therapy has been curative for lymphomas in mice. It is thought that the monoclonal antibody-toxin strategy might be most useful for eliminating residual disease that remains after the majority of the tumor cells have been killed by chemotherapy.

B. Infectious Disease

The struggle between the immune system and microbial pathogens is legendary (Marrack and Kappler, 1994). The ability to determine MHC-associated peptide sequence will undoubtedly have an impact in the battle. More effective vaccines could well be designed based on the knowledge of the microbe specific peptides presented by MHC molecules. Additionally, the potential use of costimulatory molecules (e.g., CD40L) to enhance the effectiveness of the vaccine is a significant possibility. Other possibilities include the construction of fusion proteins between the immunogen and cytokines like GM-CSF. GM-CSF increases the capacity of the macrophage to process and present antigen and can increase the antibody response to suboptimal doses of antigen. Our recent understanding that the profile of cytokines produced during an immune response can actually direct the response to an inefficient effector mechanism (e.g., predominant antibody instead of CMI), offers the potential to alter this with appropriate cytokine therapy.

C. Treatment of Autoimmune Disease

The ideal treatment of autoimmune disease would be to induce specific tolerance to the target of the immune reactivity (Sette et al., 1994). This would encompass, of course, knowing what the exact determinant is, as well as understanding the mechanisms for inducing nonresponsiveness. Certainly the technological ability exists to now determine the MHC-associated peptide that is triggering the autoimmune disease. This could lead to the design of an antagonist peptide that would bind to MHC but result in decreased binding strength by the T cell receptor, resulting in T cell "turn-off." This has already been accomplished at an experimental level (Meyers et al., 1993; Sloan-Lancaster et al., 1993; Franco et al., 1994).

The second approach is to interrupt T cell costimulation by introducing antagonists to these molecules or to cytokines involved in activation. For instance, a soluble construct of CTLA4 or administration of anti-CD40L monoclonal antibody prevents the induction of collagen-induced arthritis in mice (Jenkins and Johnson, 1993; Banchereau et al., 1994). Also, soluble constructs of cytokine receptors, such as IL-1R or TNFR, or IL-1R antagonist are being used in an attempt to reduce inflammation as a consequence of the immune response (Fanslow et al., 1990; Arend, 1993).

Finally, therapeutics are being designed that will mask the array of adhesion molecules on the vascular endothelium at the site of inflammation (Tanaka et al., 1993). Lymphocytes normally do not penetrate into tissue unless specifically invited to do so by adhesion molecules. Using antibodies to mask these molecules could diminish or prevent lymphocyte extravazation. An example where this would be useful is in multiple sclerosis where as a consequence of inflammation, lymphocytes are allowed to cross the blood-brain barrier.

D. Transplantation

As knowledge of peripheral tolerance mechanisms increase, more therapeutic strategies become possible. Disrupting costimulatory signal transduction to induce anergy in the responding T cells has become a popular strategy. The use of a soluble construct of CTLA4 to interrupt costimulatory signaling has been shown to prolong allograft life in mice (Boussiotis et al., 1994). Additionally, the use of soluble cytokine receptors might impact the rate of rejection, as well as diminish damage to the engrafted organ by the inflammation induced by chronic rejection.

XVII. PROSPECTS FOR THE FUTURE

As detailed above, there is a significant research effort aimed at manipulating immune function. It is hoped that in the near future, this effort will translate into new, effective therapeutics. This survey of immune function has revealed a complex, dynamic system that relies not on a simple "on-off switch" trigger to responsiveness, but instead on a system that integrates a variety of signals (both positive and negative) to fine tune the type and strength of the response. It has become evident that the nature and amplitude of these signals changes with time during an immune response (Yamashita and Boros, 1990; Ehlers et al., 1992). Also, there is undoubtedly significant functional redundancy designed into the system. One need only compare the activities of IL-2 and IL-15 or IL-4 and IL-13 to appreciate this. Consider as well the heterogeneity of human disease states and it is evident that the prospect of eliminating disease by administration of single cytokine is extremely simplistic. One only need examine the results of clinical trials on patients with sepsis in which different TNF antagonists were tested. In not one set of trials was there significant and beneficial improvement (Lynn and Cohen, 1995). The published preclinical data that showed efficacy of these molecules in animals only illuminates the problem of using experimental results in inbred mice as a basis to design and conduct clinical trials (Mohler et al., 1993). The humbling lesson from this experience only highlights the complexity of immune function and indicates that our knowledge, while large and formidable, is still not adequate to predict therapeutic efficacy. The only conclusion is that there is still much to understand about immune function in health and disease.

ACKNOWLEDGMENTS

The author gratefully acknowledges the efforts of Val Curnow in amassing the references used herein.

REFERENCES

Adachi, M., Watanabe-Fukunaga, R., & Nagata, S. (1993). Aberrant transcription caused by the insertion of an early transposable element in an intron of the Fas antigen gene of lpr mice. Proc. Natl. Acad. Sci. USA 90, 1756-1760.

Alderson, M. R., Armitage, R. J., Maraskovsky, E., Tough, T. W., Roux, E., Schooley, K., Ramsdell, F., & Lynch, D. H. (1993). *Fas* transduces activation signals in normal human T lymphocytes. J. Exp. Med. 178, 2231-2235.

Allison, J. P., & Havran, W. L. (1991). The immunobiology of T cells with invariant gamma/delta antigen receptors. Annu. Rev. Immunol. 9, 679-705.

Ambrus, J. L., Jr., Pippin, J., Joseph, A., Xu, C., Blumenthal, D., Tamayo, A., Claypool, K., McCourt, D., Srikiatchatochorn, A., & Ford, R. J. (1993). Identification of a cDNA for a human high-molecular-weight B-cell growth factor. Proc. Natl. Acad. Sci. USA 90, 6330-6334.

Arend, W. P. (1993). Interleukin-1 receptor antagonist. Adv. Immunol. 44, 153-174.

Armitage, R. J. (1994). Tumor necrosis factor receptor superfamily members and their ligands. Curr. Opin. Immunol. 6, 407-413.

Armitage, R. J., Fanslow, W. C., Strockbine, L., Sato, T. A., Clifford, K. N., Macduff, B. M., Andersen, D. M., Gimpel, S. D., Davis-Smith, T., Maliszewski, C. R., Clark, E. A., Smith, C. A., Grabstein, K. H., Cosman, D., & Spriggs, M. K. (1992). Molecular and biological characterization of a murine ligand for CD40. Nature 357, 80-82.

Armitage, R. J., Tough, T. W., Macduff, B. M., Fanslow, W. C., Spriggs, M. K., Ramsdell, F., & Alderson, M. R. (1993). CD40 ligand is a T cell growth factor. Eur. J. Immunol. 23, 2326-2331.

Aruffo, A., Farrington, M., Hollenbaugh, D., Li, X., Milatovich, A., Nonoyama, S., Bajorath, J., Grosmaire, L. S., Stenkamp, R., Neubauer, M., Roberts, R. L., Noelle, R. J., Ledbetter, J. A., Francke, U., & Ochs, H. D. (1993). The CD40 ligand, gp39, is defective in activated T cells from patients with X-linked hyper-IgM syndrome. Cell 72, 291-300.

Ashton-Rickardt, P. G., Bandeira, A., Delaney, J. R., Van Kaer, L., Pircher, H.-P., Zinkernagel, R. M., & Tonegawa, S. (1994). Evidence for a differential avidity model of T cell selection in the thymus. Cell 76, 651-663.

Babbitt, B. P., Allen, P. M., Matsueda, G., Haber, E., & Unanue, E. R. (1985). Binding of immunogenic peptides to Ia histocompatibility molecules. Nature 317, 359-361.

Badwey, J. A., & Karnovsky, N. L. (1980). Active oxygen species and the function of leukocytes. Ann. Rev. Biochem. 49, 695-726.

Baggioloni, M., Boulay, F., Badwey, J. A., & Curnutte, J. T. (1993). Activation of neutrophil leukocytes: Chemoattractant receptors and respiratory burst. FASEB J. 7, 1004-1010.

Banchereau, J., Bazan, F., Blanchard, D., Briere, F., Galizzi, J. P., van Kooten, C., Liu, Y. J., Rousset, F., & Saeland, S. (1994). The CD40 antigen and its ligand. Annu. Rev. Immunol. 12, 881-922.

Bancroft, G. J., Schreiber, R. D., Bosma, G. C., Bosma, M. J., & Unanue, E. R. (1987). A T-cell independent mechanism of macrophage activation by interferon gamma. J. Imm. 139, 1104-1107.

Bancroft, G. J., Sheehan, K. C. F., Schreiber, R. D., & Unanue, E. R. (1989). Tumor necrosis factor is involved in the T-cell independent pathway of macrophage activation in SCID mice. J. Imm. 143, 127-130.

Barber, L. D., & Parham, P. (1993). Peptide binding to major histocompatibility complex molecules. Annu. Rev. Cell Biol. 9, 163-206.

Beckman, E. M., Porcelli, S. A., Morita, C. T., Behar, S. M., Furlong, S. T., & Brenner, M. B. (1994). Recognition of a lipid antigen by CD1-restricted α/β T cells. Nature 372, 691-694.

Belich, M. P., Glynne, R. J., Senger, G., Sheer, D., & Trowsdale, J. (1994). Proteasome components with reciprocal expression to that of the MHC-encoded LMP proteins. Current Biology 4, 769-776.

Bendelac, A. (1995). Mouse NK1+ T cells. Current Opin. Immunol. 7, 367-374.

Bevan, M. J., Hogquist, K. A., & Jameson, S. C. (1994) Selecting the T cell repertoire. Science 264, 796-797.

Bevilacqua, M. P. (1993). Endothelial-leukocyte adhesion molecules. Annu. Rev. Immunol. 11, 767-804.

Bierer, B. E., & Hahn, W. C. (1993). T cell adhesion, avidity regulation and signaling: A molecular analysis of CD2. Immunology 5, 249-261.

Bikoff, E. K., Huang, L. Y., Episkopou, V., van Meerwijk, J., Germain, R. N., & Robertson, E. J. (1993). Defective major histocompatibility complex Class II assembly, transport peptide acquisition and CD4+ T cell selection in mice lacking invariant chain expression. J. Exp. Med. 177, 1699-1712.

Bix, M., & Locksley, R. M. (1995). Natural T cells: Cells that co-express NKRP-1 and TCR. J. Immunol. 155, 1020-1022.

Bix, M., Liao, N.-S., Zijlstra, M., Loring, J., Jaenisch, R., & Raulet, D. (1991). Rejection of class I MHC-deficient haemopoietic cells by irradiated MHC-matched mice. Nature 349, 329-331.

Bjorkman, P. J., Saper, M. A., Samraoui, B., Bennet, W. S., Strominger, J. L., & Wiley, D. C. (1987). Structure of the human Class I histocompatibility antigen HLA-A2. Nature 329, 506-512.

Boise, L. H., Gonzalez-Garcia, M., Postema, C. E., Ding, L., Lindsten, T., Turka, L. A., Mao, X., Nunez, G., & Thompson, C. B. (1993). bcl-x, a bcl-2-related gene that functions as a dominant regulator of apoptotic cell death. Cell 74, 597-608.

Boismenu, R., & Havran, W. (1994). Modulation of epithelial cell growth by intraepithelial gamma/delta T cells. Science 266, 1253-1255.

Bonomo, A., Kehn, P. J., & Shevach, E. M. (1995). Post-thymectomy autoimmunity: Abnormal T-cell homeostasis. Immunol. Today 16, 61-67.

Boussiotis, V. A., Gribben, J. G., Freemen, G. J., & Nadler, L. M. (1994). Blockade of the CD28 co-stimulatory pathway: A means to induce tolerance. Curr. Opin. Immunol. 6, 797-807.

Brandtzaeg, P. (1989). Overview of the mucosal immune system. Curr. Topics Microbiol. Immunol. 146, 13-28.

Bretscher, P., & Cohn, M. (1970). A theory of self-nonself discrimination: Paralysis and induction involve the recognition of one and two determinants on an antigen, respectively. Science 169, 1042-1049.

Brink, R., Goodnow, C. C., Crosbie, J., Adams, E., Eris, J., Mason, D. Y., Hartley, S. B., & Basten, A. (1992). Immunoglobulin M and D antigen receptors are both capable of mediating activation, deletion, or anergy after interaction with specific antigen. J. Exp. Med. 176, 991-1005.

Brown, J. H., Jardetzky, T. S., Gorga, J. C., Stern, L. J., Urban, R. G., Strominger, J. L., & Wiley, D. C. (1993). The three-dimensional structure of the human class II histocompatibility antigen HLA-DR1. Nature 364, 33-39.

Brunda, M. J. (1994). Interleukin-12. J. Leuk. Biol. 55, 280-288.

Burkly, L. C., Lo, D., Kanagawa, O., Brinster, R. L., & Flavell, R. A. (1989). T-cell tolerance by clonal anergy in transgenic mice with nonlymphoid expression of MHC class II I-E. Nature 342, 564-566.

Buus, S., Sette, A., Colon, S. M., & Grey, H. M. (1988). Autologous peptides constitutively occupy the antigen binding site on Ia. Science 242, 1045-1047.

Cambier, J. C., Pleiman, C. M., & Clark, M. R. (1994). Signal transduction by the B cell antigen receptor and its coreceptors. Annu. Rev. Immunol. 12, 457-472.

Chapman, P. B., & Houghton, A. N. (1993). Non-antibody immunotherapy of cancer. Curr. Opin. Immunol. 5, 726-731.

Chase, M. W. (1946). Inhibition of experimental drug allergy by prior feeding of the sensitizing agent. Proc. Soc. Exp. Biol. Med. 61, 257-259.

Chen, Y., Kuchroo, V. K., Inobe, J., Hafler, D. A., & Weiner, H. L. (1994). Regulatory T cell clones induced by oral tolerance: Suppression of autoimune encephalomyelitis. Science 265, 1237-1240.

Clerici, M., & Shearer, G. M. (1994). Th1-Th2 hypothesis of HIV infection: New insights. Immol. Today 15, 575-579.

Cohen, P. L., & Eisenberg, R. A. (1991). Lpr and gld: Single gene models of systemic autoimmunity and lymphoproliferative disease. Annu. Rev. Immunol. 9, 243-262.

Collins, T. L., Kassner, P. D., Bierer, B., & Burakoff, S. J. (1994). Adhesion receptors in lymphocyte activation. Curr. Opin. Immunol. 6, 385-393.

Cresswell, P. (1994). Assembly, transport, and function of MHC class II molecules. Annu. Rev. Immunol. 12, 259-293.

Cudkowicz, G., & Bennett, M. (1971). Peculiar immunobiology of bone marrow allografts. II. Rejection of parental grafts by resistant F1 hybrid mice. J. Exp. Med. 134, 1513-1522.

Curnutte, J. T. (1992). Molecular basis of the autosomal recessive forms of chronic granulomatous disease. Immunodef. Rev. 3, 149-172.

Davis, M. M., & Bjorkman, P. J. (1988). T-cell antigen receptor genes and T-cell recognition. Nature 334, 395-402.

Demotz, S., Grey, H. M., & Sette, A. (1990). The minimal number of class II MHC-antigen complexes needed for T cell activation. Science 249, 1028-1030.

Drinker, C. K., Field, M. E., & Ward, H. K. (1934). The filtering cpacity of lymph nodes. J. Exp. Med. 59, 393-412.

Du, X. X., Doerschuk, C. M., Orazi, A., & Williams, D. A. (1994). A bone marrow stromal-derived growth factor, interleukin-11, stimulates recovery of small intestinal mucosal cells after cytoablative therapy. Blood 83, 33-37.

Du, X. X., & Williams, D. A. (1994). Interleukin-11: A multifunctional growth factor derived from the hematopoietic microenvironment. Blood 83, 2023-2030.

Durie, F. H., Fava, R. A., Foy, T. M., Aruffo, A., Ledbetter, J. A., & Noelle, R. J. (1993). Prevention of collagen-induced arthritis with an antibody to gp39, the ligand for CD40. Science 261, 1328-1330.

Ehlers, S., Mielke, M. E. A., Blankenstein, T., & Hahn, H. (1992). Kinetic analysis of cytokine gene expression in the livers of naive and immune mice infected with listeria monocytogenes. J. Imm. 149, 3016-3022.

Engelhard, V. H. (1994). Structure of peptides associated with MHC class I molecules. Curr. Opin. Immunol. 6, 13-23.

Falk, K., Rotzschke, O., Stevanovic, S., Jung, G., & Rammensee, H. G. (1991). Allelle-specific motifs revealed by sequencing of self-peptides eluted from MHC molecules. Nature 351, 290-296.

Fanslow, W. C., Sims, J. E., Sassenfeld, H., Morrissey, P. J., Gillis, S., Dower, S. K., & Widmer, M. B. (1990). Regulation of alloreactivity *in vivo* by the soluble IL-1 receptor. Science 248, 739-742.

Fehling, H. J., Swat, W., Laplace, C., Kuhn, R., Rajewsky, K., Muller, U., & von Boehmer, H. (1994). MHC Class I expression in mice lacking the proteasome subunit LMP-7. Science 265, 1234-1237.

Finley, D., & Chau, V. (1991). Ubiquitination. Annu. Rev. Cell Biol. 7, 25-69.

Finn, O. J. (1993). Tumor-rejection antigens recognized by T lymphocytes. Curr. Opin. Immunol. 5, 701-708.

Foy, T. M., Durie, F. H., & Noelle, R. J. (1994). The expansive role of CD40 and its ligand, gp39, in immunity. Sem. Immunol. 6, 259-266.

Franco, A., Southwood, S., Arrhenius, T., Kuchroo, V. K., Grey, H. M., Sette, A., & Ishioka, G. Y. (1994). T cell receptor antagonist peptides are highly effective inhibitors of experimental allergic encephalomyelitis. Eur. J. Immunol. 24, 940-946.

Fujihashi, K., Taguchi, T., Aicher, W. K., McGhee, J. R., Bluestone, J. A., Eldridge, J. H., & Kiyono, H. (1992). Immunoregulatory functions for murine intraepithelial lymphocytes: γ/δ T cell receptor-positive (TCR^+) T cells abrogate oral tolerance, while α/β TCR^+ provide B cell help. J. Exp. Med. 175, 695-707.

Galli-Taliadoros, L. A., Sedgwick, J. D., Wood, S. A., & Korner, H. (1995). Gene knock-out technology: A methodological overview for the interested novice. J. Immunol. Meth. 181, 1-15.

Georgopolous, K., Bigby, M., Wang, J. H., Molnar, A., Wu, P., Winandy, S., & Sharpe, A. (1994). The Ikaros gene is required for the development of all lymphoid lineages. Cell 79, 143-156.

Germain, R. (1994). MHC-dependent antigen processing and peptide presentation: Providing ligands for T lymphocyte activation. Cell 76, 287-299.

Germain, R. N., & Margulies, D. H. (1993). The biochemistry and cell biology of antigen processing and presentation. Annu. Rev. Immunol. 11, 403-450.

Giri, J. G., Ahdieh, M., Eisenman, J., Shanebeck, K., Grabstein, K., Kumaki, S., Namen, A., Park, L. S., Cosman, D., & Anderson, D. (1994). Utilization of the beta and lambda chains of the IL-2 receptor by the novel cytokine IL-15. EMBO J. 13, 2822-2830.

Goldberg, A. L., & Rock, K. L. (1992). Proteolysis, proteasomes and antigen presentation. Nature 357, 375-379.

Goodnow, C. C. (1992). Transgenic mice and analysis of B-cell tolerance. Annu. Rev. Immunol. 10, 489-518.

Grabstein, K. H., Eisenman, J., Shanebeck, K., Rauch, C., Srinivasan, S., Fung, V., Beers, C., Richardson, J., Schoenborn, M. A., Ahdieh, M., Johnson, L., Alderson, M. R., Watson, J. D., Anderson, D. M., & Giri, J. G. (1994). Cloning of a T cell growth factor that interacts with the beta chain of the Interleukin-2 receptor. Science 264, 965-968.

Gray, D. (1993). Immunological memory. Annu. Rev. Immunol. 11, 49-77.

Gray, D. (1994). Regulation of immunological memory. Curr. Opin. Immunol. 6, 425-430.

Gruenberg, J., & Howell, K. E. (1989). Membrane traffic in endocytosis: Insights from cell-free assays. Annu. Rev. Cell Biol. 5, 453-481.

Gubler, U., Chua, A. O., Schoenhaut, D. S., Dwyer, C. M., McComas, W. Motyka, R., Nabavi, N., Wolitzky, A. G., Quinn, P. M., Familletti, P. C., & Gately, M. K. (1991). Coexpression of two distinct genes is required to generate secreted biactive cytotoxic lymphocyte maturation factor. Proc. Natl. Acad. Sci USA 88, 4143-4147.

Guerder, S., Meyerhoff, J., & Flavell, R. (1994). The role of the T cell costimulator B7-1 in autoimmunity and the induction and maintenance of tolerance to peripheral antigen. Immunity 1, 155-166.

Guy-Grand, D., Rocha, B., Mintz, P., Malassis-Seris, M., Selz, F., Malissen, B., & Vassalli, P. (1994). Different use of T cell receptor transducing modules in two populations of gut intraepithelial lymphocytes are related to distinct pathways of T cell differentiation. J. Exp. Med. 180, 673-679.

Haas, W. (1993). Gamma/delta T cells. Annu. Rev. Immunol. 11, 637-685.

Harding, C. V., & Unanue, E. R. (1990). Quantitation of antigen-presenting cell MHC Class II/peptide complexes necessary for T cell stimulation. Nature 346, 574-576.

Harding, F., McArthur, J. G., Gross, J. A., Raulet, D. H., & Allison, J. P. (1992). CD28-mediated signalling costimulates murine T cells and prevents the induction of anergy in T cell clones. Nature 356, 607-609.

Hardy, R. R., Carmack, C. E., Shinton, S. A., Kemp, D. A., & Hayakawa, K. (1991). Resolution and characterization of pro-B and pre-pro-B cell stages in normal mouse bone marrow. J. Exp. Med. 173, 1213-1225.

Hayakawa, K., & Hardy, R. R. (1988). Normal, autoimmune, and malignent $CD5^+$ B cells: The Ly1 lineage. Ann. Rev. Immunol. 6, 197-218.

Hershko, A., & Ciechanover, A. (1992). The ubiquitin system for protein degradation. Annual Review of Biochemistry 61, 761-807.

Hogquist, K. A., Jameson, S. C., Heath, W. R., Howard, J. L., Bevan, M. J., & Carbone, F. R. (1994). T cell receptor antagonist peptides induce positive selection. Cell 76, 17-27.

Hozumi, N., & Tonegawa, S. (1976). Evidence for the somatic rearragement of immunoglobulin genes coding for variable and constant regions. Proc. Natl. Acad. Sci. USA 73, 3628-3633.

Inaba, K., Steinman, R. M., Pack, M. W., Aya, H., Inaba, M., Sudo, T., Wolpe, S., & Schuler, G. (1992). Identification of proliferating dendritic cell precursors in mouse blood. J. Exp. Med. 175, 1157-1167.

Jackson, A. M., & James, K. (1994). Understanding the most successful immunotherapy of cancer. The Immunologist 2, 208-215.

Jackson, M. R., & Peterson, P. A. (1993). Assembly and intracellular transport of MHC class I molecules. Annu. Rev. Cell Biol. 9, 207-235.

Janeway, C. A., Jr. (1994). Thymic selection: Two pathways to life and two to death. Immunity 1, 3-6.

Jenkins, M. K., & Johnson, J. G. (1993). Molecules involved in T-cell costimulation. Curr. Opin. Immunol. 5, 361-367.

Joyce, S., & Nathanson, S. G. (1994). Methods to study peptides associated with MHC Class I molecules. Curr. Opin. Immunol. 6, 24-31.

June, C. H., Ledbetter, J. A., Lisely, P. S., & Thompson, C. B. (1990). Role of the CD28 receptor in T cell activation. Immunol. Today 11, 211-216.

Kantor, A. B., & Herzenberg, L. A. (1993). Origin of murine B cell lineages. Annu. Rev. Immunol. 11, 501-538.

Kappler, J. W., Roehm, N., & Marrack, P. (1987). T cell tolerance by clonal elimination in the thymus. Cell 49, 273-280.

Karjalainen, K. (1994). High sensitivity, low affinity-paradox of T cell receptor recognition. Curr. Opin. Immunol. 6, 9-12.

Kaufmann, S. H. E. (1993). Immunity to intracellular bacteria. Annu. Rev. Immunol. 11, 129-163.

Kawashima, I., Ohsumi, J., Mita-Honjo, K., Shimoda-Takano, K., Ishikawa, H., Sakakibara, S., Miyadai, K., & Takiguchi, Y. (1991). Molecular cloning of cDNA encoding adipogenesis inhibitory factor and identity with Interleukin-11. FEBS 283, 199-202.

Killeen, N., Stuart, S. G., & Littman, D. R. (1992). Development and function of T cells in mice with a disrupted CD2 gene. EMBO J. 11, 4329-4336.

Kim, K. M., Adachi, T., Nielsin, P. J., Terashima, M., Lamers, M. C., Kohler, G., & Reth, M. (1994). Two new proteins preferentially associated with membrane immunoglobulin D. EMBO J. 13, 3793-3800.

Kincade, P. W., Lee, G., Pietrangali, C. E., Hayashi, S. I., & Gimble, J. M. (1989). Cells and molecules that regulate B lymphopoiesis in bone marrow. Ann. Rev. Immunol. 7, 111-143.

Klinman, N. R. (1994). Selection in the expression of functionally distinct B-cell subsets. Curr. Opin. Immunol. 6, 420-424.

Knop, M., Schiffer, H. H., Rupp, S., & Wolf, D. H. (1993). Vaculolar/lysosomal proteolysis: Proteases, substrates, mechanisms. Curr. Opin. Cell Biol. 5, 990-996.

Korsmeyer, S. J. (1992). Bcl-2: A repressor of lymphocyte death. Immunol. Today 13, 285-288.

Korthius, R. J., Anderson, D. C., & Granger, D. N. (1994). Role of neutrophil-endothelial cell adhesion in inflammatory disorders. J. Crit. Care 9, 47-71.

Kovacsovics-Bankowski, M., & Rock, K. L. (1995). A phagosome-to-lysosome pathway for exogenous antigens presented on MHC Class I molecules. Science 267, 243-246.

Kronenburg, M. (1994). Antigens recognized by gamma/delta T cells. Curr. Opin. Immunol. 6, 64-71.

Landreth, K. S., Narayanan, R., & Dorshkind, K. (1992). Insulin-like growth factor-1 regulates pro-B cell differentiation. Blood 80, 1207-1212.

Lefrancois, L. (1991). Intraepithelial lymphocytes of the intestinal mucosa: Curiouser and curiouser. Seminars in Immunology 3, 99-108.

Linsley, P. S., & Ledbetter, J. A. (1993). The role of the CD28 receptor during T cell responses to antigen. Annu. Rev. Immunol. 11, 191-212.

Linton, P. J., Decker, D. J., & Klinman, N. R. (1989). Primary antibody forming cells and secondary B cells are generated from separate precursor subpopulations. Cell 59, 1049-1059.

Lotze, M. T., Matory, Y. L., Ettinghausen, S. E., Rayner, A. A., Sharrow, S. O., Seipp, C. A. Y., Custer, M. C., & Rosenberg, S. A. (1985). *In vivo* administration of purified human IL-2. II. Half life, immunologic effects, and expansion of peripheral lymphoid cells *in vivo* with recombinant IL-2. J. Immunol. 135, 2865-2875.

Lynn, W. A., & Cohen, J. (1995). Adjunctive therapy for septic shock: A review of experimental approaches. Clinical Infectious Diseases 20, 143-158.

Macatonia, S. E., Knight, S. C., Edwards, A. J., Griffiths, S., & Fryer, P. (1987). Localization of antigen on lymph node dendritic cells after exposure to the contact sensitizer fluorescein isothiocyanate. J. Exp. Med. 166, 1483-1494.

MacLennan, I. C. M. (1994). Germinal centers. Annu. Rev. Immunol. 12, 117-139.

Marrack, P., & Kappler, J. (1994). Subversion of the immune system by pathogens. Cell 76, 323-332.

Melchers, F., Haasner, D., Grawunder, U., Kalberer, C., Karasuyama, H., Winkler, T., & Rolink, A. G. (1994). Roles of IgH and L chains and of surrogate H and L chains in the development of cells of the B lymphocyte lineage. Annu. Rev. Immunol. 12, 209-225.

Melchers, F., Karasuyama, H., Haasner, D., Bauer, S., Kudo, A., Sakaguchi, N., Jameson, B., & Rolink, A. (1993). The surrogate light chain in B cell development. Immunol. Today 14, 60-68.

Meyers, L. K., Rosloniec, E. F., Seyer, J. M., Stuart, J. M., & Kang, A. H. (1993). A synthetic peptide analogue of a determinant of type II collagen prevents the onset of collagen-induced arthritis. J. Immunol. 150, 4652-4658.

Michalek, M. T., Benacerraf, B., & Rock, K. L. (1989). Two genetically identical antigen-presenting cell clones display heterogeneity in antigen processing. Proc. Natl. Acad. Sci. USA 86, 3316-3320.

Mizoguchi, H., O'Shea, J. J., Longo, D. L., Loeffler, C. M., McVicar, D. W., & Ochoa, A. C. (1992). Alterations in signal transduction molecules in T lymphocytes from tumor-bearing mice. Science 258, 1795-1798.

Montgomery, R. A., & Dallman, M. J. (1991). Analysis of cytokine gene expression during fetal thymic ontogeny using the polymerase chain reaction. J. Immunol. 146, 554-560.

Modlin, R. L., & Nutman, T. B. (1993). Type 2 cytokines and negative regulation in human infections. Curr. Opin. Immunol. 5, 511-517.

Mohler, K. M., Torrance, D. S., Smith, C. A., Goodwin, R. G., Stremler, K. E., Fung, V. P., Madani, H., & Widmer, M. B. (1993). Soluble tumor necrosis factor (TNF) receptors are effective therapeutic agents in lethal endotoxemia and function simultaneously as both TNF carriers and TNF antagonists. J. Immunol. 151, 1548-1561.

Momburg, F., Oritz-Navarrete, V., Neefjes, J. J., Goulmy, E., Van de Wal, Y., Spits, H., Powis, S. J., Butcher, G. W., Howard, J. C., Walden, P., & Hammerling, G. J. (1992). The proteasome subunits encoded by the major histocompatibility complex are not essential for antigen presentation. Nature 360, 174-177.

Monaco, J. J. (1992). Genes in the MHC that may affect antigen processing. Curr. Opin. Immunol. 4, 70-73.

Morahan, G., Allison, J., & Miller, J. F. A. P. (1989). Tolerance of class I histocompatibility antigens expressed extrathymically. Nature 339, 622-624.

Morrissey, P. J., Conlon, P., Charrier, K., Braddy, S., Alpert, A., Williams, D., Namen, A. E., & Mochizuki, D. (1991). Administration of IL-7 to normal mice stimulates B-lymphopoiesis and peripheral lymphadenopathy. J. Immunol. 147, 561-568.

Mosley, R. L., & Klein, J. R. (1993). Peripheral engraftment of fetal intestine into athymic mice sponsers T cell development: Direct evidence for thymopoietic function of murine small intestine. J. Exp. Med. 176, 1365-1373.

Mosman, T. R., Cherwinski, H., Bond, M. W., Giedlin, M. A., & Coffman, R. L. (1986). Two types of murine T helper clones. I. Definition according to profiles of lymphokine activities and secreted proteins. J. Immunol. 136, 2348-2357.

Musashi, M., Yang, Y.-C., Paul, S. R., Clark, S. C., Sudo, T., & Ogawa, M. (1991). Direct and synergistic effects of Interleukin 11 on murine hemopoiesis in culture. Proc. Natl. Acad. Sci. USA 88, 765-769.

Nakayama, K., Nakayama, K., Negishi, I., Kuida, K., Shinkai, Y., Louie, M. C., Fields, L. E., Lucas, P. J., Stewart, V., Alt, F. W., & Loh, D. Y. (1993). Disappearance of the lymphoid system in Bcl-2 homozygous mutant chimeric mice. Science 261, 1584-1588.

Nikolic-Zugic, J. (1991). Phenotypic and functional stages in the intrathymic development of alpha/beta T cells. Immunol. Today 12, 65-79.

Nitschke, L., Kosco, M. H., Kohler, G., & Lamers, M. C. (1993). Immunoglobulin D-deficient mice can mount normal immune responses to thymus-independent and -dependent antigens. Proc. Natl. Acad. Sci. USA 90, 1887-1891.

Nuchtern, J. G., Biddison, W. E., & Klausner, R. D. (1990). Class II MHC molecules can use the endogenous pathway of antigen presentation. Nature 343, 74-76.

Ohlen, C., Kling, G., Hoglund, P., Hansson, M., Scangos, G., Bieberich, C., Jay, G., & Karre, K. (1989). Prevention of allogeneic bone marrow graft rejection by H-2 transgene in donor mice. Science 246, 666-667.

Oltvai, Z. N., Milliman, C. L., & Korsmeyer, S. J. (1993). Bcl-2 heterodimerizes in vivo with a conserved homolog, Bax, that accelerates programed cell death. Cell 74, 609-619.

Orlowski, M. (1990). The multicatalytic proteinase complex, a major extralysosomal proteolytic system. Biochemistry 29, 10289-10297.

Owen, R. L. (1994). M cells-entryways of opportunity for enteropathogens. J. Exp. Med. 180, 7-9.

Palmitter, R. D., & Brinster, R. L. (1986). Germ-line transformation of mice. Ann. Rev. Genetics 20, 465-499.

Palombella, V. J., Rando, O. J., Goldberg, A. L., & Maniatis, T. (1994). The Ubiquitin-proteasome pathway is required for processing the NF-kB1 precursor protein and the activation of NF-kB. Cell 78, 773-785.

Pardoll, D. M. (1993). New strategies for enhancing the immunogenicity of tumors. Curr. Opin. Immunol. 5, 719-725.

Perlmutter, R. M., Levin, S. D., Appleby, M. W., Anderson, S. J., & Alberola-Ila, J. (1993). Regulation of lymphocyte function by protein phosphorylation. Annu. Rev. Immunol. 11, 451-499.

Peschon, J. J., Morrissey, P. J., Grabstein, K. H., Ramsdell, F. J., Maraskovsky, E., Gliniak, B. C., Park, L. S., Ziegler, S. F., Williams, D. E., Ware, C. B., Meyer, J. D., & Davison, B. L. (1994). Early lymphocyte expansion is severely impaired in Interleukin 7 Receptor-deficient mice. J. Exp. Med. 180, 1955-1960.

Peters, P. J., Neefjes, J. J., Oorschot, V., Ploegh, H. L., & Geuze, H. J. (1991). Segregation of MHC Class II molecules from MHC Class I molecules in the golgi complex for trasport to lysosomal compartments. Nature 349, 669-676.

Pfeffer, K., & Mak, T. W. (1994). Lymphocyte ontogeny and activation in gene targeted mutant mice. Annu. Rev. Immunol. 12, 367-411.

Pizzo, P. A. (1984). Granulocytopenia and cancer therapy: Past problems, current solutions, future challenges. Cancer 54, 2649-2661.

Poussier, P., & Julius, M. (1994). Thymus independent T cell development and selection in the intestinal epithelium. Annu. Rev. Immunol. 12, 521-553.

Powis, S. J., Deverson, E. V., Coadwell, W. J., Ciruela, A., Huskisson, N. S., Smith, H., Butcher, G. W., & Howard, J. C. (1992). Effect of polymorphism of an MHC-linked transporter on the peptides assembled in a class-I molecule. Naure 357, 211-215.

Raulet, D. H. (1992). A sense of something missing. Nature 358, 21-22.

Reed, E. (1994). The use of colony-stimulating factors as bone marrow support for systemic anticancer therapy. J. Natl. Med. Assoc. 86, 459-464.

Riethmuller, G., Schneider-Gadicke, E., & Johnson, J. P. (1993). Monoclonal antibodies in cancer therapy. Curr. Opin. Immunol. 5, 732-739.

Rock, K. L., Gramm, C., Rothstein, L., Clark, K., Stein, R., Dick, L., Hwang, D., & Goldberg, A. L. (1994). Inhibitors of the proteasome block the degradation of most cell proteins and the generation of peptides presented on MHC Class I molecules. Cell 78, 761-771.

Roes, J., & Rajewsky, K. (1993). Immunoglobulin D (IgD)-deficient mice reveal an auxiliary receptor function for IgD in antigen-mediated recruitment of B cells. J. Exp. Med. 177, 45-55.

Rolink, A., Karasuyama, H., Grawunder, U., Haasner, D., Kudo, A., & Melchers, F. (1993). B cell development in mice with a defective lambda 5 gene. Eur. J. Immunol. 23, 1284-1288.

Rosenberg, S. A. (1988). Immunotherapy of cancer using interleukin 2: Current status and future prospects. Immunol. Today 9, 58-62.

Rosenthal, A. S., & Shevach, E. M. (1973). Function of macrophages in antigen recognition by guinea pig T lymphocytes. I. Requirement for histocompatible macrophages and lymphocytes. J. Exp. Med. 138, 1194-1212.

Rotzschke, O., & Falk, K. (1994). Origin, structure and motifs of naturally processed MHC class II ligands. Curr. Opin. Immunol. 6, 45-51.

Rudensky, A. Y., Maric, M., Eastman, S., Shoemaker, L., DeRoos, P. C., & Blum, J. S. (1994). Intracellular assembly and transport of endogenous peptide-MHC class II complexes. Immunity 1, 585-594.

Saint-Ruf, C., Ungewiss, K., Groettrup, M., Bruno, L., Fehling, H. J., & von Boehmer, H. (1994). Analysis and expression of a cloned pre-T cell receptor gene. Science 266, 1208-1212.

Schild, H., Mavaddat, N., Litzenberger, C., Ehrich, E. W., Davis, M. M., Bluestone, J. A., Matis, L., Draper, R. K., & Chien, Y. (1994). The nature of major histocompatibility complex recognition by g/d T cells. Cell 76, 29-37.

Schwartz, R. H. (1990). A cell culture model for T lymphocyte clonal anergy. Science 248, 1349-1356.

Schwartz, R. H. (1992). Costimulation of T lymphocytes: The role of CD28, CTLA-4, and B7/BB1 in Interleukin-2 production and immunotherapy. Cell 71, 1065-1068.

Scott, P. (1993). IL-12: Initiation cytokine for cell-mediated immunity. Science 260, 496-497.

Seder, R. A., & Paul, W. E. (1994). Acquisition of lymphokine-producing phenotype by CD4+ T cells. Annu. Rev. Immunol. 12, 635-673.

Seidman, J. G., Leder, A., Edgell, M. H., Polsky, F., Tilghman, S. M., Tiemier, D. C., & Leder, P. (1978a). Multiple related immunoglobulin variable region genes identified by cloning and sequence analysis. Proc. Natl. Acad. Sci. USA 75, 3881-3886.

Seidman, J. G., Ledr, A., Nau, M., Norman, B., & Leder, P. (1978b). Antibody diversity. the structure of cloned immunoglobulin genes suggests a mechanism for generating anitbody diversity. Science 202, 11-22.

Sethna, M. P., van Parijs, L., Sharpe, A. H., Abbas, A. K., & Freeman, G. J. (1994). A negative regulatory function of B7 revealed in B7-1 transgenic mice. Immunity 1, 415-421.

Sette, A., Alexander, J., Ruppert, J., Snoke, K., Franco, A., Ishioka, G., & Grey, H. M. (1994). Antigen analogs/MHC complexes as specific T cell receptor antagonists. Annu. Rev. Immunol. 12, 413-431.

Shahinian, A., Pfeffer, K., Lee, K. P., Kundig, T. M., Kishihara, K., Wakeman, A., Kawai, K., Ohashi, P. S., Thompson, C. B., & Mak, T. W. (1993). Differential T cell costimulatory requirements in CD28-deficient mice. Science 261, 609-612.

Shawar, S. M., Vyas, J. M., Rodgers, J. R., & Rich, R. R. (1994). Antigen presentation by MHC ClassI-B molecules. Adv. Immunol. 12, 839-880.

Sher, A., & Coffman, R. L. (1992). Regulation of immunity to parasites by T cells and T cell derived cytokines. Annu. Rev. Immunol. 10, 385-409.

Slingluf, C. L., Hunt, D. F., & Engelhard, V. H. (1994). Direct analysis of tumor antigens. Curr. Opin. Immunol. 6, 733-740.

Sloan-Lancaster, A., Evavold, B. D., & Allen, P. M. (1993). Induction of T cell anergy by altered T cell-receptor ligand on live antigen presenting cells. Nature 363, 156-159.

Steinman, R. M. (1991). The dendritic cell system and its role in immunogenicity. Ann. Rev. Immunol. 9, 271-296.

Strasser, A., Whittingham, S., Vaux, D. L., Bath, M. L., Adams, J. M., Cory, S., & Harris, A. W. (1991). Enforced BCL2 expression in B-lymphoid cell prolongs antibody responses and elicits autoimmune disease. Proc. Natl. Acad. Sci. USA 88, 8661-8665.

Takahashi, T., Tanaka, M., Brannan, C. I., Jenkins, N. A., Copeland, N. G., Suda, T., & Nagata, S. (1994). Generalized lymphoproliferative disease in mice, caused by a point mutation in the Fas ligand. Cell 76, 969-976.

Tanaka, K., Tamura, T., Yoshimura, T., & Ichihara, A. (1992). Proteasomes: Protein and gene structures. The New Biologist 4, 173-187.

Tanaka, Y., Kobayashi, K., Takahashi, A., Arai, I., Higuchi, S., Otomo, S., Habu, S., & Nisimura, T. (1993). Inhibition of inflammatory liver injury by a monoclonal antibody against lymphocyte function-associated antigen. J. Immunol. 151, 5088-5095.

Tanaka, Y., Sano, S., Nieves, E., DeLibero, G., Rosa, D., Modlin, R. L., Brenner, M. B., Bloom, B. R., & Morita, C. T. (1994). Nonpeptide ligands for human g/d T cells. Proc. Natl. Acad. Sci. USA 91, 8175-8179.

Townsend, A., & Trowsdale, J. (1993). The transporters associated with antigen presentation. Semin. Cell Biol. 4, 53-61.

Trinchieri, G. (1994). Recognition of major histocompatibility complex Class I antigens by natural killer cells. J. Exp. Med. 180, 417-421.

Trowbridge, I. S., Collawn, J. F., & Hopkins, C. R. (1993). Signal-dependent membrane protein trafficking in the endocytic pathway. Annu. Rev. Cell Biol. 9, 129-161.

Umlauf, S. W., Beverly, B., Kang, S. Brorson, K., Tran, A., & Schwartz, R. H. (1993). Molecular regulation of the IL-2 gene: Rheostatic control of the immune system. Immunol. Rev. 133, 177-197.

Underdown, B. J., & Schiff, J. M. (1986). Immunoglobulin A: Strategic defense initiative at the mucosal surface. Annu. Rev. Immunol. 4, 389-417.

Vetrie, D., Vorechovsky, I., Sideras, P., Holland, J., Davis, A., Flinter, F., Hammarstrom, I., Kinnon, C., Levinsky, R., & Robrow, M. (1993). The gene involved in X-linked agammaglobulinemia (XLA) is a member of the src family of protein-tyrosine kinases. Nature 361, 226-233.

Vidard, L., Rock, K. L., & Benacerraf, B. (1992). Heterogeneity in antigen processing by different types of antigen-presenting cells. Effect of cell culture on antigen processing ability. J. Immunol. 149, 1905-1911.

Vitetta, E. S., & Uhr, J. W. (1994). Monoclonal antibodies as agonists: An expanded role for their therapeutic use. Cancer Research 54, 5301-5309.

von Boehmer, H., Teh, H. S., & Kiselow, P. (1989). The thymus selects the useful, neglects the useless and destroys the harmful. Immunol. Today 10, 57-61.

Walunas, T. L., Lenschow, D. J., Bakker, C. Y., Linsley, P. S., Freeman, G. J., Green, J. M., Thompson, C. B., & Bluestone, J. A. (1994). CTLA-4 can function as a negative regulator of T cell activation. Immunity 1, 405-413.

Watanabe-Fukunaga, R., Brannan, C. I., Copeland, N. G., Jenkins, N. A., & Nagata, S. (1992a). Lymphoproliferation disorder in mice explained by defects in Fas antigen that mediates apoptosis. Nature 356, 314-317.

Watanabe-Fukunaga, R., Brannan, C. I., Stoh, N., Yonchara, S., Copeland, N. G., Jenkins, N. A., & Nagata, S. (1992b). The cDNA structure, expression and chromosomal assignment of the mouse Fas antigen. J. Immunol. 148, 1274-1279.

Webb, S., Morris, C., & Sprent, J. (1990). Extrathymic tolerance of mature T cells: Clonal elimination as a consequence of immunity. Cell 63, 1249-1256.

Weiner, H. L., Friedman, A., Miller, A., Khoury, S. J., Al-Sabbagh, A., Santos, L., Sayegh, M., Nussenblatt, R. B., Trentham, D. E., & Haftler, D. A. (1994). Oral tolerance: Immunologic mechanisms and treatment of animal and human diseases by oral administration of autoantigens. Annu. Rev. Immunol. 12, 809-837.

Weiss, S., & Bogen, B. (1991). MHC Class II-restricted presentation of intracellular antigen. Cell 64, 767-776.

Wherry, J. C., Schreiber, R. D., & Unanue, E. R. (1991). Regulation of gamma interferon production by natural killer cells in SCID mice: Roles of tumor necrosis factor and bacterial stimuli. Infect. Immun. 59, 1709-1712.

Wolf, J. L., & Bye, W. A. (1984). The membranous epithelial (M) cell and the mucosal immune system. Annu. Rev. Med. 35, 95-112.

Xu, J., Foy, T. M., Laman, J. D., Elliott, E. A., Dunn, J. J., Waldschmidt, T. J., Elsemore, J., Noelle, R. J., & Flavell, R. A. (1994). Mice deficient for the CD40 ligand. Immunity 1, 423-431.

Yamashita, T., & Boros, D. L. (1990). Changing patterns of lymphocyte proliferation, IL-2 production and utilization, and IL-2 receptor expression in mice infected with Schistosoma mansoni. J. Immunol. 145, 724.

Yokoyama, W. M. (1993). Recognition structures on natural killer cells. Curr. Opin. Immunol. 5, 67-73.

Ziegler, H. K., & Unanue, E. R. (1981). Identification of a macrophage processing event required for I-region restricted antigen presentation to lymphocytes. J. Immunol. 127, 1869-1875.

Zinkernagel, R. M., & Doherty, P. C. (1974). Restriction of *in vitro* cell mediated cytotoxicity in lymphocytic choriomenengitis within a syngeneic or semiallogeneic system. Nature 248, 701-702.

Zurawski, G., & de Vries, J. E. (1994). Interleukin 13, an interleukin 4-like cytokine that acts on monocytes and B cells, but not on T cells. Immunol. Today 15, 19-26.

GROWTH FACTORS AND BONE

Anne M. Delany and Ernesto Canalis

ABSTRACT

Skeletal cells secrete a variety of growth factors that either enhance the replication of the bone cells or modify their differentiated function. The synthesis and activity of skeletal growth factors can be modified by systemic hormones, local factors, and binding proteins.

Growth Factors and Cytokines in Health and Disease
Volume 3A pages 127-155.

ISBN: 0-7623-0118-X

I. INTRODUCTION

Skeletal growth factors are a group of polypeptides synthesized by skeletal cells and known to regulate their replication and differentiated function. Skeletal growth factors act in concert with hormones that regulate bone remodeling, and systemic hormones that modify the synthesis, activation, receptor binding and the binding, proteins of growth factors secreted by skeletal cells.

The growth factors synthesized by bone cells are not specific to the skeletal system since they are synthesized by a number of nonskeletal tissues, and they are also present in the systemic circulation. Systemic growth factors may be present in plasma and are frequently synthesized by the liver or may be released after platelet aggregation. Local growth factors may be secreted by cells of the osteoblast or osteoclast lineage and act as autocrine or paracrine modulators of bone cell function. Skeletal cells synthesize fibroblast growth factors (FGF), platelet-derived growth factors (PDGF), transforming growth factors beta (TGFβ), bone morphogenetic proteins (BMP), and insulin-like growth factors (IGF). Some growth factors, such as FGF and PDGF, appear to play a major role in cell replication, whereas others, such as IGF, are weak mitogens and play a more important role in the differentiated function of the osteoblast. The reason for the existence of multiple skeletal growth factors is not completely understood. However, they have different functions, and modifications of their synthesis and activity by different hormones would suggest that they become operational under diverse physiological and pathophysiological circumstances (Tables 1 and 2).

Skeletal cells also secrete a variety of cytokines of the immune system, and they will be described in an adjoining chapter. Skeletal cytokines of the immune system appear to play a more important role in the bone resorptive process. We will limit this chapter to the presentation of more classic growth factors that play a role in

Table 1. Effects of Skeletal Growth Factors on Cells of the Osteoblastic Lineage

Growth Factor Tested	*Cell Replication*	*Collagen Synthesis*	*Other Differentiated Function*
FGF	↑	↑ cell number dependent ↓ independent	↓
PDGF	↑	↑ cell number dependent ↓ independent	↓
TGFβ	biphasic	↑ cell number independent	↓
BMP	↑	↑ cell dependency unknown	↑
IGF	Small ↑	↑ cell number independent	↑

Table 2. Effects of Skeletal Growth Factors on Osteoclast Recruitment, Bone Resorption, Matrix Degradation, and Matrix Metalloproteinase (MMP)-13 Expression

Growth Factor Tested	*Osteoclast Recruitment*	*Bone Resorption*	*Collagen Degradation*	*Osteoblastic MMP-13*
FGF	N.T.*	↑	→	↑
PDGF	Probably ↑	↑	↑	↑
TGFβ	↓	Biphasic	→	↓
BMP	N.T.	N.T.	N.T.	↓
IGF I	↑	↑	↓	↓
IGF II	→	→	↓	↓

***Note*:** N.T.: Not tested.

bone formation, although this does not exclude effects of other cytokines on bone formation.

II. FIBROBLAST GROWTH FACTORS

Fibroblast growth factors or heparin-binding growth factors are a family of structurally related polypeptides (Burgess and Maciag, 1989; Baird, 1993). FGFs were initially isolated from the central nervous system and are now known to be synthesized by a variety of tissues (Esch et al., 1985; Abraham et al., 1986). Acidic and basic FGF are synthesized by skeletal cells, and they regulate the replication and differentiated function of the osteoblast (Globus, 1989). Acidic and basic FGF have similar effects on bone cell function, although basic FGF appears to be somewhat more potent than acidic FGF as a bone cell mitogen (Canalis et al., 1987, 1988a; McCarthy et al., 1989a). Heparin enhances the effects of acidic FGF but not those of basic FGF on DNA synthesis rates in primary cultures of osteoblasts (McCarthy et al., 1989a). This differential sensitivity to heparin is found in other cell systems, and it has been postulated that heparin modifies the binding of acidic FGF to cell surface receptors. The stimulatory effects of acidic and basic FGF on cell replication are observed in cultures of intact calvariae and of isolated bone cells. The effect is not specific for osteoblasts since both forms of FGF stimulate the replication of skeletal fibroblasts/preosteoblasts as well as osteoblasts, suggesting a generalized increase in cell replication. Although the mitogenic effects of FGFs are not specific, they target cells of the osteoblastic lineage, and exposure of calvariae to FGF for short periods of time causes an increase in a cell population, which is subsequently able to synthesize type I collagen, a marker of osteoblastic function. The stimulatory effect of FGFs on type I collagen synthesis is dependent on cell number and it is prevented by DNA synthesis inhibitors, indicating that FGFs do not have a direct stimulatory effect on the differentiated function of the osteoblast (Canalis et al., 1988a).

Investigations conducted in primary cultures of osteoblasts, osteosarcoma cell lines, and osteoblastic MC3T3 cells indicate that FGFs directly inhibit the differentiated function of the osteoblast. Basic FGF has been studied in more detail than acidic FGF, and has been found to cause a dose-dependent inhibition of alkaline phosphatase, type I collagen, and osteocalcin mRNA levels, independently of its growth stimulatory effects (Rodan et al., 1989; Hurley et al., 1993). In addition, basic FGF blocked the stimulatory effects of parathyroid hormone (PTH) on cyclic AMP production confirming its inhibitory actions on the differentiated function of the osteoblast. The inhibition of type I collagen mRNA expression by basic FGF is due to transcriptional mechanisms.

The activity of FGFs in skeletal cells can be modified by changes in the expression and affinity of FGF receptors. There are four FGF receptor genes described and termed FGFR1, 2, 3, and 4. The various FGF receptors are structurally related protein tyrosine kinases that contain immunoglobulin-like domains (Wang et al., 1994; Klagsbrun and Baird, 1991). FGFs also bind to cell surface associated proteoglycans and these proteoglycans regulate the binding of FGF to high affinity receptors (Elenius et al., 1992; Mali et al. 1993). Studies on the expression and modulation of FGF receptors and cell surface binding proteoglycans in bone cells have not been reported, and at present, there is no information regarding the regulation of FGF binding to cells of the osteoblastic lineage. However, this level of regulation is likely to exist in skeletal cells and to be a critical step modulating FGF actions on bone cell function.

Basic FGF may affect osteoblastic cellular functions either directly or indirectly since it modifies the synthesis of other growth factors secreted by skeletal cells (Table 3). In osteoblast-like cells, basic FGF increases TGFβ1 mRNA levels by increasing transcription and not transcript stability (Noda and Vogel, 1989). Basic FGF also increases TGFβ activity in the medium of osteoblast cultures, and TGFβ might mediate selected actions of basic FGF in bone. Basic FGF inhibits the synthesis of IGF I and II in primary cultures of rat osteoblasts, an effect that is unrelated to its mitogenic properties for these cells. IGF I and II enhance the differentiated function of the osteoblast and the decrease in their expression by basic FGF correlates with the inhibitory actions of FGF on the osteoblastic phenotype (Canalis et al., 1993). Basic FGF also decreases the expression of insulin-like growth factor binding protein (IGFBP) 5, a binding protein consistently shown to increase bone cell growth (Canalis and Gabbitas, unpublished observations).

Acidic and basic FGF are synthesized by osteoblasts and stored in the extracellular matrix (Globus et al., 1989), however, the mechanisms by which FGFs are released from bone cells and matrix are currently unknown. Since neither acidic

Table 3. Regulation of Growth Factor Synthesis by Skeletal Growth Factors in Osteoblasts

	Synthesis								
	FGF		*PDGF*		*TGFβ*			*IFG*	
Growth Factor Tested	*Acidic*	*Basic*	*A*	*B*	*1*	*2*	*3*	*I*	*II*
FGF	NT*	↑	NT		↑	NT		↓	↓
PDGF	NT	P ↑	↑	→	NT			↓	↓
TGF β	NT	↑	↑	↑	↑	NT		↓	↓
IGF	NT		NT		NT			NT	
BMP	NT		NT		↑	NT		↑	↑

Notes: NT*: Not tested.
P ↑: Probable increase.

nor basic FGF has a peptide leader sequence, they are not secreted following classic mechanisms. Recent experiments performed in NIH 3T3 fibroblasts indicate that heat shock induces the release of acidic FGF, suggesting that it is secreted during periods of cell stress (Jackson et al., 1992). This would suggest a role for FGFs in tissue repair, and it would be in agreement with their stimulatory effects on angiogenesis (Montesano et al., 1986). Studies on the synthesis of acidic FGF in bone cells have not been reported. However, the synthesis of basic FGF in skeletal cells is regulated by other growth factors, as it has been reported in nonskeletal cells (Goldsmith et al., 1991; Hurley et al., 1994). PDGF induces the synthesis of basic FGF in non-skeletal cells and similar effects may occur in osteoblasts although this has not been reported. Basic FGF itself and TGF β induce the expression of basic FGF transcripts in osteoblastic cells, whereas hormones appear not to have a stimulatory effect on FGF expression (Table 4). Autoregulation of growth factor synthesis is not selective to basic FGF and it occurs with multiple growth factors in skeletal and nonskeletal cells. Growth factor autoregulation would suggest that following the initial induction of growth factor synthesis, the presence and therefore effect of a growth factor may persist for an extended period of time. Studies designed to determine the level of regulation of basic FGF synthesis in skeletal cells have not been completed. The basic FGF gene promoter contains GC rich elements but does not contain typical TATA or CAAT boxes. The basic FGF promoter contains TPAresponse elements, the binding sites for the transcription factor AP-1, suggesting that agents that modulate components of the AP-1 complex may regulate basic FGF synthesis in skeletal and nonskeletal cells. PDGF and FGF regulate Fos and Jun in skeletal cells; consequently, they have the potential to regulate basic FGF expression by transcriptional mechanisms (Shibata et al., 1991).

Table 4. Regulation of Growth Factor Synthesis by Selected Hormones in Osteoblasts*

	Synthesis								
	FGF		*PDGF*		*TGF β*			*IFG*	
Hormone Tested	*Acidic*	*Basic*	*A*	*B*	*1*	*2*	*3*	*I*	*II*
Parathormone	NT*	→	→	NT	→°	NT		↑	→
Vitamin D	NT	NT	→	NT	→°	NT		→	→
Glucocorticoids	NT	NT	→	NT	→°	NT		↓	→
Estrogens	NT	NT	→	NT	↑	NT		↑	→
Growth Hormone	NT	NT	→	NT		NT		Small ↑	→
Thyroid Hormones	NT	NT	→	NT		NT		Small ↑	→

Notes: * No studies reported on BMPs.
NT = Not tested.
° Glucocorticoids and bone resorbers activate TGF β.

The function of basic and acidic FGF in skeletal tissue is probably related to their mitogenic properties. As such, they may be important in the restoration of a cell population during specific physiological or pathophysiological conditions. Recombinant human basic FGF enhances fracture repair and cartilage repair in experimental animal models, suggesting a role in healing and fracture repair (Kawaguchi et al., 1994; Cuevas et al., 1988). Since both acidic and basic FGF inhibit the differentiated function of the osteoblast, it is unlikely that they are important in the maintenance of bone mass. There are no skeletal disorders where abnormalities in the synthesis of FGFs have been reported. However, mutations in FGFR1, 2, and 3 genes cause a variety of skeletal disorders. Mutations in FGFR1 and 2 genes cause the craniosynostosis and craniofacial abnormalities found in Pfeiffer, Jackson-Weiss, and Crouzon syndromes (Muenke et al., 1994; Jabs et al., 1994; Reardon et al., 1994). Mutations in the transmembrane domain of the FGFR3 gene causes achondroplasia, a common form of dwarfism (Shiang et al., 1994; Rousseau et al., 1994).

III. PLATELET-DERIVED GROWTH FACTORS

Platelet-derived growth factors are polypeptides which were initially isolated from circulating platelets and were subsequently found to be synthesized by a number of cells, including those of the skeletal tissue (Heldin and Westermark 1987, 1989; Westermark and Heldin 1993; Canalis et al., 1992). There are two PDGF genes, termed A and B, and both are expressed by normal osteoblasts and osteosarcoma cell lines (Rydzield and Canalis, 1996). The PDGF A and B genes encode two different PDGF chains, which associate in either homodimeric or heterodimeric form. Therefore, skeletal cells may synthesize PDGF AA, BB, or AB dimers.

The three PDGF dimers, AA, BB, and AB, have been tested for their effects on bone cell function. Whereas PDGF AA, BB, and AB are all biologically active, PDGF BB is a more potent mitogen than PDGF AA, and AB has intermediate effects on bone cell replication (Canalis et al., 1992). The potency of the three PDGF dimers is comparable in skeletal and nonskeletal cells and most investigators reporting the effects of PDGF in bone have tested PDGF AB or BB dimers. PDGF stimulates DNA synthesis and cell replication in intact calvariae and in cultures of osteoblast-enriched cells (Centrella et al., 1991d; Canalis et al., 1989). However, the effect on DNA synthesis is not specific for cells of the osteoblastic lineage and it is also observed in skeletal fibroblasts (Canalis et al., 1989). PDGF increases collagen and non-collagen protein synthesis in intact calvariae and osteoblast cultures, a modest effect that is directly related to an increase in cell number. On a cell-to-cell basis, however, PDGF does not stimulate collagen synthesis. Studies using histomorphometric analysis of intact calvariae treated with PDGF revealed a generalized effect on cell replication and a modest inhibition of bone matrix apposition rates, confirming that PDGF does not increase bone matrix formation (Hock and Canalis, 1994). Furthermore, the decrease in matrix apposition rates would suggest that PDGF causes a direct inhibition of the differentiated function of the osteoblast, but it does not preclude that cells dividing under the influence of PDGF eventually differentiate into mature functioning osteoblasts. In accordance with the inhibition of the osteoblastic differentiated phenotype, PDGF decreases the expression of IGF I and IGF II, the two skeletal growth factors consistently shown to increase osteoblastic function, and opposes the stimulatory effects of IGF I on bone collagen synthesis (Canalis et al., 1989). PDGF BB increases bone resorption and collagen degradation, effects that appear secondary to an increase in osteoclast number and an increase in interstitial collagenase expression by the osteoblast (Canalis et al., 1989). Interstitial collagenase is the major collagenase that initiates collagen degradation at neutral pH, and its induction by PDGF is probably relevant to bone matrix degradation. PDGF AA does not stimulate bone resorption and does not modify interstitial collagenase expression by the osteoblast. This may be due to the weaker activity of PDGF AA or to the fact that it plays no role in the process of bone matrix degradation. The effect of PDGF BB on interstitial collagenase expression by the osteoblast is likely due to transcriptional and post-transcriptional mechanisms. The collagenase gene contains AP-1 binding sites, and PDGF BB induces Fos and Jun, components of the AP-1 transcription factor. Studies on the regulation of the collagenase gene by PDGF BB in osteoblasts are currently underway in our laboratory.

There are two types of PDGF receptors, type A or α, which binds PDGF A and B chains and type B or β, which binds only PDGF B chains with high affinity. The PDGF receptors contain ligand activated tyrosine-kinase activity, and are structurally related to receptors for other cytokines such as colony stimulating factor 1 (Seifert et al., 1989; Yarden et al., 1986; Gronwald et al., 1988; Claesson-Welsh et al., 1989). Skeletal cells, like other cells, express PDGF α and β receptors and the biological actions of PDGF can be regulated by modifications in receptor number or affinity. Interleukin 1 and tumor necrosis factor α are the major regulators of PDGF binding to osteoblasts, although their effects vary within the cell line examined. In primary cultures of rat osteoblasts interleukin 1 and tumor necrosis factor α selectively increase PDGF AA binding, whereas in murine MC3T3 cell lines they increase the binding of PDGF AA and BB dimers (Centrella et al., 1992; Tsukamoto et al., 1991). This is not surprising since in MC3T3 cells interleukin 1 and tumor necrosis factor α increase PDGF α receptor transcripts and PDGF α receptor binds both PDGF A and B chains. The increase in PDGF AA binding to rat osteoblasts caused by interleukin 1 and tumor necrosis factor α results in an enhanced PDGF AA mitogenic activity. In contrast, in human osteoblasts interleukin 1 decreases PDGF AA binding and activity, and the differences in the results obtained do not have an obvious explanation (Gilardetti et al., 1991). In addition to cytokines, growth factors regulate PDGF binding to its receptor. TGF β decreases PDGF AA binding in skeletal and nonskeletal cells and FGF increases PDGF α receptor transcripts (Centrella et al., 1992; Gronwald et al., 1989; Schollman et al., 1992). PDGF itself increases PDGF α receptors by transcriptional mechanisms in fibroblasts, but similar studies have not been reported in osteoblasts (Eriksson et al., 1991). An increase in PDGF receptors and autoregulation of PDGF synthesis are probably important regulatory steps in the autocrine function of PDGF in bone cells. PDGF B chains bind to SPARC (secreted protein acidic enriched cysteine) or osteonectin, an extracellular matrix protein expressed at high levels in skeletal tissue (Lane and Sage, 1994). Interactions of SPARC with PDGF B chains have the capability of inhibiting the binding of PDGF B to its receptors and preventing its biological effects (Raines et al., 1992). Consequently, changes in the expression of SPARC by osteoblasts could modulate the activity of PDGF B. Glucocorticoids induce osteonectin expression in skeletal cells, therefore they may decrease the mitogenic activity of PDGF BB in cells of the osteoblastic lineage through this mechanism.

PDGF likely acts as a systemic regulator of skeletal cell function following platelet aggregation, and as a local regulator of bone cells since malignant and normal osteoblasts express PDGF A and B genes (Graves et al., 1984; Betsholtz et al., 1986; Heldin et al., 1986; Rydziel et al., 1994). Growth factors, but not hormones, enhance the expression of PDGF AA synthesis by osteoblasts. PDGF AA and BB increase PDGF A transcripts and polypeptide levels in osteoblast cultures (Rydziel et al., 1994). This effect is mimicked by phorbol esters suggesting that protein kinase C dependent mechanisms are involved in the regulation of PDGF

A expression by bone cells. Autoregulation of PDGF synthesis is not specific to the skeletal tissue, and is also observed in nonskeletal cells (Bhandari et al., 1994). TGF β increases PDGF A expression in osteoblasts and microvascular endothelial cells (Rydziel et al., 1992; Kavanaugh et al., 1988). The level of regulation of PDGF A and B synthesis in skeletal cells has not been established, but in nonskeletal cell lines, growth factors regulate PDGF expression by transcriptional mechanisms (Kavanaugh et al., 1988). The human PDGF A gene promoter region has been characterized and contains serum response elements, which are probably responsible for the induction of PDGF A by PDGF (Lin et al., 1992b). The PDGF A gene contains two alternative promoters, and at the present time it is not certain if one or both are operational in osteoblastic cells (Rorsman et al., 1988). Studies on the regulation of PDGF B in osteoblasts have been limited. Recent work from our laboratory indicated that the PDGF B gene is expressed by osteoblasts, and it is regulated by other skeletal growth factors. The PDGF B promoter contains common regulatory elements including SP-1 and AP-1 binding sites, and it is responsive to phorbol esters suggesting that agents that induce protein kinase C activity in osteoblasts may modify PDGF B expression (Pech et al., 1989).

Since PDGF has potent mitogenic activity and does not simulate the differentiated function of the osteoblast, it is likely that its main function in bone is related to the process of cell restoration and possibly fracture repair. This would be analogous to the effect of PDGF on wound healing (Pierce et al., 1991; Lynch et al., 1987). It is not likely that PDGF is relevant to the maintenance of bone mass since it decreases bone matrix apposition rates. The level of expression of PDGF in bone is low, and there is no information about its role in the pathogenesis or treatment of various skeletal disorders.

IV. TRANSFORMING GROWTH FACTORS BETA

Transforming growth factors beta are members of a family of closely related polypeptides with various degrees of structural homology and important effects on cell function (Kingsley, 1994; Barnard et al., 1990). There are five TGF β genes, termed TGF β 1-5, and skeletal cells like other mammalian cells, express TGF β1, 2, and 3 (Pelton et al., 1990). Bone matrix contains TGF β 1, 2, and 3 homodimers as well as 1.2, and 2.3 heterodimers (Ogawa et al., 1992). TGF β1, 2, and 3 have analogous effects on bone cell function with minor differences in potency (Centrella et al., 1991c). In cultures of intact calvariae, TGF β stimulates DNA synthesis and cell replication and has a modest stimulatory effect on collagen synthesis when it is transiently administered but not when calvariae are continuously exposed to the growth factor (Centrella et al., 1986). Bone histomorphometric analysis of intact rat calvariae treated with TGF β demonstrates a stimulatory effect of TGF β on matrix apposition rates (Hock et al., 1990). Similarly, *in vivo* studies have demonstrated a stimulatory effect of TGF β on bone formation (Rosen et al., 1994; Marcelli et al., 1990). The effects of TGF β on cell replication and the differentiated function of the osteoblast vary with the osteoblastic cell line used. In primary cultures of rat osteoblasts, TGFβ has a biphasic stimulatory effect on DNA synthesis, whereas in rat osteosarcoma cell lines, it inhibits cell growth (Rosen et al., 1988; Centrella et al., 1987; Noda and Rodan, 1987). TGF β causes a uniform inhibition of alkaline phosphatase activity in all osteoblast cell models tested suggesting that it inhibits the differentiated function of the osteoblast. Furthermore, TGF β inhibits osteocalcin production in rat osteosarcoma cells by transcriptional mechanisms (Noda, 1988). Although TGF β inhibits classic phenotypic markers of the osteoblast, it enhances type I collagen, fibronectin, osteonectin, and osteopontin expression in osteosarcoma cell lines (Noda and Rodan, 1987; Bonewald et al., 1992). The changes in fibronectin synthesis are cell line dependent, and may be responsible for phenotypic changes dependent on extracellular matrix composition, whereas the stimulation of type I collagen synthesis may be responsible for changes in bone formation (Ignotz and Massague, 1986). The stimulation of type I collagen transcripts in osteoblasts by TGF β is dependent on the degree of cell differentiation, and tends to decline as cells become more differentiated (Shibata et al., 1993).

The actions of TGF β on bone resorption have also been controversial. TGF β has a biphasic effect on the production of osteoclast-like cells, and at low concentrations it enhances osteoclast-like cell formation, whereas at high concentrations it is inhibitory. The stimulatory effect on osteoclast formation seems to be related to the production of prostaglandins (Shinar and Rodan, 1990). The inhibitory effect of TGF β on osteoclast-like cell formation is observed in human bone marrow cultures and appears secondary to a decrease in the differentiation of early stem cells into cells of the osteoclast lineage with a shift to cells of the granulocyte pathway (Chenu et al., 1988). The subsequent decrease in mature osteoclasts is probably responsible for the inhibitory actions of TGF β on bone resorption (Pfeilschifter et al, 1988).

TGF β type I and II receptors were recently identified and cloned, and both appear necessary for complete cellular responses to TGF β (Ebner et al., 1993; Lin et al., 1992a; Chen et al., 1993b; Laiho et al., 1990; Bassing et al., 1994). Osteoblasts express type I and type II TGFβ receptors and it is reasonable to assume that, like in other cell types, they require the expression of the two receptors for full cellular responses. However, studies to prove this hypothesis in skeletal cells have not been conducted. Parathyroid hormone and glucocorticoids modify the binding of TGF β to its receptors on osteoblasts (Centrella et al., 1988, 1991b). Glucocorticoids shift the binding of TGF β from type I and II receptors to betaglycan, formerly known as TGF β type III receptor, and increase betaglycan mRNA expression by the osteoblast (Nakayama et al., 1994). Betaglycan is not signal transducing, but it may modulate the binding of TGF β to its receptors (Wang et al., 1991; Lopez-Casillas et al., 1994). The increased expression of betaglycan, secondary to glucocorticoid exposure, may be in part responsible for the inhibitory effects of glucocorticoid on bone cell replication, since glucocorticoids oppose the stimulatory effect of TGF β on DNA synthesis in osteoblast cultures. This does not preclude a direct action of glucocorticoids on specific osteoblastic genes or modifications in the synthesis or activity of other growth factors (Delany et al.,1995).

TGF β is secreted as a latent high molecular weight complex consisting of the carboxy terminal remnant of the TGF β precursor and a TGF β binding protein (Derynck et al., 1986; Wakefield et al., 1988). The biologically active levels of TGF β depend on changes in its synthesis or in the activation of its latent form. Glucocorticoids and bone resorbing agents increase the levels of biologically active TGF β in cultures of osteoblasts and intact bone, possibly by inducing the production of lysosomal proteases (Oursler et al., 1991; 1993). Activation is an important control mechanism for TGF β. For instance, glucocorticoids do not regulate TGF β synthesis, but increase its active form. Basic FGF enhances TGF β1 gene expression in osteoblasts and studies using TGF β1 gene promoter constructs reveal TGF β1 autoregulation, which is likely to occur in skeletal cells as it has been reported in nonskeletal cells (Kim et al., 1989a, 1989b). Estrogens increase TGF β1 polypeptide and mRNA levels in cultures of human osteoblasts and ovariectomy reduces the concentrations of TGF β in rat bone (Oursler et al., 1991; Finkelman et al., 1992). There have been limited studies on the regulation of TGF β2 and TGF β3 synthesis in osteoblasts. Examination of the TGF β2 and TGF β3 gene promoters reveals the existence of cyclic AMP responsive elements (CREs) suggesting that agents that induce cyclic AMP may regulate TGF β2 and β3 at the transcriptional level in osteoblast cultures (Noma et al., 1991; Lafyatis et al., 1991).

While there has been controversy about the actual role of TGF β in bone cell function, *in vivo* studies demonstrating its osteogenic potential would suggest a possible role in fracture healing, analogous to that demonstrated in wound healing (Joyce et al., 1990; Beck et al., 1990). Studies involving targeted gene disruption of the mouse TGF β1 gene did not result in changes in embryonic development or skeletal structure (Shull et al., 1992). This may have been due to early death of the TGF β1 deficient mice secondary to severe inflammatory disease. It is, however, possible that TGF β plays a role in the maintenance of bone formation and bone mass. Decreased skeletal TGF β concentrations in ovariectomized rats suggests a possible role in bone abnormalities in the estrogen deficient state. Similarly, changes in TGFβ receptor binding and activity induced by glucocorticoids would suggest a role for TGF β in the mechanism of action of glucocorticoids in bone. From a therapeutic point of view, it is likely that the local administration of TGF β will be useful in wound and fracture healing, but it is unlikely that the systemic administration of TGF β will be a therapeutic option because of possible side effects including generalized stem cell growth inhibition (Migdalska et al., 1991).

V. BONE MORPHOGENETIC PROTEINS

Investigations performed during the past several decades revealed that, in addition to growth factor activity, bone matrix extracts have osteoinductive properties.

These properties are attributed to a family of BMPs or osteoinductive factors which have the capability of inducing new endochondral bone formation when implanted with a carrier in subcutaneous tissue (Canalis et al., 1988b; Celeste et al., 1990; Wozney et al., 1988). BMPs are members of the TGF β superfamily of polypeptides, which have a conserved carboxy terminal region containing seven cysteine repeats. Members of the TGF β superfamily also include the products of genes involved in the development of *Drosophila* melanogaster embryos, activins, and inhibins (Table 5; Lyons et al., 1991; Doctor et al., 1992; Wharton et al., 1991; Jones et al., 1992; Wall et al., 1993). BMPs induce bone formation and the differentiation of mesenchymal cells into the osteogenic pathway (Reddi, 1983; Sampath et al., 1993).

There have been a number of studies assessing the biological activities of BMP 2, BMP 3 or osteogenin, BMP 4, and BMPs 7 and 8 or osteogenic proteins 1 and 2. These investigations indicate that BMPs stimulate DNA and collagen synthesis in osteoblast cultures, and this effect is similar to that of TGF β (Takuwa et al., 1991; Chen et al., 1991a). However, BMPs have unique activities, since they stimulate parameters of the differentiated function of the osteoblast, such as alkaline phosphatase activity, cyclic AMP responsiveness to parathyroid hormone, and osteocalcin synthesis. These effects are different from those observed with TGF β and demonstrate that BMPs stimulate the osteoblastic differentiated phenotype (Luyten et al., 1989; Vukicevic et al., 1989; Sampath et al., 1992; Knutsen et al., 1993; Ozkaynak et al., 1992). In addition to its direct effects on osteoblastic function, BMP 2 induces the differentiation of stromal cells into cells of the osteoblastic lineage and inhibits myogenic differentiation (Thies et al., 1992; Yamaguchi et al., 1991). This observation was confirmed by demonstrating that BMP inhibits myogenesis and the expression of known myogenic determination

Table 5. Members of the Transforming Growth Factor β Superfamily

Transforming growth factor beta (TGF β) 1, 2, 3, 4, 5
Activins, Inhibins
Bone morphogenetic protein (BMP) 2 (Formerly 2A)
Bone morphogenetic protein 3 or osteogenin
Bone morphogenetic protein 4 (Formerly 2B)
Bone morphogenetic protein 5
Bone morphogenetic protein 6 or Vgr-1
Bone morphogenetic protein 7or osteogenic protein 1 (OP-1)
Bone morphogenetic protein 8or osteogenic protein 2 (OP-2)
Drosophila decapentaplegic gene (DPP)
Drosophila 60 A gene
Xenopus vegetal pole derived transcripts (Vg-1)
Vg-related (Vgr) 1 and 2
Muellerian Inhibiting Substance (MIS)
Growth and Differentiation Factor (GDF)

genes, including myogenin, myo D, and myf 5 (Murray et al., 1993). Expression of inhibitors of differentiation which contain helix-loop-helix domains was surprisingly enhanced by BMP in osteoblast-like cells (Ogata et al., 1993). Some of the activities of BMPs in bone may be mediated by the expression of other cytokines and BMP 2 enhances interleukin 6 and TGF β1 mRNA levels in skeletal cells (Zheng et al., 1994). Interleukin 6 is important in the recruitment of osteoclasts, suggesting that BMP 2 may play a role in the process of bone resorption and bone remodeling.

Studies on expression of BMP receptors in bone cells have been limited. Recent work in *Drosophila* demonstrated the existence of two BMP type I receptors, encoded by the genes saxophone (sax) and thick veins (tkv) (Nellen et al., 1994; Brummel et al., 1994; Penton et al., 1994; ten Dijke et al. 1994). These receptors possess serine/threonine kinase domains, and are implicated in the effects of *Drosophila* decapentaplegic on embryonic development. Studies on the regulation of BMP expression in skeletal cells have been limited, and it is important to note that BMPs are mostly expressed in nonskeletal tissues (Ozkaynak et al., 1990, 1992; Chen et al., 1993a).

A number of *in vivo* studies have shown that BMPs enhance endochondral bone formation, and local application of BMPs heal bone defects (Reddi and Cunningham, 1991). Consequently, BMPs could play an important function in bone repair and in the treatment of non-union fractures. Their role in the pathogenesis of bone disorders and their therapeutic potential as systemic agents is currently not known. Activin has similar effects to TGF β in osteoblast cultures, although it probably acts through independent receptors (Centrella et al., 1991a). The effects of activin on bone collagen synthesis may result in increased bone formation, but its role in bone remodeling has not been clarified. It is also unknown if activin is expressed by skeletal cells or if it alters the expression of other growth factors in bone tissue.

VI. INSULIN-LIKE GROWTH FACTORS

Insulin-like growth factors I and II are among the most abundant growth factors secreted by bone cells. They have a modest effect on cell replication while stimulating the differentiated function of the osteoblast (Table 1). These two 7500 dalton growth factors are structurally related, however, in bone IGF I is a more potent modulator of cell function than IGF II (Daughaday and Rotwein, 1989; McCarthy et al., 1989b). The IGF I receptor mediates most of the biological actions of IGF I and II. This receptor is a transmembrane glycoprotein tetramer ($\alpha_2\beta_2$) with ligand-activated tyrosine kinase activity. Insulin receptor substrate I (IRS-1) is a well characterized substrate for the IGF receptor tyrosine kinase, mediating the effects of IGF binding (Chuang et al., 1993). IRS-I can associate with and phosphorylate proteins containing Src homology (SH) domains, and activate diverse signal transduction cascades which include mitogen-associated kinase and phosphatidyl inositol-3-kinase (Giorgetti et al., 1994; Way and Mooney, 1993). Further downstream regulators which play a role in IGF I signal transduction include protein kinase A and protein kinase C (Kachra et al., 1994). A recent study showed that IRS-1 can associate with $\alpha_v\beta_3$ integrin (a vitronectin receptor) in insulin-stimulated fibroblasts, suggesting an additional potential mechanism by which IGFs can modify osteoblast function (Vuori and Ruoslahti, 1994). IGF I receptor number in osteoblasts can be modulated by a number of agents which regulate bone cell metabolism, including PDGF, beta 2 microglobulin, prostaglandins, and 1, 25 dihydroxyvitamin D_3 (Rubini et al., 1994; Hakeda et al., 1991; Kurose et al., 1990; Centrella et al., 1989).

IGF I stimulates the differentiated function of the osteoblast. It increases trabecular bone formation in ovariectomized rats, however, at high doses IGF I increases osteoclast number and activity (Mueller et al., 1994). The effect of IGF I on bone formation is mediated in part by its ability to increase matrix apposition rates (Hock et al., 1988). Type I collagen is the most important structural protein in bone, and IGF I increases the synthesis of α1(I) collagen at least in part by increasing gene transcription (Delany and Canalis, unpublished observations). In osteoblasts IGFs decrease the synthesis of interstitial collagenase the only enzyme able to initiate the degradation of type I collagen at neutral pH (Canalis et al., 1995; Matrisian and Hogan, 1990). Although osteoclasts are responsible for bone degradation, there is increasing evidence for the coupling of osteoclastic and osteoblastic function, and osteoblast collagenase may play a role in the initiation of bone collagen degradation and bone resorption (Rodan and Martin, 1981; Martin and Ng, 1994). Due to its ability to increase collagen synthesis while decreasing collagenase levels, IGF I plays a key role in the maintenance of bone mass.

The mechanisms by which growth factors and hormones regulate IGF synthesis are currently under investigation. The genes for IGF I and IGF II are well conserved across species (Daughaday and Rotwein, 1989), allowing for the development of animal and cell culture models for studying IGF gene regulation. However, the genes for IGF I and IGF II are complex. The IGF II gene gives rise to multiple species of transcripts due to the use of four alternative promoters, splicing polyadenylation signals (van Dijk et al., 1991). Similarly, the IGF I gene gives rise to multiple transcripts due to the use of alternative promoters located on exons 1 and 2, alternate splicing, and differences in the length of the 3' untranslated region (Hall et al., 1992). Little is known about the mechanisms by which IGF II synthesis is regulated in bone. Hormones do not modify IGF II expression in osteoblasts, but growth factors are inhibitory (Tables 2 and 3). IGF I gene regulation has been studied more extensively than IGF II. The rat gene has six exons and the 70 amino acid mature protein is encoded by sequences found within exons 3 and 4 (Daughaday and Rotwein, 1989). In bone, as in other extrahepatic tissues, transcripts initiated within the exon 1 promoter predominate (Pash et al., 1995). The exon 1 promoter does not contain classical eukaryotic promoter elements, such as a CAAT or TATA box, and it has four transcription initiation sites (Hall et al., 1992;

Adamo et al., 1993). In osteoblasts, start sites 2 and 3 are the most prominently used, and current knowledge indicates that agents which modulate IGF I gene transcription do not alter start site preference (Pash et al., 1995; Delany and Canalis, 1995). IGF I transcripts range in size from seven kilobases (kb) to 0.9 kb, and the 6.5 kb mRNA species is most abundant in rat osteoblasts. This transcript has > 5 kb of 3' untranslated sequence, which contains well conserved AU-rich regions and potential stem loop structures that may play a role in the regulation of mRNA stability (Lund et al., 1989; Hoyt et al., 1992; Carter and Malter, 1991). Indeed, in rat liver the largest IGF I transcript is ~3 times less stable than the 0.9 kb species (Hepler et al., 1990). It is not yet known if the stability of the IGF I transcripts is differentially regulated in osteoblasts.

Agents which decrease IGF I expression are often associated with bone loss. For example, glucocorticoids decrease IGF I expression by transcriptional mechanisms, and this may play a role in the development of glucocorticoid-induced osteoporosis (Delany and Canalis, 1995). In osteoblasts, IGF I and IGF II expression is decreased by TGF β, PDGF BB, and bFGF, growth factors which are found in serum and are synthesized by bone cells and mononuclear cells within the bone compartment (Canalis et al., 1993; Gabbitas et al., 1994). TGF β, bFGF, and PDGF BB are potent osteoblast mitogens that decrease the differentiated function of the osteoblast (Table 1). Acute inhibition of IGF expression by these growth factors may be necessary during periods of mitogenesis and may play a role in coordinating bone remodeling. *In vitro*, IGF I expression increases in rat osteoblastic cells as the cultures progress from the proliferative to the differentiating stage, and it is then decreased as the cultured osteoblasts begin to form nodules and calcify (Gabbitas and Canalis, unpublished observations). Developmental regulation of IGF I may be mediated in part by autocrine and paracrine skeletal growth factors.

There are few agents which increase IGF I expression in bone. However, BMP 2, an important locally produced factor that induces osteoblast proliferation and differentiated function, increases IGF I and IGF II expression (Thies et al., 1992; Canalis and Gabbitas, 1994). This probably plays a role in the ability of BMP 2 to promote bone cell differentiated function, but the mechanisms by which BMP 2 regulates IGF expression are not yet known. Estrogens increase IGF I by stimulating gene transcription, an effect mediated through an AP-1 like element within the IGF I promoter (Umayahara et al., 1994). Since estrogens increase the expression of IGF I, decreased IGF I as a result of estrogen deficiency may play a role in the pathogenesis of post-menopausal osteoporosis (Ernst et al., 1989). Prostaglandins are potent inducers of IGF I gene transcription in osteoblasts (Pash et al., 1995). Stimulation of prostaglandins by inflammatory mediators, such as IL-1, is probably involved in the bone remodeling which occurs during fracture repair. The systemic hormone PTH increases osteoblast IGF I gene transcription, but continuous exposure of calvariae to PTH decreases collagen synthesis without decreasing IGF I expression, suggesting additional direct inhibitory actions of PTH on bone formation (Canalis et al., 1994). This would favor the use of PTH intermittently for the treatment of bone disorders involving decreased bone mass. In fact, intermittent PTH administration was shown to decrease bone loss in estrogen deficient women, suggesting that it may be beneficial for the prevention of post-menopausal osteoporosis (Finkelstein et al., 1994).

The importance of IGF I and IGF II for normal growth and development was highlighted by a study in which targeted gene disruption was used to create mice that lacked functional IGF I receptors, making the animals insensitive to both IGF I and IGF II. These animals died at birth from respiratory failure, were only ~45% of normal size, and showed developmental delays in ossification (Liu et al., 1993). This suggests that IGFs play a key role in bone development and maintenance. In aging, serum levels of IGF I and the IGF I content in cortical, but not trabecular, bone decreases and it has been suggested that this decrease in IGF I may play a role in the bone loss that occurs with this process (Nicolas et al., 1994; Canalis, 1994; Clemmons and Van Wyk 1994). The potential for using IGF I to treat osteoporosis is limited by its side effects, and the growth factor affects multiple organ systems. A study testing the effects of short-term administration of IGF I to normal post-menopausal women showed that it increased markers for both collagen synthesis and collagen breakdown in a dose-dependent manner, suggesting that it increased bone remodeling. Higher doses of IGF I were associated with side effects, while lower doses caused minimal side effects yet still increased indices of bone remodeling (Ebeling et al., 1993). Studies in ovariectomized rats suggest that IGF I treatment can increase cancellous bone mass (Mueller et al., 1994). Determining the mechanisms by which hormones and growth factors modify IGF I and IGF II synthesis in bone will be important for the development of new compounds that regulate skeletal IGF and that may be used for the treatment of osteoporosis and other metabolic bone disorders.

VII. INSULIN-LIKE GROWTH FACTOR BINDING PROTEINS

In addition to the regulation of IGF synthesis, the activity and availability of these growth factors are regulated by a family of insulin-like growth factor binding proteins (IGFBPs). Presently, six IGFBPs have been characterized, and the genes for these proteins have been cloned from a number of tissues (Rechler, 1993). These proteins range in size from 24 to 45 kD, and can exist in glycosylated and unglycosylated forms. Rat osteoblasts express transcripts for IGFBPs 1-6, while human osteoblasts express transcripts for all the binding proteins except for IGFBP-2 (Okazaki et al., 1994). IGFBPs can potentiate or inhibit the effects of IGF I and IGF II on cell function; IGFBP-5 is stimulatory while IGFBPs -2, -3, and -4 are inhibitory (Andress and Birnbaum, 1992; Bautista et al., 1991; Mohan et al., 1989; Feyen et al., 1991; Campbell and Novak, 1991). IGFBPs vary in their affinity for IGF I and IGF II, and this may play a role in their modulation of IGF bioavailability (Bach et al., 1993). The binding of IGFs to high affinity IGFBPs may sequester the growth factor and keep it from interacting with the IGF I receptor, decreasing the effect of IGFs (Cohick and Clemmons, 1993). On the other hand, binding of IGFs to lower affinity IGFBPs associated with the cell surface or with the extracellular matrix may increase the local effective concentration of the growth factor and potentiate its effects (Andress and Birnbaum, 1992; Jones et al., 1993). In addition to playing a role in the regulation IGF bioavailability and protecting the growth factor from degradation, it was recently suggested that IGFBPs may help IGF I achieve and maintain proper folding and disulfide bonding in the extracellular space (Hober et al., 1994).

Only IGFBP-5 has been shown to stimulate osteoblast mitogenesis and to potentiate IGF activity (Andress and Birnbaum, 1992; Jones et al., 1993). The ability of IGFBP-5 to stimulate osteoblast mitogenesis in the absence of IGF I may be mediated by a cell surface receptor for IGFBP-5. Such receptors have been found for IGFBP-3 on a human breast cancer cell line, and it is possible that these cell surface receptors exist for other IGFBPs (Oh et al., 1993). IGFBP-5 can be associated with the extracellular matrix and the interaction of IGF I with IGFBP-5 can be inhibited by O-linked glycosaminoglycans found within the matrix, including heparin, heparin sulfate, and dermatan sulfate (Arai et al., 1994). By inhibiting the interaction between IGF I and IGFBP-5, these matrix components may increase the availability of IGF I to its receptor.

In vitro studies suggest that the pattern of IGFBP expression is modulated during the developmental stages of the osteoblast. IGFBP-2 and -5 expression is highest in the proliferative phase of rat osteoblastic cell cultures. During the matrix maturation phase of these cells, IGFBP-3, -4, and -6 expression peaks, and during the mineralization phase IGFBP-4 and -6 expression is maintained (Birnbaum and Wiren, 1994). The regulation of IGFBP expression during osteoblastic differentiation may be related to the time course of expression of autocrine and paracrine skeletal growth factors such as those which regulate the expression of the IGFs themselves. For example, IGFs increase osteoblast IGFBP-5 expression, whereas growth factors with potent mitogenic activity are inhibitory (McCarthy et al., 1994; Conover and Kiefer, 1993; Canalis and Gabbitas, 1995). In addition, PGE_2, which can mediate the effects of other growth factors and cytokines, increases expression of IGFBP-3, -4, and -5 (McCarthy et al., 1994). The expression of IGFBP-4, which inhibits the effects of IGFs, is stimulated by PDGF BB and bFGF, two osteoblast mitogens that inhibit differentiated function, as well as by cyclic AMP inducers (Chen et al., 1993c; LaTour et al., 1990).

In addition to local autocrine and paracrine factors, systemic hormones and growth factors can modulate IGFBP synthesis, however, the effects appear to be cell line and culture condition dependent (Hassager et al., 1992). For example, growth hormone increases IGFBP-3 and -5 transcripts in normal rat osteoblasts, but fails to do so in human osteosarcoma cells (Conover and Kiefer, 1993). In rat osteosarcoma cells, PTH increases IGFBP-5 (Conover et al., 1993a; Chen et al., 1993c). Glucocorticoids also evoke differential regulation of IGFBPs in normal and malignant osteoblastic cells. In normal human osteoblasts, glucocorticoids decrease IGFBP-3, -4, and -5 expression while increasing IGFBP-1 (Okazaki et al., 1994). In contrast, IGFBP-5 is not decreased by glucocorticoids in the human osteosarcoma cell line U-2 (Conover and Kiefer, 1993). Differences in the regulation of IGFBPs in normal and malignant osteoblastic cell cultures may suggest that altered IGFBP expression contributes to the malignant phenotype. The systemic hormone 1, 25-dihydroxyvitamin D_3 increases osteoblast IGFBP-3 and -4 expression, but does not alter the synthesis of IGFBP-5 (Chen et al., 1991b; Moriwake et al., 1992; Scharla et al., 1991; Conover and Kiefer, 1993). IGFBP-3 and-4 are increased during the matrix maturation phase of osteoblast development, which is also the time when osteocalcin, a 1,25 dihydroxyvitamin D_3 dependent protein, is expressed (Birnbaum and Wiren, 1994; Owen et al., 1993). Thus, differential regulation of IGFBPs by hormones and growth factors allows for another level of regulation of IGF activity within the bone compartment, and the mechanisms by which these agents affect IGFBP synthesis are currently under investigation.

The abundance of IGFBPs in the extracellular space can also be regulated at the level of proteolytic degradation. IGFBP proteases have been characterized from diverse sources, including cultured fibroblasts, osteoblasts, and neuronal cell lines, as well as late gestation pregnancy serum (Fowlkes et al., 1994; Conover et al., 1993b; Cheung et al., 1994; Kanzaki et al., 1994; Nam et al., 1994). Normal human osteoblasts and osteosarcoma cells produce proteases for IGFBP-4, the major inhibitory IGFBP, and for IGFBP-5, the stimulatory IGFBP for bone cells. Interestingly, the protease activity is modulated by IGFs, which promote the degradation of IGFBP-4 and inhibit the degradation of IGFBP-5, suggesting a mechanism by which IGF activity can be potentiated in bone (Kanzaki et al., 1994). Studies in other cell systems showed that the activity or synthesis of IGFBP proteases is modulated by agents which can regulate bone cell function, including glucocorticoids and the protein kinase C agonist phorbol myristate acetate (Cheung et al., 1994; Conover et al., 1993b). The IGFBP protease activity present in bone cell cultures has not been characterized, but in other systems IGFBP proteolytic activity has been attributed to serine proteases and metalloproteinases (Cheung et al., 1994; Fowlkes et al., 1994; Nam et al., 1994). Osteoblasts produce matrix metalloproteinases as well as serine proteases which may be candidates for the IGFBP proteolytic activity observed in bone cell cultures. The regulation of IGFBPs at multiple levels and their differential regulation by hormones and growth factors suggest that they play a role in fine- tuning the activity of IGFs and osteoblast function in the bone compartment.

ACKNOWLEDGMENTS

Work carried out in the authors' laboratory was supported by grants from the National Institutes of Health, AR 21707, DK 42424, DK 45227, and DK 09038. The authors thank Ms. Beverly Faulds for valuable secretrial assistance.

REFERENCES

Abraham, J.A., Whang, J.L., Tumolo, A., Mergia, A., Friedman, J., Gospodarowicz J., & Fiddes, J.C. (1986). Human basic fibroblast growth factor: Nucleotide sequence and genomic organization. EMBO J. 5, 2523-2528.

Adamo, M.L., Lanau, F., Neuenschwander, S., Werner, H., LeRoith, D., & Roberts, C.T., Jr. (1993). Distinct promoters in the rat insulin-like growth factor-I (IGF-I) gene are active in CHO cells. Endocrinology 132, 935-937.

Andress, D.L., & Birnbaum, R.S. (1992). Human osteoblast-derived insulin-like growth factor (IGF) binding protein-5 stimulates osteoblast mitogenesis and potentiates IGF action. J. Biol. Chem. 267, 22467-22472.

Arai, T., Parker, A. Busby, Jr., W., & Clemmons, D.R. (1994). Heparin, heparan sulfate, and dermatan sulfate regulate formation of the insulin-like growth factor-I and insulin-like growth factor-binding protein complexes. J. Biol. Chem. 269, 20388-20393.

Bach, L.A., Hsieh, S., Sakano, K., Fujiwara, H., Perdue, J.F., & Rechler, M.M. (1993). Binding of mutants of human insulin-like growth factor II to insulin-like growth factor binding proteins 1-6. J. Biol. Chem. 268, 9246-9254.

Baird, A. (1993). Editorial: Fibroblast growth factors: What's in a name? Endocrinology 132, 487-488.

Barnard, J.A., Lyons, R.M., & Moses, H.L. (1990). The cell biology of transforming growth factor β Biochim. Biophys. Acta 1032, 79-87.

Bassing, C.H., Howe, D.J., Segarin, P.R., Donahoe P.K., & Wang, X-F. (1994). A single heteromeric receptor complex is sufficient to mediate biological effects of transforming growth factor-β ligands. J. Biol. Chem. 269, 14861-14864.

Bautista, C.M., Baylink, D.J., & Mohan, S. (1991). Isolation of a novel insulin-like growth factor (IGF) binding protein from human bone: A potential candidate for fixing IGF II in human bone. Biochem. Biophys. Res. Commun. 176, 756-763.

Beck, L.S., Chen, T.L., Mikalauski, P., & Ammann, A.J. (1990). Recombinant human transforming growth factor-beta 1 (rhTGF-β1) enhances healing and strength of granulation skin wounds. Growth Factors 3, 267-275.

Betsholtz, C., Johnsson, A., Heldin, C-H., Westermark, B., Lind, P., Urdea, M.S., Eddy, R., Shows, T.B., Philpott, K., Mellor, A.L., Knott, T.J., & Scott, J. (1986). cDNA sequence and chromosomal localization of human platelet-derived growth factor A-chain and its expression in tumour cell lines. Nature 320, 695-699.

Bhandari, B., Grandaliano, G., & Abboud, H.E. (1994). Platelet-derived growth factor (PDGF) BB homodimer regulates PDGF A- and PDGF B-chain gene transcription in human mesangial cells. Biochem. J. 297, 385-388.

Birnbaum, R.S., & Wiren, K.M. (1994). Changes in insulin-like growth factor-binding protein expression and secretion during the proliferation, differentiation, and mineralization of primary cultures of rat osteoblasts. Endocrinology 135, 223-230.

Bonewald, L.F., Kester, M.B., Schwartz, Z., Swain, L.D., Khare, A., Johnson, T.L., Leach, R.J., & Boyan, B.D. (1992). Effects of combining transforming growth factor β and 1,25-Dihydroxyvitamin D_3 on differentiation of a human osteosarcoma (MG-63). J. Biol. Chem. 267, 8943-8949.

Brummel, T.J., Twombly, V., Marques, G., Wrana, J.L., Newfeld, S.J., Attisano, L., Massague, J., O'Connor, M.B., & Gelbart, W.M. (1994). Characterization and relationship of dpp receptors encoded by the saxophone and thick veins genes in drosophila. Cell 78, 251-261.

Burgess, W.H., & Maciag, T. (1989). The heparin-binding (fibroblast) growth factor family of proteins. Annu. Rev. Biochem. 58, 575-606.

Campbell, G., & Novak, J.F. (1991). Insulin-like growth factor binding protein (IGFBP) inhibits IGF action on human osteosarcoma cells. J. Cell. Physiol. 149, 293-300.

Canalis, E. (1994). Editorial: Skeletal growth factors and aging. J. Clin. Endocrinol. Metab 78, 1009-1010.

Canalis, E., & Gabbitas, B. (1994). Bone morphogenetic protein 2 increases insulin-like growth factor I and II transcripts and polypeptide levels in bone cell cultures. J. Bone Miner. Res. 9, 1999-2005.

Canalis, E., & Gabbitas, B. (1995). Skeletal growth factors regulate the synthesis of insulin-like growth factor binding protein-5 in bone cell cultures. J. Biol. Chem. 270, 10771-10776.

Canalis, E., Lorenzo, J., Burgess, W.H., & Maciag, T. (1987). Effects of endothelial cell growth factor on bone remodeling *in vitro*. J. Clin. Invest. 79, 52-58.

Canalis, E., Centrella, M., & McCarthy, T. (1988a). Effects of basic fibroblast growth factor on bone formation *in vitro*. J. Clin. Invest. 81, 1572-1577.

Canalis, E., McCarthy, T., & Centrella, M. (1988b). Isolation of growth factors from adult bovine bone. Calcif. Tissue Int. 43, 346-351.

Canalis, E., McCarthy, T.L., & Centrella, M. (1989). Effects of platelet-derived growth factor on bone formation *in vitro*. J. Cell. Physiol. 140, 530-537.

Canalis, E., Varghese, S., McCarthy T.L., & Centrella, M. (1992). Role of platelet-derived growth factor in bone cell function. Growth Regul. 2, 151-155.

Canalis, E., Pash, J., Gabbitas, G., Rydziel S., & Varghese, S., (1993). Growth factors regulate the synthesis of insulin-like growth factor-I in bone cell cultures. Endocrinology 133, 33-38.

Canalis, E., Hock, J.M., & Raisz, L.G. (1994). Anabolic and catabolic effects of parathyroid hormone on bone and interactions with growth factors. In: The Parathyroids (Bilezikian, J.P., Levine, M.A., & Marcus, R., Eds.) pp. 65-82. Raven Press, Ltd., New York.

Canalis, E., Rydziel, S., Delany, A.M., Varghese, S., & Jeffrey, J.J. (1995). Insulin-like growth factors inhibit interstitial collagenase synthesis in bone cell cultures. Endocrinology 136, 1348-1354.

Carter, B.Z., & Malter, J.S. (1991). Biology of disease. Regulation of mRNA stability and its relevance to disease. Lab. Invest. 65, 610-621.

Celeste, A.J., Iannazzi, J.A., Taylor, R.C., Hewick, R.M., Rosen, V., Wang, E.A., & Wozney, J.M. (1990). Identification of transforming growth factor β family members present in bone-inductive protein purified from bovine bone. Proc. Natl. Acad. Sci, USA 87, 9843-9847.

Centrella, M., Massague, J., & Canalis, E. (1986). Human platelet-derived transforming growth factor-β stimulates parameters of bone growth in fetal rat calvariae. Endocrinology 119, 2306-2312.

Centrella, M., McCarthy, T.L., & Canalis, E. (1987). Transforming growth factor β is a bifunctional regulator of replication and collagen synthesis in osteoblast-enriched cell cultures from fetal rat bone. J. Biol Chem. 262, 2869-2874.

Centrella, M., McCarthy, T.L., & Canalis, E. (1988). Parathyroid hormone modulates transforming growth factor β activity and binding in osteoblast-enriched cell cultures from fetal rat parietal bone. Proc. Natl. Acad. Sci. USA 85, 5889-5893.

Centrella, M., McCarthy, T.L., & Canalis, E. (1989). b2-microglobulin enhances insulin-like growth factor I receptor levels and synthesis in bone cell cultures. J. Biol. Chem. 264, 18268-18271.

Centrella, M., McCarthy, T.L., & Canalis, E. (1991a). Activin-A binding and biochemical effects in osteoblast-enriched cultures from fetal-rat parietal bone. Mol. Cell. Biol. 11, 250-258.

Centrella, M., McCarthy, T.L., & Canalis, E. (1991b). Glucocorticoid regulation of transforming growth factor β1 (TGFβ1) activity and binding in osteoblast-enriched cultures from fetal rat bone. Mol. Cell. Biol. 11, 4490-4496.

Centrella, M., McCarthy, T.L., & Canalis, E. (1991c). Transforming growth factor-beta and remodeling of bone. J. Bone Joint Surg. 73-A, 1418-1428.

Centrella, M., McCarthy, T.L., Kusmik, W.F., & Canalis, E. (1991d). Relative binding and biochemical effects of heterodimeric and homodimeric isoforms of platelet-derived growth factors in osteoblast-enriched cultures from fetal rat bone. J. Cell. Physiol. 147, 420-426.

Centrella, M., McCarthy, T.L., Kusmik, W.F., & Canalis, E. (1992). Isoform specific regulation of platelet-derived growth factor activity and binding in osteoblast-enriched cultures from fetal rat bone, J. Clin. Invest. 89, 1076-1084.

Chen, T.L., Bates, R.L., Dudley, A., Hammonds, R.G., & Amento E.P. (1991a). Bone morphogenetic protein-2b stimulation of growth and osteogenic phenotypes in rat osteoblast-like cells: comparison with TGFβ1. J. Bone Miner. Res. 6, 1387-1393.

Chen, T.L., Chang, L.Y., Bates, R.L., & Perlman, A.J. (1991b). Dexamethasone and 1,25-dihydroxyvitamin D3 modulation of insulin-like growth factor-binding proteins in rat osteoblast-like cell cultures. Endocrinology 128, 73-80.

Chen, D., Feng, J.Q., Feng, M., Harris, M.A., Mundy, G.R., & Harris, S.E. (1993a). Cloning and sequence of bone morphogenetic protein 4 cDNA from fetal rat calvarial cell. Biochim. Biophys. Acta 1174, 289-292.

Chen, R-H., Ebner, R. & Derynck, R. (1993b). Inactivation of the type II receptor reveals two receptor pathways for the diverse TGF-β activities. Science 260, 1335-1338.

Chen, T.L., Chang, L.Y., DiGregorio, D.A., Perlman, A.J., & Huang, Y.F. (1993c). Growth factor modulation of insulin-like growth factor-binding proteins in rat osteoblast-like cells. Endocrinology 133, 1382-1389.

Chenu, C., Pfeilschifter, J., Mundy, G.R., & Roodman, G.D. (1988). Transforming growth factor β inhibits formation of osteoclast-like cells in long-term human marrow cultures. Proc. Natl. Acad. Sci. USA 85, 5683-5687.

Cheung, P-T., Wu, J., Banach, W., & Chemausek, S.D. (1994). Glucocorticoid regulation of an insulin-like growth factor-binding protein-4 protease produced by a rat neuronal cell line. Endocrinology 135, 1328-1335.

Chuang, L-M., Myers, Jr., M.G., Seidner, G.A., Bimbaum, M.J., White, M.F., & Kahn, C.R. (1993). Insulin receptor substrate 1 mediates insulin and insulin-like growth factor 1-stimulated maturation of *Xenopus* oocytes. Proc. Natl. Acad. Sci. 90, 5172-5175.

Claesson-Welsh, L., Eriksson, A., Westermark, B., & Heldin, C-H. (1989). cDNA cloning and expression of the human A-type platelet-derived growth factor (PDGF) receptor establishes structural similarity to the B-type PDGF receptor. Proc. Natl. Acad. Sci. USA 86,4917-4921.

Clemmons, D.R., & Van Wyk, J.J. (1984). Factors controlling blood concentrations of somatomedin C. Clin. Endocrinol. Metab. 13, 113-143.

Cohick, W.S., & Clemmons, D.R. (1993). The insulin-like growth factors. Annu. Rev. Physiol. 55, 131-153.

Conover, C.A., & Kiefer, M.C. (1993). Regulation and biological effect of endogenous insulin-like growth factor binding protein-5 in human osteoblastic cells. J. Clin. Endocrinol. Metab. 76, 1153-1159.

Conover, C.A., Bale, L.K., Clarkson, J.T., & Torring, O. (1993a). Regulation of insulin-like growth factor binding protein-5 messenger ribonucleic acid expression and protein availability in rat osteoblast-like cells. Endocrinology 132, 2525-2530.

Conover, C.A., Clarkson, J.T., & Bale, L.K. (1993b). Phorbol ester tumor promoters regulate insulin-like growth factor-binding protein-4 proteolysis. Endocrinology 133, 1347-1351.

Cuevas, P., Burgos, J., & Baird, A. (1988). Basic fibroblast growth factor (FGF) promotes cartilage repair *in vivo*. Biochem. Biophys. Res. Commun. 156, 611-618.

Daughaday, W.H., & Rotwein, P. (1989). Insulin-like growth factor I and II. Peptide, messenger ribonucleic acid and gene structures, serum, and tissue concentrations. Endocr. Rev. 10, 68-91.

Delany, A.M., & Canalis, E. (1995). Transcriptional repression of insulin-like growth factor I by glucocorticoids in rat bone cells. Endocrinology 136, 4776-4781.

Delany, A.M., Gabbitas, B.Y., & Canalis, E. (1995). Cortisol down-regulates osteoblast α1(I) procollagen mRNA by transcriptional and post-transcriptional mechanisms. J. Cell. Biochem. 57, 488-494.

Derynck, R., Jarrett, J.A., Chen, E.Y., & Goeddel, D.V. (1986). The murine transforming growth factor-β precursor. J. Biol. Chem. 261, 4377-4379.

Doctor, J.S., Jackson, P.D., Rashka, K.E., Visalli, M., & Hoffmann, F.M. (1992). Sequence, biochemical characterization, and developmental expression of a new member of the TGF-β superfamily in *Drosophila melanogaster*. Dev. Biol. 151, 491-505.

Ebeling, P.R., Jones, J.D., O'Fallon, W.M., Janes, C.H., & Riggs, B.L. (1993). Short-term effects of recombinant human insulin-like growth factor I on bone turnover in normal women. J. Clin. Endocrinol. Metab. 77, 1384-1387.

Ebner, R., Chen, R-H., Shum, L., Lawler, S., Zioncheck, T.F., Lee, A., Lopez, A.R., & Derynck, R. (1993). Cloning of a type I TGF-β receptor and its effect on TGF-β binding to the type II receptor. Science 260, 1344-1348.

Elenius, K., Maatta, A. Salmivirta, M., & Jalkanen, M. (1992). Growth factors induce 3T3 cells to express bFGF-binding syndecan. J. Biol. Chem. 267, 6435-6441.

Eriksson, A., Nister, M., Leveen, P., Westermark, B., Heldin, C-H., & Claesson-Welsh, L. (1991). Induction of platelet-derived growth factor α and β receptor mRNA and protein by platelet-derived growth factor BB. J. Biol. Chem. 266, 21138-21144.

Ernst, M., Heath, J.K., & Rodan, G.A. (1989). Estradiol effects on proliferation, messenger ribonucleic acid for collagen and insulin-like growth factor-I, and parathyroid hormone-stimulated adenylate cyclase activity in osteoblastic cells from calvariae and long bones. Endocrinology 125, 825-833.

Esch, F., Baird, A. Ling, N., Ueno, N., Hill, F., Denoroy, L., Klepper, R. Gospodarowicz, D., Bohlen, P., & Guillemin, R. (1985). Primary structure of bovine pituitary basic fibroblast growth factor (FGF) and comparison with the amino-terminal sequence of bovine brain acidic FGF. Proc. Natl. Acad. Sci. USA 82, 6507-6511.

Feyen, J.H.M., Evans, D.B., Binkert, C., Heinrich, G.F., Geisse S., & Kocher, H.P. (1991). Recombinant human [Cys^{28}] insulin-like growth factor-I-stimulated proliferation and collagen synthesis in fetal rat calvariae. J. Biol. Chem. 266, 19469-19474.

Finkelman, R.D., Bell, N.H., Strong, D.D., Demers, L.M., & Baylink, D.J. (1992). Ovariectomy selectively reduces the concentration of transforming growth factor β in rat bone: Implications for estrogen deficiency-associated bone loss. Proc. Natl. Acad. Sci. USA 89, 12190-12193.

Finkelstein, J.S., Klibanski, A., Schaefer, E.H., Hornstein, M.D., Schiff, I., & Neer, R.M. (1994). Parathyroid hormone for the prevention of bone loss induced by estrogen deficiency. N. Engl. J. Med. 331, 1618-1623.

Fowlkes, J.L., Suzuki, K., Nagase, H., & Thrailkill, K.M. (1994). Proteolysis of insulin-like growth factor binding protein-3 during rat pregnancy: A role for matrix metalloproteinases. Endocrinology 135, 2810-2813.

Gabbitas, B., Pash, J., & Canalis, E. (1994). Regulation of insulin-like growth factor-II synthesis in bone cell cultures by skeletal growth factors. Endocrinology 135, 284-289.

Gilardetti, R.S., Chaibi, M.S., Stroumza, J., Williams, S.R., Antoniades, H.N., Carnes, D.C., & Graves, D.T. (1991). High-affinity binding of PDGF-AA and PDGF-BB to normal human osteoblastic cells and modulation by interluekin-1. Am. J. Physiol. 261, C980-C985.

Giorgetti, S., Pelicci, P.G., Pelicci, G., & Van Obberghen, E. (1994). Involvement of Src-homology/collagen (SHC) proteins in signaling through the insulin receptor and the insulin-like-growth-factor-I-receptor. Eur. J. Biochem. 223, 195-202.

Globus, R.K., Plouet, J., & Gospodarowicz, D. (1989). Cultured bovine bone cells synthesize basic fibroblast growth factor and store it in their extracellular matrix. Endocrinology 124, 1539-1547.

Goldsmith, K.T., Gammon, R.B., & Garver, R.I., Jr. (1991). Modulation of bFGF in lung fibroblasts by TGF-β and PDGF. Am. J. Physiol. 261, L378-385.

Graves, D.T., Owen, A.J., Barth, R.K., Tempst, P., Winoto, A., Fors, L., Hood, L.E., & Antoniades, H.N. (1984). Detection of *c-sis* transcripts and synthesis of PDGF-like proteins by human osteosarcoma cells. Science 226, 972-974.

Gronwald, R.G.K., Grant, F.J., Haldeman, B.A., Hart, C.E., O'Hara, P.J., Hagen, F.S., Ross, R., Bowen-Pope, D.F., & Murray, M.J. (1988). Cloning and expression of a cDNA coding for the human platelet-derived growth factor receptor: Evidence for more than one receptor class. Proc. Natl. Acad. Sci. USA 85, 3435-3439.

Gronwald, R.G.K., Seifert, R.A., & Bowen-Pope, D.R. (1989). Differential regulation of expression of two platelet-derived growth factor receptor subunits by transforming growth factor-β. J. Biol. Chem. 264, 8120-8125.

Hakeda, Y., Harada, S-i., Matsumoto, T., Ken-ichi, T., Higashino, K., Kodama, H., Hashimoto-Goto, T., Ogata, E., & Kumegawa, M. (1991). Prostaglandin $F_{2\alpha}$ stimulates proliferation of clonal osteoblastic MC3T3-E1 cells by up-regulation of insulin-like growth factor I receptors. J. Biol. Chem. 266, 21044-21050.

Hall, L.J., Kajimoto, Y., Bichell, D., Kim, S-W., James, P.L., Counts, D., Nixon, L.J., Tobin, G., & Rotwein, P. (1992). Functional analysis of the rat insulin-like growth factor I gene and identification of an IGF-I gene promoter. DNA Cell Biol. 11, 301-313.

Hassager, C., Fitzpatrick, L.A., Spencer, E.M., Riggs, B.L., & Conover, C.A. (1992). Basal and regulated secretion of insulin-like growth factor binding proteins in osteoblast-like cells is cell line specific. J. Clin. Endocrinol. Metab. 75, 228-233.

Heldin, C-H., & Westermark, B. (1987). PDGF-like growth factors in autocrine stimulation of growth. J. Cell. Physiol. Suppl. 5, 31-34.

Heldin, C-H., & Westermark, B., (1989). Platelet-derived growth factors: A family of isoforms that bind to two distinct receptors. Br. Med. Bull. 45, 453-464.

Heldin, C-H., Johnsson, A., Wennergren, S., Wernstedt, C., Betsholtz, C., & Westermark, B. (1986). A human osteosarcoma cell line secretes a growth factor structurally related to a homodimer of PDGF A-chains. Nature 319, 511-514.

Hepler, J.E., Van Wyk, J.J., Lund, P.K. (1990). Different half-lives of insulin-like growth factor I mRNAs that differ in length of 3' untranslated sequence. Endocrinology 127, 15509-1552.

Hober, S., Hansson, A., Uhlen, M., & Nilsson, B. (1994). Folding of insulin-like growth factor I is thermodynamically controlled by insulin-like growth factor binding protein. Biochemistry 33, 6758-6761.

Hock, J.M., & Canalis, E. (1994). Platelet-derived growth factor enhances bone cell replication but not differentiated function of osteoblasts. Endocrinology 134, 1423-1428.

Hock, J., Centrella, M., & Canalis, E. (1988). Insulin-like growth factor I (IGF-I) has independent effects on bone matrix formation and cell replication. Endocrinology 122, 254-260.

Hock, J., Canalis, E., & Centrella, M. (1990). Transforming growth factor beta (TGF-beta-1) stimulates bone matrix apposition and bone cell replication in cultured fetal rat calvariae. Endocrinology 126, 421-426.

Hoyt, E.C., Hepler, J.E., VanWyk, J.J., & Lund, P. K. (1992). Structural characterization of exon 6 of the rat IGF-I gene. DNA Cell Biol. 11, 433-441.

Hurley, M.M., Abreu, C., Harrison, J.R., Lichtler, A.C., Raisz, L.G., & Kream, B.E. (1993). Basic fibroblast growth factor inhibits type I collagen gene expression in osteoblastic MC3T3-E1 cells, J. Biol. Chem. 268, 5588-5593.

Hurley, M.M., Abreu, C., Gronowicz, G., Kawaguchi, H., & Lorenzo, J. (1994) Expression and regulation of basic fibroblast growth factor mRNA levels in mouse osteoblastic MC3T3-E1 cells. J. Biol. Chem. 269, 9392-9396.

Ignotz, R.A., & Massague, J. (1986). Transforming growth factor-β stimulates the expression of fibronectin and collagen and their incorporation into the extracellular matrix. J. Biol. Chem. 261, 4337-4345.

Jabs, E.W., Li, X., Scott, A.F., Meyers, G., Chen, W., Eccles, M., Mao, J-i., Charnas, L.R., Jackson, C.E., & Jaye, M. (1994). Jackson-Weiss and Crouzon syndromes are allelic with mutations in fibroblast growth factor receptor 2. Nature 8, 275-279.

Jackson, A., Friedman, S., Zhan, X., Engleka, K.A., Forough, R., & Maciag, T. (1992). Heat shock induces the release of fibroblast growth factor I from NIH 3T3 cells. Proc. Natl. Acad. Sci, USA 89, 10691-10695.

Jones, C.M., Simon-Chazottes, D., Guenet, J-L., & Hogan, B.L.M. (1992). Isolation of Vgr-2, a novel member of the transforming growth factor-β-related gene family. Mol. Endocrinol. 6, 1961-1968.

Jones, J.I., Gockerman, A., Busby, Jr., W.H., Camacho-Hubner, C., & Clemmons, D.R. (1993). Extracellular matrix contains insulin-like growth factor binding protein-5: Potentiation of the effects of IGF-I. J. Cell Biol. 121, 679-687.

Joyce, M.L., Roberts, A.B., Sporn, M.B., & Bolander, M.E. (1990). Transforming growth factor-β and the initiation of chondrogenesis and osteogenesis in the rat femur. J. Cell Biol. 110, 2195-2207.

Kachra, Z., Yang, C-R., & Posner, B.I. (1994). The augmentation of insulin-like growth factor-I messenger ribonucleic acid in cultured rat hepatocytes: Activation of protein kinase-A and -C is necessary, but not sufficient. Endocrinology 134, 702-708.

Kanzaki, S., Hilliker, S., Baylink, D.J., & Mohan, S. (1994). Evidence that human bone cells in culture produce insulin-like growth factor-binding protein-4 and -5 proteases. Endocrinology 134, 383-392.

Kavanaugh, W.M., Harsh IV, G.R., Starksen, N.S., Rocco, C.M., & Williams, L.T. (1988). Transcriptional regulation of the A and B chain genes of platelet-derived growth factor in microvascular endothelial cells. J. Biol. Chem. 263, 8470-8472.

Kawaguchi, H., Kurokawa, T., Hanada, K. Hiyama, Y., Tamura, M., Ogata, E., & Matsumoto, T. (1994). Stimulation of fracture repair by recombinant human basic fibroblast growth factor in normal and streptozotocin-diabetic rats. Endocrinology 135, 774-781.

Kim, S-J, Glick, A., Sporn, M.B., & Roberts, A.B. (1989a). Characterization of the promoter region of the human transforming growth factor-β1 gene. J. Biol. Chem. 264, 402-408

Kim, S-J., Jeang, K.T., Glick, A.B., Sporn, M.B., & Roberts, A.B. (1989b). Promoter sequences of the human transforming growth factor-β1 gene responsive to transforming growth factor-β1 autoinduction. J. Biol. Chem. 264, 7041-7045.

Kingsley, D.M. (1994). The TGF-β superfamily: New members, new receptors, and new genetic tests of function in different organisms. Genes Dev. 8, 133-146.

Klagsbrun, M., & Baird, A. (1991). A dual receptor system is required for basic fibroblast growth factor activity. Cell 67, 229-231.

Knutsen, R., Wergedal, J.E., Sampath, T.K., Baylink, D.J., & Mohan, S. (1993). Osteogenic protein-1 stimulates proliferation and differentiation of human bone cell *in vitro*. Biochem. Biophys. Res. Commun. 194, 1352-1358.

Kurose, H., Yamaoka, K., Okada, S., Nakajima, S., & Seino, Y. (1990). 1,25-Dihydroxyvitamin D_3 [1,25-$(OH)_2D_3$] increases insulin-like growth factor I (IGF-I) receptors in clonal osteoblastic cells. Study on interaction of IGF-I and 1,25-$(OH)_2D_3$. Endocrinology 126, 2088-2094.

Lafyatis, R., Lechleider, R., Kim, S-J., Jakowlew, S., Roberts, A.B., & Sporn, M.B. (1991). Structural and functional characterization of the transforming growth factor β3 promoter. J. Biol. Chem. 265, 19128-19136.

Laiho, M., Weis, F.M.B., & Massague, J. (1990). Concomitant loss of transforming growth factor (TGF)-β receptor types I and II in TGFβ-resistant cell mutants implicates both receptor types in signal transduction. J. Biol. Chem. 265, 18518-18524.

Lane, T.L., & Sage, E.H. (1994). The biology of SPARC, a protein that modulates cell-matrix interactions. FASEB J. 8, 163-173.

LaTour, D., Mohan, S., Linkhart, T.A., Baylink, D.J., & Strong, D.D. (1990). Inhibitory insulin-like growth factor-binding protein: Cloning, complete sequence, and physiological regulation. Mol. Endocrinol. 4, 1806-1814.

Lin, H.Y., Wang, X-F., Ng-Eaton, E., Weinberg, R., & Lodish, H. (1992a). Expression cloning of the TGF-β type II receptor, a functional transmembrane serine/threonine kinase. Cell 68, 775-785.

Lin, X., Wang, Z., Gu, L., & Deuel, T.F. (1992b). Functional analysis of the human platelet-derived growth factor A-chain promoter region. J. Biol. Chem. 267, 25614-25619.

Liu, J-P, Baker, J, Perkins, A.S., Robertson, E.J., & Efstratiadis, A. (1993). Mice carrying null mutations of the genes encoding insulin-like growth factor I (Igf-1) and type 1 IGF receptor (Igf1r). Cell 75, 59-72.

Lopez-Casillas, F., Payne, H.M., Andres, J.L., & Massague, J. (1994). Betaglycan can act as a dual modulator of TGFβ access to signaling receptors: Mapping of ligand binding and GAG attachment sites. J. Cell Biol. 124, 557-568.

Lund, P.K., Hoyt, E.C., & Van Wyk, J.J. (1989). The size heterogeneity of rat insulin-like growth factor-I mRNAs is due primarily to differences in the length of the 3' end. Mol. Endocrinol. 3, 2054-2061.

Luyten, F.P., Cunningham, N.S., Ma, S., Muthukumaran, N., Hammonds, R.G., Nevins, W.B., Wood, W.I., & Reddi, A.H. (1989). Purification and partial amino acid sequence of osteogenin, a protein initiating bone differentiation. J. Biol. Chem. 264, 13377-13380.

Lynch, S.E., Nixon, J.C., Colvin, R.B., & Antoniades, H.N. (1987). Role of platelet-derived growth factor in wound healing: Synergistic effects with other growth factors. Proc. Natl. Acad. Sci. USA 84, 7696-7700.

Lyons, K.M., Jones, C.M., & Hogan, B.L. (1991). The DVR gene family in embryonic development. Trends Genet. 7, 408-412.

Mali, M., Elenius, K. Miettinen, H.M., & Jalkanen, M. (1993). Inhibition of basic fibroblast growth factor-induced growth promotion by overexpression of syndecan-1. J. Biol. Chem. 268, 24215-24222.

Marcelli, C., Yates, A.J., & Mundy, G.R. (1990). *In vivo* effects of human recombinant transforming growth factor β on bone turnover in normal mice. J.Bone Miner. Res. 5, 1087-1096.

Martin, T.J., & Ng, K.W. (1994). Mechanisms by which cells of the osteoblast lineage control osteoclast formation and activity. J. Cell. Biochem. 56, 357-366.

Matrisian, L.M., & Hogan, B.L.M. (1990). Growth factor-regulated proteases and extracellular matrix remodeling during mammalian development. Curr. Top. Dev. Biol. 24, 219-259.

McCarthy, T.L., Centrella, M., & Canalis, E. (1989a). Effects of fibroblast growth factors on deoxyribonucleic acid and collagen synthesis in rat parietal bone cells. Endocrinology 125, 2118-2126.

McCarthy, T.L., Centrella, M., & Canalis, E. (1989b). Regulatory effects of insulin-like growth factor I and II on bone collagen synthesis in rat calvarial cultures. Endocrinology, 124, 301-309.

McCarthy, T.L., Casinghino, S., Centrella, M., & Canalis, E. (1994). Complex pattern of insulin-like growth factor binding protein expression in primary rat osteoblast enriched cultures: Regulation by prostaglandin E2, growth hormone, and the insulin-like growth factors. J. Cell. Physiol. 160, 163-175.

Migdalska, A., Molineux, G., Demuynck, H., Evans, G.S., Ruscetti, F., & Dexter, T.M. (1991). Growth inhibitory effects of transforming growth factor-β1 *in vivo*. Growth Factors 4, 239-245.

Mohan, S., Bautista, C.M., Wergedal, J., & Baylink, D.J. (1989). Isolation of an inhibitory insulin-like growth factor (IGF) binding protein from bone cell-conditioned medium: A potential local regulator of IGF action. Proc. Natl. Acad. Sci. USA 86, 8338-8342.

Montesano, R., Vassalli, J.D., Baird, A., Guillemin, R., & Orci, L. (1986). Basic fibroblast growth factor induces angiogenesis *in vitro*. Proc. Natl. Acad. Sci. USA 83, 7297-7301.

Moriwake, T., Tanaka, H., Kanzaki, S., Higuchi, J., & Seino, Y. (1992). 1,25-dihydroxyvitamin D_3 stimulates the secretion of insulin-like growth factor binding protein 3 (IGFBP-3) by cultured human osteosarcoma cells. Endocrinology 130, 1071-1073.

Mueller, K., Cortesi, R., Modrowski, D., & Marie, P.J. (1994). Stimulation of trabecular bone formation by insulin-like growth factor I in adult ovariectomized rats. Endocrinol. Metab. 30, E1-E6.

Muenke, M., Schell, U., Hehr, A., Robin, N.H., Losken, H.W., Schinzel, A. Pulleryn, L.J., Rutland, P., Reardon, W., Malcolm, S., & Winter, R.M. (1994). A common mutation in the fibroblast growth factor receptor 1 gene in Pfeiffer syndrome. Nature Genetics 8, 269-274.

Murray, S.S., Murray, E.J.B., Glackin, C.A., & Urist, M.R. (1993). Bone morphogenetic protein inhibits differentiation and affects expression of helix-loop-helix regulatory molecules in myoblastic cells. J. Cell. Biochem. 53, 51-60.

Nakayama, H. Ichikawa, F., Andres, J.L., Massague, J., & Noda, M. (1994). Dexamethasone enhancement of betaglycan (TGF-β type III receptor) gene expression in osteoblast-like cells. Exp. Cell Res. 211, 301-306.

Nam, T.J., Busby, W.H., Jr., & Clemmons, D.R. (1994). Human fibroblasts secrete a serine protease that cleaves insulin-like growth factor-binding protein-5. Endocrinology 135, 1385-1391.

Nellen, D., Affolter, M., & Basler, K. (1994). Receptor serine/threonine kinases implicated in the control of drosophila body pattern by decapentaplegic. Cell 78, 225-237.

Nicolas, V., Prewett, A., Bettica, P., Mohan, S., Finkelman, R.D., Baylink, D.J., & Farley, J.R. (1994). Age-related decreases of insulin-like growth factor-I and transforming growth factor-β in femoral cortical bone from both men and women: Implications for bone loss with aging. J. Clin. Endocrinol. Metab. 78, 1011-1016.

Noda, M., (1989). Transcriptional regulation of osteocalcin production by transforming growth factor-β in rat osteoblast-like cells. Endocrinology 124, 612-617.

Noda, M. & Rodan, G.A. (1987). Type β transforming growth factor (TGF β) regulation of alkaline phosphatase expression and other phenotype-related mRNAs in osteoblastic rat osteosarcoma cells. J. Cell. Physiol. 133, 426-437.

Noda, M., & Vogel, R. (1989). Fibroblast growth factor enhances type beta 1 transforming growth factor gene expression in osteoblast-like cells. J. Cell Biol. 109, 2529-2535.

Noda, M., Yoon, K., Prince, C.W., Butler, W.T., & Rodan, G.A. (1988). Transcriptional regulation of osteopontin production in rat osteosarcoma cells by type β transforming growth factor. J. Biol. Chem. 263, 13916-13921.

Noma, T., Glick, A.B., Geiser, A.G., O'Reilly, M.A., Miller, J., Roberts, A.B., & Sporn, M.B. (1991). Molecular cloning and structure of the human transforming growth factor-β2 gene promoter. Growth Factors 4, 247-255.

Ogawa, Y., Schmidt, D.K., Dasch, J.R., Chang, R-J., & Glaser, C.B. (1992). Purification and characterization of transforming growth factor-β2.3 and -β1.2 heterodimers from bovine bone. J. Biol. Chem. 267, 2325-2328.

Ogata, T., Wozney, J.M., Benezra, R., & Noda, M. (1993). Bone morphogenetic protein 2 transiently enhances expression of a gene, Id (inhibitor of differentiation), encoding a helix-loop-helix molecule in osteoblast-like cells. Proc. Natl. Acad. Sci. USA 90, 9219-9222.

Oh, Y., Muller, H.L., Lamson, G., & Rosenfeld, R.G. (1993). Insulin-like growth factor (IGF)-independent action of IGF-binding protein-3 in Hs578T human breast cancer cells. J. Biol. Chem. 268, 14964-14971.

Okazaki, R., Riggs, B.L., & Conover, C.A. (1994). Glucocorticoid regulation of insulin-like growth factor-binding protein expression in normal human osteoblast-like cells. Endocrinology 134, 126-132.

Oursler, M.J., Cortese, C., Keeting, P., Anderson, M.A., Bonde, S.K., Riggs, B.L., & Spelsberg, T.C. (1991). Modulation of transforming growth factor-β production in normal human osteoblast-like cells by 17β-estradiol and parathyroid hormone. Endocrinology 129, 3313-3320.

Oursler, M.J., Riggs, B.L., & Spelsberg, T.C. (1993). Glucocorticoid-induced activation of latent transforming growth factor-β by normal human osteoblast-like cells. Endocrinology 133, 2187-2196.

Owen, T.A., Bortell, R., Shalhoub, V., Heinrichs, A., Stein, J.L., Stein, G.S., & Lian, J.B. (1993). Postproliferative transcription of the rat osteocalcin gene is reflected by vitamin D-responsive developmental modifications in protein-DNA interactions at basal and enhancer promoter elements. Proc. Natl. Acad. Sci. USA 90, 1503-1507

Ozkaynak, E., Rueger, D.C., Drier, E.A., Corbett, C., Ridge, R.J., Sampath, T.K., & Oppermann, H. (1990). OP-1 cDNA encodes an osteogenic protein in the TGF-β family. EMBO J. 9, 2085-2093.

Ozkaynak, E., Schnegelsberg, P.N.J., Jin, D.F., Clifford, G.M., Warren, F.D., Drier, E.A., & Oppermann, H. (1992). Osteogenic protein-2. A new member of the transforming growth factor-β superfamily expressed early in embryogenesis. J. Biol. Chem. 267, 25220-25227.

Pash, J.M., Delany, A.M., Adamo, M.L., Roberts, C.T., Jr., LeRoith, D., & Canalis, E. (1995). Regulation of insulin-like growth factor I transcription by prostaglandin E_2 in osteoblast cells. Endocrinology.

Pech, M., Rao, C.D., Robbins, K., & Aaronson, S.A. (1989). Functional identification of regulatory elements within the promoter region of platelet-derived growth factor 2. Mol. Cell. Biol. 9, 396-405.

Pelton, R.W., Dickinson, M.E., Moses H.L., & Hogan, B.L.M. (1990). *In situ* hybridization analysis of TGFβ3 RNA expression during mouse development: Comparative studies with TGF β1 and β2. Development 110, 609-620.

Penton, A., Chen, Y., Staehling-Hampton, K., Wrana, J.L., Attisano, L., Szidonya, J., Cassill, J.A., Massague, J., & Hoffmann, F.M. (1994). Identification of two bone morphogenetic protein type I receptors in *Drosophila* and evidence that Brk25D is a decapentaplegic receptor. Cell 78, 239-250.

Pfeilschifter, J., Seyedin, S.M., & Mundy, G.R. (1988). Transforming growth factor beta inhibits bone resorption in fetal rat long bone cultures. J. Clin. Invest. 82, 680-685.

Pierce, G.F., Mustoe, T.A., Altrock, B.W., Deuel, T.F., & Thomason, A. (1991). Role of platelet-derived growth factor in wound healing. J. Cell. Biochem. 45, 319-326.

Raines, E.W., Lane, T.F., Iruela-Arispe, M.L., Ross, R., & Sage, E.H. (1992). The extracellular glycoprotein SPARC interacts with platelet-derived growth factor (PDGF)-AB and -BB and inhibits the binding of PDGF to its receptors. Proc. Natl. Acad, Sci. USA 89, 1281-1285.

Reardon, W., Winter, R.M., Rutland, P., Pulleryn, L.J., Jones, B.M., & Malcolm, S. (1994) Mutations in the fibroblast growth factor receptor 2 gene cause Crouzon syndrome. Nature Genetics 8, 98-103.

Rechler, M.M. (1993). Insulin-like growth factor binding proteins. Vitam. Horm. 47, 1-117.

Reddi, A.H. (1983). Regulation of local differentiation of cartilage and bone by extracellular matrix: C cascade type mechanism. Prog. Clin. Biol. Res. 261-268.

Reddi, A.H., & Cunningham, N.S. (1991). Recent progress in bone induction by osteogenin and bone morphogenetic proteins: Challenges for biomechanical and tissue engineering. J. Biomech. Eng. 113,189-190.

Rodan, G.A., & Martin, T.J. (1981). Role of osteoblasts in hormonal control of bone resorption—a hypothesis. Calcif. Tissue Int. 33, 349-351.

Rodan, S.B., Wesolowski, G., Yoon, K., & Rodan, G. A. (1989). Opposing effects of fibroblast growth factor and pertussis toxin on alkaline phosphatase, osteopontin, osteocalcin and type I collagen mRNA levels in ROS 17/2.8 cells. J. Biol. Chem. 264, 19934-19941.

Rorsman, F., Bywater, M., Knott, T.J., Scott, J., & Betsholtz, E. (1988). Structural characterization of the human platelet-derived growth factor A-chain cDNA and gene: Alternative exon usage predicts two different precursor proteins. Mol. Cell. Biol. 8, 571-577.

Rosen, D.M., Stempien, S.A., Thompson, A.Y., & Seyedin, S.M., (1988). Transforming growth factor-beta modulates the expression of osteoblast and chondroblast phenotypes *in vitro*. J. Cell. Physiol. 134, 337-346.

Rosen, D., Miller, SC., DeLeon, E., Thompson, A.Y., Bentz, H., Mathews, M., & Adams, S. (1994). Systemic administration of recombinant transforming growth factor beta 2 (rTGF-β2) stimulates parameters of cancellous bone formation in juvenile and adult rats. Bone 15, 355-359.

Rousseau, F., Bonaventure, J., Legeal-Mallet, L., Pelet, A., Rozet, J-M., Maroteaux, P., LeMerrer, M., & Munnich, A. (1994). Mutations in the gene encoding fibroblast growth factor receptor-3 in achondroplasia. Nature 371, 252-254.

Rubini, M., Werner, H., Gandini, E., Roberts, C.T., Jr., LeRoith, D., & Baserga, R. (1994). Platelet-derived growth factor increases the activity of the promoter of the insulin-like growth factor-1 (IGF-1) receptor gene. Exp. Cell Res. 211, 374-379.

Rydziel, S., & Canalis, E. (1996). Expression and growth factor regulation of platelet-derived growth factor B transcripts in primary osteoblast cell cultures. Endocrinology 137. In press.

Rydziel, S., Ladd, C., McCarthy, T.L., Centrella, M., & Canalis, E. (1992). Determination and expression of platelet-derived growth factor-AA in bone cell cultures. Endocrinology 130, 1916-1922.

Rydziel, S. Shaikh, S., & Canalis, E. (1994). Platelet-derived growth factors AA and BB enhance the synthesis of platelet-derived growth factor AA in bone cell cultures. Endocrinology 134, 2541-2546.

Sampath, T.K., Maliakal, J.C., Hauschka, P.V., Jones, W.K., Sasak H., Tucker, R.F., White, K.H., Coughlin, J.E, Tucker, M.M., Pang, R.H.L., Corbett, C., Ozkaynak E., Opperman, H., & Rueger, D.C. (1992). Recombinant human osteogenic protein-1 (hOP-1) induces new bone formation *in vivo* with a specific activity comparable with natural bovine osteogenic protein and stimulates osteoblast proliferation and differentiation *in vitro*. J. Biol. Chem. 267, 20352-20362.

Sampath, T.K., Rashka, K.E., Doctor, J.S., Tucker, R.F., & Hoffman, F.M. (1993). *Drosophila* transforming growth factor β superfamily proteins induce endochondral bone formation in mammals. Proc. Natl. Acad. Sci. USA 90, 6004-6008.

Scharla, S.H., Strong, D.D., Mohan, S., Baylink, D.J., & Linkhart, T.A. (1991). 1,25-dihydroxyvitamin D_3 differentially regulates the production of insulin-like growth factor I (IGF-I) and IGF-binding protein-4 in mouse osteoblasts. Endocrinology 129, 3139-3146.

Schollman, C., Grugel, R., Tatje, D., Hoppe, J., Folkman, J., Marme, D., & Weich, H.A. (1992). Basic fibroblast growth factor modulates the mitogenic potency of the platelet-derived growth factor (PDGF) isoforms by specific up-regulation of the PDGF α receptor in vascular smooth muscle cells. J. Biol. Chem. 267, 18032-18039.

Seifert, R.A., Hart, C.E., Phillips, P.E., Forstrom, J.W., Ross, R., Murray, M.J., & Bowen-Pope, D.W. (1989). Two different subunits associate to create isoform-specific platelet-derived growth factor receptors. J. Biol. Chem. 264, 8771-8778.

Shiang, R., Thompson, L.M., Zhu, Y-Z., Church, D.M., Fielder, T.J., Bocian, M., Winokur, S.R., & Wasmuth, J.J. (1994). Mutations in the transmembrane domain of FGFR3 cause the most common genetic form of dwarfism, achondroplasia. Cell 78, 335-342.

Shibata, F., Baird, A., & Florkiewicz, R.Z. (1991). Functional characterization of the human basic fibroblast growth factor gene promoter. Growth Factors 4, 277-287.

Shibata, Y. Abiko, Y. Moriya, Y., Yoshida, W., & Takiguchi, H. (1993). Effects of transforming growth factor-β on collagen gene expression and collagen synthesis level in mineralizing cultures of osteoblast-like cell line, MC3T3-E1. Int. J. Biochem. 25, 239-245.

Shinar, D.M., & Rodan, G.A. (1990). Biphasic effects of transforming growth factor-β on the production of osteoclast-like cells in mouse bone marrow cultures: The role of prostaglandins in the generation of these cells. Endocrinology 126, 3153-3158.

Shull, M.M., Ormsby, I., Kier, A.B., Pawlowski, S., Diebold, R.J., Yin, M., Allen, R., Sidman, C., Proetzel, G., Calvin, D., Annunziata, N., & Doetschman, T. (1992). Targeted disruption of the mouse transforming growth factor-β1 gene results in multifocal inflammatory disease. Nature 359, 693-699.

Takuwa, Y, Ohse, C., Wang, E.A., Wozney, J.M., & Yamashita, K. (1991). Bone morphogenetic protein-2 stimulates alkaline phosphatase activity and collagen synthesis in cultured osteoblastic cells, MC3T3-E1. Biochem. Biophys. Res. Commun. 174, 96-101.

ten Dijke, P., Yamashita, H, Sampath, T.K., Reddi, A. H., Estevez, M., Riddle, D.L., Ichijo, H., Heldin, C-H., & Miyazono, K. (1994). Identification of type I receptors for osteogenic protein-1 and bone morphogenetic protein-4. J. Biol. Chem. 269, 16985-16988.

Thies, R.S., Bauduy, M., Ashton, B.A., Kurtzberg, L., Wozney, J.M., & Rosen, V. (1992). Recombinant human bone morphogenetic protein-2 induces osteoblastic differentiation in W-20-17 stromal cells. Endocrinology 130, 1318-1324.

Tsukamoto, T., Matsui, T., Nakata, H., Ito, M., Natazuka, T., Fukase, M., & Fujita, T. (1991). Interleukin-1 Enhances the response of osteoblasts to platelet-derived growth factor through the a receptor-specific up-regulation. J. Biol. Chem. 16, 10143-10147.

Umayahara, Y., Kawamori, R., Watada, H., Imano, E., Iwama, N., Morishima, T., Yamasaki, Y., Kajimoto, Y., & Kamada, T. (1994). Estrogen regulation of the insulin-like growth factor I gene transcription involves an AP-1 enhancer. J. Biol. Chem. 269, 16433-16442

van Dijk, M.A., van Schaik, F.M.A., Bootsma, H.J., Holthuizen, P., & Sussenbach, J.S. (1991). Initial characterization of the four promoters of the human insulin-like growth factor II gene. Mol. Cell. Endocrinol. 81, 81-94

Vukicevic, S., Luyten, F.P., & Reddi, A.H. (1989). Stimulation of the expression of osteogenic and chondrogenic phenotypes *in vitro* by osteogenin. Proc. Natl. Acad. Sci. USA 86, 8793-8797.

Vuori, K., & Ruoslahti, E. (1994). Association of insulin receptor substrate-1 with integrins. Science 266, 1576-1578.

Wakefield, L.M., Smith, D.M., Flanders, K.C., & Sporn, M.B. (1988). Latent transforming growth factor-β from human platelets. J. Biol. Chem. 263, 7646-7654.

Wall, N.A., Blessing, M., Wright, C.V.E., & Hogan, B.L.M. (1993). Biosynthesis and *in vivo* localization of the decapentaplegic-vg-related protein, DVR-6 (bone morphogenetic protein-6). J. Cell. Biol. 120, 493-502.

Wang, J., Gao, G., & Goldfarb, M. (1994). Fibroblast growth factor receptors have different signaling and mitogenic potentials. Mol. Cell. Biol. 14, 181-188.

Wang, X-F., Lin, H.Y., Ng-Eaton, E., Downward, J., Lodish, H.F., & Weinberg, R.A. (1991). Expression cloning and characterization of the TGF-β type III receptor. Cell 67, 797-805.

Way, B.A., & Mooney, R.A. (1993). Activation of phosphatidylinositol-3-kinase by platelet-derived growth factor and insulin-like growth factor-1 is inhibited by a transmembrane phosphotyrosine phosphatase. J. Biol. Chem. 268, 26409-26415.

Westermark, B., & Heldin, C-H., (1993). Platelet-derived growth factor. Acta Oncol. 32, 101-105.

Wharton, K.A., Thomsen, G.H., & Gelbart, W.M. (1991). *Drosophila 60A* gene, another transforming growth factor beta family member, is closely related to human bone morphogenetic proteins. Proc. Natl. Acad. Sci. USA 88, 9214-9218.

Wozney, J.M., Rosen, V., Celeste, A.J., Mitsock, L.M., Whitters, M.J., Kriz, R.W., Hewick, R.M., & Wang, E.A. (1988). Novel regulators of bone formation: Molecular clones and activities. Science 242, 1528-1534.

Yamaguchi, A., Katagiri, T., Ikeda, T., Wozney, J.M., Rosen, V., Wang, E.A., Kahn, A.J., Suda, T., & Yoshiki, S. (1991). Recombinant human bone morphogenetic protein-2 stimulates osteoblastic maturation and inhibits myogenic differentiation *in vitro*. J. Cell Biol. 113, 681-687.

Yarden, Y., Escobedo, J.A., Kuang, W-J., Yang-Feng, T.L., Daniel, T.O., Tremble, P.M., Chen, E.Y., Ando, M.E., Harkins, R.N., Francke, U., Fried, V.A., Ullrich, A., & Williams, L.T. (1986). Structure of the receptor for platelet-derived growth factor helps define a family of closely related growth factor receptors. Nature 323, 226-232.

Zheng, M.H., Wood, D.J., Wysocki, S., Papadimitriou, J.M., & Wang, E.A. (1994). Recombinant human bone morphogenetic protein-2 enhances expression of interleukin-6 and transforming growth factor-β1 genes in normal human osteoblast-like cells. J. Cell. Physiol. 159, 76-82.

HEMATOPOIESIS:

CLINICAL APPLICATION OF COLONY-STIMULATING FACTORS

John E. Janik and Langdon L. Miller

Growth Factors and Cytokines in Health and Disease
Volume 3A pages 157-189.

ISBN: 0-7623-0118-X

ABSTRACT

The deleterious effects of chemotherapy on normal hematopoietic tissues may produce severe or protracted deficiencies in red blood cells, neutrophils, or platelets, and result in life-threatening infection or bleeding. Hematopoietic growth factors have had a profound influence on the management of patients undergoing cancer chemotherapy. Three recombinant hematopoietic growth factors have been licensed for clinical use in the United States: erythropoietin, G-CSF, and GM-CSF. Erythropoietin reduces the transfusion requirements of chemotherapy-induced anemia, and G-CSF and GM-CSF prevent infections due to chemotherapy-induced neutropenia, shorten the duration of neutropenia following high-dose chemotherapy, and stem cell transplantation and mobilize stem cells for transplantation. This chapter reviews randomized placebo-controlled trials of the licensed colony-stimulating factors in management of patients with cancer and briefly examines new agents that stimulate megakaryocyte and platelet development.

I. INTRODUCTION

Blood is composed of a complex mixture of plasma proteins and cellular elements. The cells circulating in the blood are essential for survival and have various effector functions. Blood cells have a rapid turnover and a continuous production is required for maintenance of stable blood cell counts. Bone marrow is the primary tissue responsible for blood cell production in adult life and is capable of responding to the increased production needs associated with anemia or infection. The stress induced by these stimuli causes the marrow to increase its production of red blood cells or white blood cells respectively. Hematopoietic growth factors are responsible for this increase in blood cell production by the bone marrow.

The complex process of blood cell production has been intensely investigated and many of the colony-stimulating factors (CSFS) responsible for the production of specific blood cell populations have been cloned, sequenced, and expressed in yeast, bacteria, and mammalian cells using recombinant DNA techniques. Table 1 provides a partial list of CSFS that have been approved for clinical use and those that are under investigation. These hematopoietic growth factors have been employed extensively in the management of patients receiving radiation or chemotherapy and their use has helped to reduce the morbidity associated with the administration of cancer therapy. Important questions about the ability of these agents to permit administration of dose-intense chemotherapy and improve cancer response and cure rates can now be evaluated.

It is critically important to define the expectations that need to be fulfilled before CSFS can be accepted as effective therapeutic agents. It is clear that not all patients treated with these cytokines will experience a clinical benefit and that, in fact, a significant number of patients do not achieve the desired therapeutic endpoint. Is

Table 1. Human Recombinant Hematopoietic Growth Factors

Factor	Molecular Weight	Chromosomal Location	Site of Synthesis[1]	Hematopoietic Activities
Erythropoietin	34,000	7q11-22	Kidney, Liver	Stimulates erythroid maturation.
G-CSF	19,600	17q11-22	MFE	Stimulates granulocyte maturation and activation.
GM-CSF	22,000	5q23-31	MFET	Stimulates granulocyte, monocyte, dendritic cell, erythroid and megakaryocyte maturation. Activates granulocytes.
Interleukin-1 alpha	17,500	2q14	MaEp	Sensitizes early hematopoietic progenitors to later acting factors. Stimulates lymphoid cells.
Interleukin-1 beta	17,500	2q14	EF Others	Sensitizes early hematopoietic to later acting factors. Stimulates lymphoid cells.
Interleukin-3	25,000	5q23-31	T	Stimulates early growth of granulocytes, monocytes, erythroid, and megakaryocytes.
Interleukin-6	26,000	7q15	FMaT Others	Stimulates megakaryocyte maturation. Stimulates B cell maturtion.
Interleukin-11	23,000	19q13.3-4	Stromal	Synergizes with IL-3 to stimlate megakaryocyte colonies. Shortens cell cycle duration of early hematopoietic progenitors.
PIXY-321	35,000	NA	NA	Stimulates megakayocyte, granulocyte maturation.
Thrombopoietin	25,000-31,000	3q26 or 27	Liver, kidney, Other	Stimulates megakaryocyte maturation, sensitizes platelets to aggregating agents.

Note: [1] M = Monocyte, F = Fibroblast, T = T cell, Ma = Macrophage, Ep = Epithelial cell.

it necessary that the use of a CSF result in an improvement in survival before its use can be recommended? Is a reduction in the cost associated with the administration of cancer therapy important in justifying the administration of a CSF? Are other endpoints, such as a decrease in neutropenia or neutropenic fevers, sufficient criteria to warrant their use? Finally, what effects on quality of life do these hematopoietic growth factors have for patients undergoing cancer therapy? In addressing these questions this chapter will focus primarily on placebo-controlled randomized trials testing the ability of the CSF to stimulate red cell and neutrophil maturation in support of patients undergoing cancer treatment. Randomized trials of agents that stimulate platelet production have been reported and a brief review of agents that stimulate thrombopoiesis will be included. A number of reviews of

CSF and the recommendations for their use have been published and the interested reader is referred to these publications for further information. (Boogaerts et al.,1995; American Society of Clinical Oncology, 1994; Vose and Armitage, 1995).

II. THE USE OF ERYTHROPOIETIN IN PATIENTS WITH CANCER

Anemia is a common problem in patients with cancer and frequently requires therapeutic intervention. Nutritional deficiencies, the myelotoxic effects of radiation and chemotherapy, and the production of negative regulators of hematopoiesis in response to cancer may all participate in producing anemia. Erythropoietin levels in anemic cancer patients are significantly lower than those of patients with iron-deficiency anemia with comparable levels of hemoglobin (Miller et al.,1990). This relative deficiency of erythropoietin production in cancer patients is often worsened by chemotherapy, particularly platinum-based regimens. This latter effect may be a result of the nephrotoxicity of these agents.

The use of erythropoietin to treat cancer-associated anemia is thus based on the rationale of its relative deficiency and the known efficacy of erythropoietin in the treatment of anemia in patients with end stage renal failure who are undergoing hemodialysis (Eschbach et al., 1987, 1989a, 1989b). Three clinical settings can be considered for clinical evaluation of erythropoietin: (1), in the treatment of anemia due solely to cancer, (2), in chemotherapy-related anemia, and (3), after bone marrow transplantation. Randomized trials of erythropoietin have been conducted in all three settings.

A single major study has been performed to evaluate the use of erythropoietin in cancer patients who were receiving no therapy (Abels,1993). In this study, 124 anemic (hematocrite 32%) patients were randomized to recombinant human erythropoietin (rHuEPO) 100 U/kg or placebo given subcutaneously three times per week. There was no change in hematocrit in the placebo group but a 2.8% increase in hematocrit was seen in patients receiving rHuEPO. An increase in hematocrit to 38% was observed in 21% of rHuEPO treatment patients and 32% of the group had an increase in hematocrit of 6% or greater. However, transfusion requirements were not significantly reduced during the period of study. Administration of higher doses of rHuEPO for longer periods of time may have improved these therapeutic endpoints.

In the second and more common circumstance, patients with cancer are receiving chemotherapy that will further exacerbate cancer-associated anemia. In one study, 157 anemic (hematocrit 32%) patients were randomized to rHuEPO, at a dose of 150 U/kg, or placebo subcutaneously three times weekly for 12 weeks or until the hematocrit reached 38-40% (Case et al.,1993). No significant change in the hematocrit was demonstrated for the placebo group whereas 41% of the rHuEPO-treated patients attained a hematocrit of 38% and 58% of patients had a 6% or greater increase in hematocrit. Transfusion requirements for the 12-week period were not altered by rHuEPO therapy. This result may mask a therapeutic benefit; no change in transfusion requirement was noted in the first month of treatment but transfusion requirements were reduced in the rHuEPO group in the second and third months of treatment. This result indicates that a lag time needed to produce responsive erythroid precursors is required before erythropoietin can be effective.

It is not likely that rHuEPO would have an impact on survival in any cancer setting, therefore cost-benefit analysis and quality of life assessments play major deciding roles in the use of this cytokine in the management of patients with cancer. In this study, a statistically significant improvement in energy level and ability to perform the activities of daily living as well as a trend toward improvements in overall quality of life were noted in the rHuEPO group.

Cisplatinum-based chemotherapy may represent a particularly suitable setting for the evaluation of rHuEPO therapy in anemic cancer patients. Erythropoiesis is impaired in most patients receiving platinum-based chemotherapy regimens; the anemia is normochromic and normocytic and associated with a low erythropoietin level (Doll and Weiss, 1983). There appears to be a correlation between the cumulative platinum dose and decline in hematocrit. Two trials have evaluated rHuEPO in the management of anemic cancer patients receiving cisplatinum-based chemotherapy. In both cases patients received rHuEPO at a dose of 150 U/kg three time weekly. In the larger trial, 132 patients were randomized to treatment or placebo (Abels, 1993) The mean increase in hematocrit was 6% in the rHuEPO group and 1.3% in the placebo group: 35.9% of patients in the rHuEPO group achieved a hematocrit of 38% and 48.4% had an increase in hematocrit of 6% or greater. In the placebo group 1.6% of patients achieved a hematocrit of 38%. The transfusion requirements during the trial were not altered by rHuEPO although a trend toward a lesser number of transfusions was again noted in the second and third month of therapy.

In the smaller study, 57 patients were randomized to rHuEPO with protoferrin or protoferrin alone (Gamucci et al.,1993). Hemoglobin values rose by 0.9 g/dl in erythropoietin-treated patients whereas they fell by 1.5 g/dl in the control group. rHuEPO was not universally effective; an increase in hemoglobin of 2.1 g/dl was seen in 76% of the rHuEPO group but in 24% of patients declined by 2.8 g/dl. In the control group 53% of patients maintained a stable hemoglobin and the remaining 47% had a drop of 3.5 g/dl. No quality of life assessment was performed in this study.

Erythropoietin has been tested after allogeneic and autologous bone marrow transplantation to evaluate its effectiveness in reducing red cell transfusion requirements and time to transfusion independence. In the largest study, 329 patients undergoing allogeneic or autologous bone marrow transplantation were randomized to rHuEPO or placebo after marrow reinfusion (Link et al.,1992). rHuEPO was administered at a daily dose of 150 IU/kg as a continuous intravenous infusion until the red cell count was stable or for a maximum of 42 days. In the allogeneic transplant group, 106 patients received rHuEPO and 109 were given a placebo. The median time to red cell transfusion independence was significantly reduced from

27 days in patients receiving placebo to 19 days in those randomized to rHuEPO. There was no difference in the mean number of erythrocyte transfusion in the first 20 days after transplant, nor was there a reduction in the number of transfusions for the first 100 days after transplantation. There was a significantly reduced erythrocyte transfusion requirement from day 21 to 41 of the study; patients randomized to placebo received a mean of 2.7 units of red cells and those randomized to rHuEPO received 1.4 units during this period. A near significant reduction in transfusion requirement was also noted in the followup period off rHuEPO. The major factors influencing transfusion requirements in this study were bleeding events, acute graft-versus-host disease (GVHD), and major ABO blood group incompatibilities. The greatest effect for rHuEPO was observed in patients with severe GVHD. Transfusion requirements were reduced from a mean of 18.4 units to 8.5 units.

In contrast to this minimal benefit seen in patients undergoing allogeneic transplantation no difference in any parameter was detectable in patients undergoing autologous transplantation. The median time to transfusion independence and the number of units of red cells transfused was not reduced. The incidence of severe transplantation related toxicities including hepatic veno-occlusive disease were not adversely affected by rHuEPO treatment in either setting.

The results of this study were confirmed by other investigators in both the allogeneic and autologous setting. In the allogeneic transplantation study, 50 patients were randomized to rHuEPO 200 U/kg daily for four weeks and then twice weekly for an additional four weeks or to placebo (Klaesson et al.,1994). The time to red cell transfusion independence was significantly reduced from 24 days in the control group to 14 days in the rHuEPO group and the red blood cell transfusion requirements were decreased. The mean number of units of red cells transfused was reduced from 10 units in the placebo group to five in the rHuEPO cohort. There was no difference in transplantation-related mortality or toxicity.

The lack of efficacy of rHuEPO in the autologous transplant setting was confirmed in a second trial (Chao et al.,1994). In this study 35 patients with lymphoma were randomized to rHuEPO (600 U/kg as an intravenous bolus three times weekly) or placebo. Therapy with rHuEPO was initiated three weeks before transplantation because of the lag time for red cell response seen in previous trials. rHuEPO was held during the preparatory chemotherapy regimen and resumed on the day of marrow reinfusion. All patients received G-CSF following marrow reinfusion. There was no reduction in the median number of units of red cells transfused; no comment concerning the time to erythrocyte transfusion independence was made. rHuEPO treatment did not impair or improve granulocyte or platelet engraftment.

A recently published study evaluated rHuEPO in the management of patients with a normal hematocrit undergoing chemotherapy for small cell lung cancer (de Campos et al.,1995). In previous studies, the target populations were anemic cancer patients (hematocrit 32%) and although rHuEPO improved red cell counts, no reduction in transfusion requirements was demonstrated. In this small study, 36 patients were treated with carboplatin, etoposide, ifosfamide, and vincristine chemotherapy on a 28-day schedule for up to six cycles. Twelve patients were randomized to each group; one group received no additional treatment and the others received rHuEPO at one of two dose levels (150 or 300 U/kg). Patients were transfused to maintain a hemoglobin of 9 g/dL. Hemoglobin values declined in all groups but the decrease was delayed in groups receiving rHuEPO. Fewer patients required red cell transfusions and the total number of units of red cells transfused was more than halved by rHuEPO treatment. There was no difference in transfusion requirements for the two groups treated with low and high dose rHuEPO. Interestingly, the number of units of platelets transfused was also decreased in the rHuEPO group. Homology between thrombopoietin and erythropoietin may account for the beneficial effect on platelet counts seen with this relatively high dose rHuEPO regimen (Bartley et al., 1994). This was the first study to demonstrate a reduction in red cell transfusion requirement in patients undergoing standard outpatient cancer chemotherapy. No improvement in response rate or survival was described in the study. It will be important to confirm these observations in larger trials and to evaluate the cost-effectiveness of this approach.

Algorithms have been developed to identify patients with a high probability of response or unresponsiveness to rHuEPO. Therapy can be discontinued in patients with a low probability of response thereby improving cost-effectiveness. An algorithm was developed with a group of 40 patients and verified with a subsequent group of 40 patients (Ludwig et al., 1994). A positive or negative response could be predicted using two parameters, hemoglobin and serum erythropoietin. Patients who have shown an increase in hemoglobin of 0.5 g/dL and have a serum erythropoietin level of less than 100 mU/mL after two weeks of treatment have a high probability of response (predictive power, 95%) whereas patients who have a serum EPO level of greater than 100 mU/mL and an increase in hemoglobin of less than 0.5 g/dL are very unlikely to respond (predictive power, 93%). The serum ferritin can also be used to predict response. Patients with a serum ferritin of greater than 400 ng/mL after two weeks of therapy are unlikely to respond (predictive power, 88%) whereas patients with a ferritin less than that value are likely to respond (predictive power, 72%).

On the basis of the trials described above the FDA approved the use of rHuEPO in cancer patients who will be receiving chemotherapy for a minimum of two months. This approval was based on a significantly decreased transfusion requirement after the first month of treatment. It remains to be shown in large randomized studies that the use of rHuEPO will decrease transfusion requirements and make its use more cost-effective than red blood cell transfusion. The toxicity of rHuEPO in patients with cancer has been minimal. Hypertension, seen in patients with renal failure treated with rHuEPO, has not occurred in cancer patients with any significant frequency. It is important to consider the administration of iron to patients who show an initial response to therapy but then fail to respond. A functional iron deficiency can be produced in patients treated with rHuEPO.

One additional rationale for using rHuEPO in cancer patients who need red blood cell transfusion is the induction of an immunosuppressed state following allogeneic red blood cell infusions. Renal transplant recipients who have undergone allogeneic red blood cell transfusions have an improved allograft survival compared with those untransfused before transplant (Blajchman and Singal, 1989; Opelz and Terasaki, 1978). It has been postulated that the transfusion induces a state of immunosuppression that enhances allograft survival. Patients with colon cancer frequently undergo red blood cell transfusion and a large number of retrospective studies have demonstrated an inferior outcome in transfused patients. These studies are counterbalanced by other studies that show equivalent survival in transfused and non-transfused patients. Two recent meta-analyses of the question found an increased risk of cancer recurrence in transfused patients (Chung et al., 1993; Vamvakas and Moore, 1993). If allogeneic transfusion does produce a state of immunosuppression that results in higher recurrence rates then the rationale for use of rHuEPO in patients with cancer will be enhanced and the potential for an improvement in survival with rHuEPO may be attained.

III. USE OF COLONY-STIMULATING FACTORS TO STIMULATE NEUTROPHIL MATURATION

Three general strategies for use of CSF have been applied in the management of patients who have developed or are at risk for the development of neutropenic fevers or infections. These approaches have primarily been developed in patients undergoing chemotherapy for the treatment of cancer but can be applied to other patients at risk for neutropenic complications.

Primary prophylaxis is the administration of a CSF before the onset of neutropenia. In this setting the CSF is typically administered in the first cycle immediately after the completion of chemotherapy. Most randomized trials of CSF have tested the benefit of CSF use in this setting.

Secondary prophylaxis is the administration of a CSF to prevent subsequent episodes of neutropenia or neutropenic fever in a patient who has already experienced this complication. This approach would appear to be the most rational because it targets an expensive therapy to the patient population most likely to benefit. This avoids the administration of CSF to many patients who do not need treatment but does have the disadvantage that it allows some patients to experience neutropenia or a neutropenic fever that could have been prevented.

CSF therapy is the use of CSF for the treatment of neutropenia or neutropenic fever. This approach specifically targets the population known to have developed a neutropenic event and prevents the unnecessary administration of CSF to patients who do not need to be treated. However, therapeutic CSF use after the occurrence of neutropenia lacks the ability to prevent complications; its utility is limited to the theoretical enhancement of neutrophil function and a reduction in the duration of neutropenia, fever, or infection.

Two myeloid growth factors have been licensed for use in the United States by the Food and Drug Administration (FDA) and the development of others is ongoing or has been halted (Table 2). The next sections will review the use of G-CSF and GM-CSF as primary prophylaxis, secondary prophylaxis, and treatment for established neutropenia.

A G-CSF: Primary Prophylaxis

Outpatient Chemotherapy

Several trials have evaluated filgrastim G-CSF as primary prophylaxis following standard dose chemotherapy in the outpatient setting. In the trial conducted by Crawford et al. (1991) 199 patients with small cell lung cancer (SCLC) were treated with a chemotherapy regimen that consisted of cyclophosphamide, doxorubicin, and etoposide (CAE) followed by randomization to treatment with G-CSF at a dose of 230 μg/m^2/day or placebo. Patients who developed neutropenic fevers were crossed over to receive open-label G-CSF. This design unfortunately compromised the ability to determine the efficacy of G-CSF on survival and dose-intensity. In

Table 2. CSFs that Stimulate Neutrophil Maturation

CSF Glycosylation	*Generic/ Trade Name*	*Synthetic Method*	*Primary Indication*	*Developmental Status*
G-CSF No	Filgrastim Neupogen	*E. coli*	Prevention of chemotherapy-induced neutropenic complication	Commercially available worldwide
G-CSF Yes	Lenograstim Neutrogin	CHO cells	Prevention of chemotherapy-induced neutropenic complication	Commercilly available outside the United States
GM-CSF Yes	Sargramostim Leukine	*S. cerevisiae*	Stimulation of neutrophil engraftment following autologous bone marrow transplantaion	Commercially available in the United States
GM-CSF No	Molgramostim Leucomax	*E. coli*	Prevention of chemotherapy-induced neutropenic complication	Investigational in the United States Commercially available outside the United States
GM-CSF Yes	Regramostim Not applicable	CHO cells	None	Development halted
GM-CSF No	Ecogamostim Not Applicable	*E. coli*	None	Development halted

the other trial, conducted in Europe (Trillet-Lenoir et al.1993), 129 patients with SCLC were treated with an identical regimen but cross-over to G-CSF was not permitted. This trial allowed an analysis of G-CSF on treatment outcome and chemotherapy dose intensity.

The analysis of both trials focused on the results in the first cycle of therapy because of the comparability of chemotherapy doses in the initial cycles. In both trials, G-CSF reduced the depth and duration of neutropenia; a greater then 20% decrease in the incidence of febrile neutropenia was similarly observed. As a result of these effects significantly fewer hospitalizations for antibiotic therapy were needed in the patients treated with G-CSF. It is of interest that the majority of the benefit from G-CSF therapy was apparent after the first cycle of chemotherapy. In subsequent cycles, few patients in either treatment group developed febrile neutropenia, perhaps as a consequence of chemotherapy dose reductions.

The European study permits an analysis of chemotherapy dose-intensity and treatment outcome. Patients treated with G-CSF had fewer chemotherapy dose-reductions and received an 8% more dose-intense treatment. Unfortunately, this increment in dose-intensity was not associated with an improvement in response rate or survival for patients with limited or extensive disease.

The filgrastim form of G-CSF was also evaluated in 80 patients with intermediate grade lymphoma treated with combination chemotherapy on a weekly basis (Pettengell et al., 1992). Patients were randomized to chemotherapy alone or with G-CSF given continuously except on the days preceding and during therapy with cyclophosphamide, doxorubicin, and etoposide. Significantly fewer episodes of neutropenia (ANC < 1,000) and neutropenic fever were associated with G-CSF use but because of the low threshold for hospitalization (temperature 37.5 C and ANC < 1,000/µl) there was no benefit in rate of hospitalization or antibiotic use. Similar to the study of G-CSF in SCLC, chemotherapy dose-intensity was slightly greater (approximately 8-11%) in patients treated with G-CSF but without a concomitant increase in response rate or survival.

No effect of G-CSF on platelet counts was noted in these trials with the exception of the European study of CAE which was associated with more thrombocytopenia in the later cycles of chemotherapy (Trillet-Lenoir et al., 1993). This may be a reflection of the higher chemotherapy dose-intensity permitted by G-CSF. Mucositis was less frequent in G-CSF-treated patients undergoing CAE chemotherapy (Crawford et al., 1994) but prompted more treatment delays in G-CSF-treated patients with lymphoma (Pettengell et al., 1992). No effect of G-CSF on mortality was noted in any trial.

Induction Therapy in Acute Leukemia

One trial has evaluated the use of G-CSF after induction therapy for acute leukemia (Ohno et al., 1990). In this study a heterogeneous population of patients with relapsed or refractory leukemia were treated. The administration of G-CSF was associated with a more rapid neutrophil recovery and fewer days of fever and documented infections. The rates of infectious mortality in the study were not reduced by G-CSF therapy nor was the duration of hospitalization or antibiotic usage reduced. The complete response rates and relapse rates were not affected by G-CSF administration. There was no clinically detectable adverse effect of cytokine therapy on stimulation of leukemia regrowth (Ohno et al., 1990, 1993).

High-Dose Chemotherapy and Stem Cell Transplantation

The use of a CSF after high-dose chemotherapy and stem cell transplantation represents a special case in the primary prophylactic use of CSF. All patients treated in this way are expected to have a protracted period of neutropenia and many will develop neutropenic fevers and documented infections despite the use of a CSF. Clinical benefit must therefore be measured in terms of a reduction in the number of days of neutropenia, days of hospitalization, and mortality. Randomized trials of both filgrastim and lenograstim have been conducted primarily in patients with lymphoma and leukemia (Gisselbrecht et al., 1994; Blaise et al., 1992; Schmitz et al., 1992; Linch et al., 1993; Stahel et al., 1994). The use of G-CSF is associated with a more rapid neutrophil recovery and the duration of fevers was significantly reduced although mortality due to infection was not reduced in any trial. The effects of G-CSF on antibiotic usage and the number of days in the hospital varied, depending on the study.

Myelodysplastic Syndromes

G-CSF has been evaluated in patients with myelodysplastic syndromes (MDS) in an attempt to reduce infections and mortality, based on *in vitro* studies that show an improvement in neutrophil superoxide anion production and neutrophil alkaline phosphatase activity in patients with refractory anemia with excess blasts (RAEB) or refractory anemia with excess blasts in transformation (RAEB-T; Yuo et al., 1987). A preliminary report randomized 102 patients with MDS to chronic administration of G-CSF or observation. Entry criteria on the study required a neutrophil count of less than 800/µl. All patients receiving G-CSF had an improvement in neutrophil count within 2-3 weeks of study entry compared with the untreated controls (Greenberg et al., 1993). Survival in RAEB-T patients randomized to G-CSF therapy was comparable but mortality in the patients with RAEB was higher with G-CSF therapy. A subsequent analysis suggested that the G-CSF treated cohort had a greater proportion of patients with poor prognostic features (number of marrow blasts, platelet count and age) than the observation arm (Sanz et al., 1989). When patients were stratified by risk category, decreased survival was again observed in association with G-CSF therapy but only in the high-risk subgroup. No analysis of the clinical benefit of G-CSF in terms of time to first infection, relative frequency of infectious complications and infectious mortality has been reported.

B. GM-CSF: Primary Prophylaxis

Outpatient Chemotherapy

The randomized trials of GM-CSF primary prophylaxis have not produced results of sufficient magnitude to warrant approval by the FDA for use after standard dose chemotherapy in the outpatient setting. GM-CSF is licensed for administration after bone marrow transplantation and was recently recommended for administration after induction chemotherapy for acute leukemia in older patients.

The most positive trials of primary GM-CSF prophylaxis have involved the treatment of patients with non-Hodgkin's lymphoma. In one trial, 172 patients were randomized to GM-CSF at a dose of 400 μg per day after COP-BLAM chemotherapy (Gerhartz et al., 1993a). Among the patients who were able to tolerate GM-CSF administration through six courses, there was a reduction in the number of days of neutropenia, fever, hospitalization, and antibiotic use. The response rate was the same or higher in the GM-CSF-treated group. However, a high dropout rate in the GM-CSF-treated patients prevented an adequate assessment of GM-CSF efficacy in this trial if all patients who received study medication were included in the analysis, statistically significant benefit from 6m-CSF prophylaxis was lost. Positive results were also reported in a small trial of GM-CSF in HIV-related NHL (Kaplan et al., 1991). In the 21 randomized patients, GM-CSF use was associated with a reduction in neutropenia, febrile neutropenia, and hospitalization. Another study evaluated the use of GM-CSF after a dose-intense chemotherapy regimen of cyclophosphamide, etoposide, and cisplatinum in patients with breast cancer or lymphoma (Neidhart et al., 1994). Forty-two of the 56 patients enrolled in the study received two cycles of therapy and were considered evaluable. The duration of neutropenia, febrile neutropenia, and hospitalization for fever was significantly reduced in the first cycle in the GM-CSF-treated cohort with trends favoring GM-CSF in later cycles. A reduction in the frequency of neutropenic fevers was not observed perhaps because of the high chemotherapy dose-intensity of the regimen.

GM-CSF has also been evaluated as primary prophylaxis in a randomized study of 148 patients with SCLC receiving CAE chemotherapy. Patients were randomized to 0, 10, or 20 μg/kg/d of molgramostim GM-CSF administered subcutaneously after chemotherapy (Hamm et al., 1994). There was no change in the frequency of febrile neutropenia; the incidence of such occurrences, however, was only 10% per cycle, much lower than the frequency in the control group or the G-CSF-treated group in the Crawford study (Crawford et al., 1991) because lower doses of chemotherapy were employed in the 6M-CSF trial. There was a trend toward a shorter duration of Grade 4 neutropenia (ANC < 500/μl) and a significantly higher median nadir ANC. GM-CSF did not allow dose increases in CAE chemotherapy and there was no benefit associated with the higher dose of GM-CSF.

Two additional trials of molgramostim GM-CSF, one in testicular cancer (Bajorin et al., 1995) and another in sarcomas in children (Wexler et al., 1994) showed no significant effects on neutrophil recovery or febrile neutropenia. This apparent failure of GM-CSF to show clinical benefit in primary prophylaxis following outpatient chemotherapy may be due to its lesser biological activity (Lord et al., 1989, 1992) or to the induction of negative regulators of hematopoiesis that impair GM-CSF's positive effects (Stehle et al., 1990). Alternatively, the ability of GM-CSF to produce fevers may obscure its effects on hematopoiesis and prevent observation of a beneficial decrease in the incidence of febrile neutropenia.

Acute Leukemia

Two studies evaluated the use of GM-CSF after induction chemotherapy for acute leukemia. In a study conducted by the Eastern Cooperative Oncology Group 117 patients between the ages of 55 and 70 were randomized to GM-CSF or placebo after initial induction chemotherapy for acute leukemia (Rowe et al., 1993). This study formed the basis for FDA recommendation of GM-CSF in primary prophylaxis in elderly patients with acute leukemia. The trial showed that 6M-CSF prophylaxis was associated with significant reductions in the duration of Grade 4 neutropenia and in the incidence of Grades 3, 4, and 5 infectious toxicities among patients receiving GM-CSF. This trial also showed a significant improvement in median survival of patients treated with GM-CSF. This trial was the first to show an improvement in survival associated with CSF administration. A second study was reported by the Cancer and Leukemia Group-B (CALGB) in patients over the age of 60 with newly diagnosed AML (Stone et al., 1994). Molgramostim GM-CSF or placebo was administered to 347 evaluable patients. This study failed to confirm the results of the ECOG study; there was no statistically significant reduction in the duration of neutropenia, infectious complication, or mortality. These differences may have been obscured by the high rate of removal of patients from the study for toxicity. More than 30% of the patients in both arms of the study were removed from treatment because of the perception of excessive toxicity, on the part of the treating physicians thus only two-thirds of the patients received the planned course of treatment.

High-dose Chemotherapy and Stem Cell Transplant

Sargramostim GM-CSF is licensed by the FDA for primary prophylaxis following high-dose chemotherapy and bone marrow transplant in patients with non-Hodgkin's and Hodgkin's lymphoma and acute lymphocytic leukemia (Immunex Corporation, 1991). In the licensing trial, 128 patients with lymphoma or leukemia were randomized to GM-CSF or placebo. The primary value of GM-CSF was to accelerate neutrophil recovery to an ANC > 500/µl following stem cell infusion (Nemunaitis et al., 1991). There was a high incidence of neutropenic fevers in both groups and no difference in the duration of fevers. The mean time to achieve an ANC of 100/µl was not significantly different. The number of documented infections was reduced in the GM-CSF group but the difference did not achieve statistical significance. The important clinical parameters of duration of hospitalization and antibiotic use were reduced by GM-CSF. Several additional studies of GM-CSF have documented a more rapid neutrophil recovery in lymphoma patients undergoing high-dose chemotherapy and stem cell infusion (Gulati and Bennett, 1992; Advani et al., 1992; Gorin et al., 1992; Link et al., 1992; Khwaja et al., 1992).

Myelodysplastic Syndrome

GM-CSF was shown in phase I trials to increase neutrophil counts within two or three days but the response was not maintained when CSF administration was discontinued (Willemze et al., 1992; Schuster et al., 1990; Rose et al., 1994). GM-CSF administration can also improve neutrophil function (Boogaerts et al., 1990) in patients with myelodysplastic syndromes, another potential benefit of therapy. Neutrophil chemotaxis is reduced by GM-CSF therapy, a potentially negative effect of this therapy and an effect that has also been seen in patients after receiving chemotherapy and GM-CSF (Kaplan et al., 1989). Monocyte and eosinophil counts increase but consistent effects on red cell and platelet production were not observed.

A randomized trial of molgramostim GM-CSF, 3 μg/kg/d versus observation was conducted in 133 myelodysplastic patients with neutropenia, ANC 1500/μl (Schuster et al., 1990). Statistically significant increases in neutrophil counts were observed at day 30, 60, and 90 after starting therapy. Eosinophil, monocyte, and lymphocyte counts also increased in response to treatment. No change in transfusion requirements of red blood cells or platelets was reported. The incidence of major infections (neutropenic fever or IV antibiotics) was reduced from 33% in the control arm to 15% in those receiving GM-CSF. There was no evidence that GM-CSF caused an increased incidence of acute leukemia or RAEB-T; four patients in each arm progressed during the 90-day trial. Responding patients were offered continued therapy during the first 14 days of each month. Improvements in neutrophil counts were noted only during GM-CSF administration in these patients. The effects of marrow cytogenetics, *in vitro* hematopoietic progenitor culture, marrow blast percentage, platelet count, and age on response to GM-CSF have not yielded consistent results and cannot be used to select patients for treatment (Willemze et al., 1992; Estey et al., 1991).

C. Secondary CSF Prophylaxis of Infection

There have been no randomized trials to determine whether CSF prophylaxis will prevent new episodes of neutropenic fevers in patients with a prior episode during an earlier cycle of chemotherapy. However, in the Crawford study of G-CSF primary prophylaxis, patients in both arms of the trial were crossed over to open label G-CSF after an episode of febrile neutropenia (Crawford et al., 1991). The

52 patients who were treated with placebo and had an episode of neutropenic fever in the first cycle of therapy had a reduction in the duration of Grade 4 neutropenia from six days in Cycle 1 to three days in Cycle 2. In addition, as compared to the 100% incidence of neutropenic fever in Cycle 1, only 23% of these patients had an episode of neutropenic fever in Cycle 2. Assuming that all patients would experience a second episode of neutropenic fever in the second cycle of therapy had they not gotten G-CSF, these results indicate a 77% absolute reduction in incidence of neutropenic fevers, an improvement that is substantially better than the 20-30% absolute benefit with primary prophylaxis. Patients who develop an episode of neutropenic fever appear to be at greater risk than those who do not.

Secondary prophylaxis with sargramostim GM-CSF was evaluated in a preliminary fashion in a phase I-II trial (Vadhan-Raj et al., 1992). Patients with neutropenia in an earlier cycle of therapy with cyclophosphamide, doxorubicin, and dacarbazine received various doses of GM-CSF. A reduction in the duration of neutropenia (mean of 6 vs. 3 days) and a trend toward fewer episodes of neutropenic fevers were observed. The small sample size and the different dose levels of GM-CSF hinder interpretation.

These trials, although not randomized, suggest that secondary prophylaxis is of benefit to patients with an earlier episode of neutropenic fever. The degree of benefit may be even greater than that associated with the use of CSF primary prophylaxis. Secondary application of the CSF may also provide economic advantages by limiting use of the support only in those situations where the neutropenic risk has been documented.

D. CSF Therapy of Established Neutropenia and Neutropenic Fevers

CSF therapy has commonly been instituted in patients after the onset of neutropenia and particularly in patients with febrile neutropenia admitted to the hospital for a febrile intravenous antibiotics who have been as part of standard medical care.

A small randomized trial evaluated the efficacy of G-CSF in neutropenic (ANC 1000/μl) patients with non-small cell lung cancer receiving cisplatin, vindesine, and mitomycin-C (Fukuda et al., 1993). CSF therapy did not shorten the duration of neutropenia or produce clinically meaningful benefit. The small sample size and the apparent administration of G-CSF with chemotherapy complicate interpretation of the results of this study. Regramostim and molgramostim GM-CSF were evaluated in afebrile leukopenic (white blood count 2000/μl) patients without clinical benefit (Gerhartz et al., 1993b). Large prospective trials are ongoing to better define the utility of CSF therapy in this setting.

The use of CSF in patients with chemotherapy-induced febrile neutropenia has been evaluated in five randomized clinical trials (Maher et al., 1994; Anaissie et al., 1994; Riikonen et al., 1994; Biesma et al., 1990; Mayordomo et al., 1995). The largest study was a multicentered trial conducted in Australia and involved the treatment of 216 patients with fever (temperature 38.2) and neutropenia, ANC

1000/μl, (Maher et al., 1994). The antibiotic regimen used consisted of tobramycin and piperacillin. Randomization and initiation of CSF therapy was required within 12 hours of instituting empiric antibiotic coverage. Filgrastim G-CSF, 12 μg/kg/day was administered until an ANC > 500/μl was achieved and the patient was without fever for four days. Patients with hematologic malignancy (lymphoma and acute lymphocytic leukemia) or solid tumors were eligible and patients with myeloid leukemia were excluded. Placebo was administered to 107 patients and G-CSF to 109 patients. G-CSF significantly reduced the median number of days of neutropenia (3 compared with 4 days of ANC < 500/μl) and the time to resolution of neutropenic fever (5 compared to 6 days) but the clinically relevant parameters, duration of antibiotic treatment and the time of hospitalization, were not reduced (median of 8 days in both groups). The need for alternative antibiotics was also equivalent in both cohorts. Fewer patients required empiric antifungal therapy (6% compared with 11%) but the difference was not significant. Subset analysis showed that G-CSF reduced the need for prolonged hospitalization by one-half but it was not possible to identify these patients before institution of therapy. Although G-CSF was effective at increasing neutrophil counts regardless of baseline values, a reduction in the time to resolution of fever was only evident in the subgroup with initial neutrophil counts lower than 100/μl. Other subgroups benefitting from G-CSF were patients with solid tumors, patients for whom more than 10 days had elapsed between completion of chemotherapy and study entry, and in patients with documented infection. No difference in mortality was seen between patients treated with G-CSF and placebo.

A single-institution study evaluated molgramostim GM-CSF in patients with fever (temperature 38.3) and neutropenia, ANC < 1,000/microliter, (Anaissie et al., 1994). The initial empiric antibiotic regimen consisted of ticarcillin-clavulanate and netilmicin. Patients were randomized to GM-CSF (3μg/kg/day as a four-hour infusion) or placebo. The majority of patients were being treated for acute or chronic leukemia, lymphoma, or solid tumors. One hundred episodes of febrile neutropenia were studied. The time required to achieve a neutrophil count of 500/μl was reduced (7 days vs. 8 days) but the difference was not significant. The median duration of fever (4 days) and the duration of antibiotic treatment (7 days) was comparable in both groups. Subset analysis suggested that patients with tissue infections, leukemic patients and those with a baseline ANC < 100/μl were more likely to benefit from CSF. Mortality due to infection was not affected by CSF treatment. A second smaller study of regramostim GM-CSF conducted in Europe similarly failed to document a reduction in duration of hospitalization or antibiotic therapy (Biesma et al., 1990).

Two trials have claimed demonstrated a clinical benefit with CSF therapy (Riikonen et al., 1994; Mayordomo et al., 1995). In one of these trials pediatric patients with neutropenic fevers were randomized to receive 5 microgram/kg/day of GM-CSF or placebo. Because these children were receiving intensive chemotherapy, there was a mean duration of neutropenia of 11.9 days in the placebo group.

The addition of GM-CSF reduced the mean duration of neutropenia to 7.7 days. The duration of hospitalization was significantly reduced in this study, although the distribution was not affected. The intensive nature of the chemotherapy and the prolonged neutropenia favored the identification of a beneficial result with CSF therapy, similar to that observed in the Australian study of G-CSF that suggested a clinical benefit in patients with prolonged neutropenia.

In the other trial repeated 121 nonleukemic patients with fever (temperature 38 C) and neutropenia, ANC < 500/µl, (Mayordomo et al., 1995) were given an empiric antibiotic combination of ceftazadime and amikacin. Patients were randomized to G-CSF 5 µg/kg/day (39), GM-CSF 5 µg/kg/day (39) or placebo (43). The median duration of neutropenia (ANC 500 and 1,000) was significantly reduced in both CSF arms although the median duration of fever was similar. A significant reduction in the median duration of hospitalization was seen in both CSF arms of the trial (5 days compared with 7 days for placebo) although this result may have been biased in favor of CSF-therapy because unrealistically sustained recovery of ANC (> 100 for 2 days) was required for hospital discharge.

E. Mobilization of Peripheral Blood Progenitors Cells (PBPC)

An important use of CSF is in the mobilization of peripheral blood progenitors for autologous or allogeneic stem cell transplants. These cells can be collected by leukopheresis and stored for reinfusion after high dose chemotherapy. The initial studies of peripheral blood progenitors for hematologic reconstitution collected cells from patients without any mobilizing agents (CSF or chemotherapy). Many apheresis procedures were needed to acquire sufficient cells for hematologic reconstitution because of the small number of stem cells present in unstimulated blood (Kessinger and Armitage, 1991). Chemotherapy was subsequently found to stimulate release of hematopoietic progenitors into the blood probably as a result of CSF release in response to bone marrow hypoplasia. Later it was noted that both G- (Dührsen et al., 1988) and GM-CSF (Socinski et al., 1986; Mangan et al., 1993) can stimulate release of PBPC from the bone marrow and their use has significantly eased the collection of these cells. The yield of PBPC peaks at four to eight days after CSF alone (Vredenburgh et al., 1992; Bensinger et al., 1993; Shpall et al., 1994; Bishop et al., 1994) and is typically greatest during the logarithmic phase of neutrophil recovery when CSF are administered after chemotherapy (Van Hoef et al., 1994; Gianni et al., 1989; Fukuda et al., 1992; Demuynck et al., 1992; Dreger et al., 1993; Pettengell et al., 1993b; Jagannath et al., 1992; Ho et al., 1993; Tarella et al., 1991; Patrone et al., 1992). A recent study evaluated the ability of G-, GM-CSF, or the combination to mobilize progenitors from normal volunteers (Lane et al., 1995). Both G-CSF and the combination of G- and GM-CSF efficiently mobilized PBPC for hematologic reconstitution in the absence of chemotherapy; one or two phereses would be sufficient to perform allogeneic transplantation with cells mobilized by CSF alone. The yield of PBPC with GM-CSF alone, however, was about 10-fold lower than that seen with G-CSF or the combination of both cytokines. Although few randomized trials have compared the yield of PBPC harvested from the steady state or after hematologic recovery from chemotherapy alone the accumulated data from historically-controlled trials (Dührsen et al., 1988; Socinski et al., 1986; Sheridan et al., 1992; Mangan et al., 1993; Gianni et al., 1989; Chao et al., 1993; Demuynck et al., 1992; Kessinger and Armitage, 1991; Schwartzberg et al., 1992; Jagannath et al., 1992; Kessinger et al., 1992; Haas et al., 1990, 1992; Tarella et al., 1991; Bishop et al., 1994) provide convincing evidence that CSF mobilization produced stem cells that achieve more reproducible and reliable hematologic reconstitution.

The use of PBPC to achieve hematologic reconstitution has been compared to that of autologous bone marrow transplantation. In historically-controlled trials, PBPC transplantation appears to provoke more rapid hematologic recovery particularly in platelet reconstitution (Sheridan et al., 1992; Bensinger et al., 1993; Pettengell et al., 1993a; Shpall et al., 1994). The results of randomized trials of PBPC and bone marrow for hematologic reconstitution are now being reported. High-dose chemotherapy with carboplatin, ifosfamide, and etoposide was administered to 47 consecutive patients with relapsed or refractory germ cell tumors followed by infusion of PBPC or bone marrow (Beyer et al., 1995). All patients received G-CSF from the day following stem cell infusion until the ANC was greater than 1,000/μl for three consecutive days. The results confirmed the historically-controlled finding; the time to ANC > 500 and 1,000/μl and platelet count greater than 20,000/μl were significantly shorter in the group treated with PBPC. The time to platelet and red cell transfusion independence and the number of days of intravenous antibiotic treatment were significantly shortened but there was no statistically significant decrease in the number of units of red cells or platelets transfused, number of febrile days, or in the duration of hospitalization. There was no difference in overall or event-free survival. A preliminary report of a comparison of CSF-stimulated bone marrow and PBPC suggests that the two sources are equivalent in hematologic reconstitution (Janssen et al., 1994).

F. Myeloprotective Effects of CSF

Studies of hematopoietic growth factors have primarily focused on their myelorestorative properties when used after administration of chemotherapy. Several cytokines have myeloprotective activity and can prevent the lethal effects of radiation or chemotherapy on the bone marrow. Administration of interleukin-1, tumor necrosis factor alpha, stem cell factor, or interleukin-12 before a lethal dose of chemotherapy or radiation can prevent mortality in animals. Clinical trials in man have either failed to document such activity or have not been performed. GM-CSF has potential activity as a myelorestorative agent and enhances hematopoietic reconstitution following bone marrow transplantation. In addition it has significant potential as a myeloprotective agent when administered before chemotherapy. Aglietta first demonstrated that heightened progenitor cell cycling during GM-CSF administration is followed by a profound and rapid decline in the number of bone marrow stem cells undergoing DNA synthesis (Aglietta et al., 1989). Subsequent studies showed that within 24-48 hours of stopping GM-CSF the rate of marrow DNA synthesis islower then baseline and this state of bone marrow quiescence persists for at least seven days (Vadhan-Raj et al., 1992). The biological basis for this repression remains poorly understood, but it is conceivable that negative regulators of hematopoiesis (e.g., TNFα, heavy chain ferritin, or MIP-1α) may be responsible for decreased cell cycling following GM-CSF withdrawal (Broxmeyer et al., 1988; Falk et al., 1991). Another possibility is down-regulation of expression of the GM-CSF receptor. The hypothesis that might possible to administer cell cycle active chemotherapy during this refractory period without producing significant myelosuppression has led to several clinical trials of prechemotherapy GM-CSF priming.

In a non-randomized study, GM-CSF priming was evaluated in patients with sarcoma receiving cyclophosphamide, doxorubicin, and dacarbazine chemotherapy (Vadhan-Raj et al., 1992). Patients who experienced an episode of neutropenia were treated in subsequent cycles with a 14-day course of GM-CSF by continuous intravenous infusion before the next cycle of chemotherapy. Both the degree and duration of neutropenia were reduced by GM-CSF priming. In addition to preventing severe myelosuppression, GM-CSF priming allowed dose escalation of chemotherapy in 41% of patients in later cycles of chemotherapy. Twelve percent of patients experienced severe myelosuppression despite GM-CSF priming.

A short course of GM-CSF was administerred before adjuvant chemotherapy in women with node-positive breast cancer. Patients were randomized to three days of GM-CSF administered from day four to six before chemotherapy or to chemotherapy alone. Administration of chemotherapy was delayed in 22% of the cycles in patients treated with chemotherapy alone but no chemotherapy delays were necessary in the GM-CSF primed group (Aglietta et al., 1993).

GM-CSF priming was tested in the extreme setting of high dose chemotherapy and autologous blood cell transplant (Kritz et al., 1993). In this study, patients received GM-CSF priming alone or GM-CSF priming and reinfusion of GM-CSF primed peripheral blood progenitors after high-dose chemotherapy. All patients treated with GM-CSF priming alone required reinfusion of back-up bone marrow progenitors because of persistent marrow aplasia. Delayed neutrophil and platelet recovery, an increased number of episodes of febrile neutropenic sepsis, and platelet transfusion were needed in the group treated with GM-CSF priming alone as compared with those who were primed but also received PBPC. Similarly discouraging was a randomized study in which prechemotherapy GM-CSF was administered before very high-dose cyclophosphamide, etoposide, and cisplatin and followed by G-CSF prophylaxis. The extreme hematologic toxicity associated with the regimen may have overwhelmed any myelopprotection afforded by GM-CSF (Schwartzberg et al., 1993).

The most positive study to examine the effects of prechemotherapy GM-CSF has been a compound trial of sargramostim GM-CSF given before and after topotecan with only postchemotherapy GM-CSF (Janik et al., 1993). The preliminary results of the trial showed a significant reduction in the incidence of Grade 4 neutropenia (ANC < 500/μl) in GM-CSF primed patients (27%) compared with those treated only with postchemotherapy GM-CSF (77%). A trend toward a reduction in the incidence of febrile neutropenia was also observed but the difference did not acheive statistical significance.

G. Toxicity of CSF

The toxicities associated with CSF therapy are listed in Tables 3 and 4. There are differences in the toxicity profiles associated with the different forms of GM-CSF. Non-glycosylated GM-CSF is thought to produce more severe side effects because of its higher receptor affinity and rapid *in vivo* distribution (Dorr, 1993; Donahue et al., 1986; Mayer et al., 1987). Side-effects seen with both forms of GM-CSF administered at high doses include pericarditis, atrial fibrillation, pleural effusion, thrombosis, and capillary leak (Antman et al., 1988; Edmonson et al., 1992; Ho et al., 1990; Lieschke et al., 1989, 1990; Logothetis et al., 1990; Steward et al., 1989; Brandt et al., 1988). A first-dose reaction characterized by symptoms of flushing, tachycardia, hypotension, musculoskeletal pain, dyspnesa, nausea, rigors, and leg spasms has been seen infrequently with the non-glycosylated forms of GM-CSF. This has rarely been seen with the glycosylated form. This reaction may not only be seen in association with the first cycle of therapy but can

***Table 3*.** G-CSF Toxicity

Clinical Side Effects:
Medullary bone pain
Sweet syndrome
Injection site inflammation
Rash
Allergic reaction
Alopecia
Splenomegaly
Exacerbation of eczema,psoriasis
Laboratory Side Effects:
Neutrophilia*
Monocytosis
Lymphocytosis
Elevation Lactate dehydrogenase
Elevation of Uric acid
Elevation of Alkaline Phosphatase

Note: * Neutrophils show toxic granulation, Dohle bodies.

Table 4. GM-CSF Toxcity

Fever
Nausea
Fatigue
Headache
Bone pain
Chills
Myalgia
Injection site reaction
Diarrhea
Anorexia
Arthralgia
Skin rash
Facial flushing
Capillary leak
Dyspnea
Thrombosis
Hypotension
Conjunctivitis
First dose reaction

occur in later cycles; for this reason it is important to observe patients treated with non-glycosylated GM-CSF for several hours after the first administration in each cycle of therapy. Because of the toxicity seen with the non-glycosylated form of GM-CSF, there is a perception that GM-CSF is more toxic than G-CSF, However, a randomized, double-blind comparison of glycosylated (yeast-derived) GM-CSF and G-CSF for prevention of febrile neutropenia showed no statistical difference of CSF toxicity (Beveridge and Miller, 1993). The incidence of injection site reactions, significant fever, chills, or joint pain in the 137 assessable patients were comparable. Neutralizing antibodies have been seen to sargramostim in about 4% of patients tested but their clinical significance remains to be defined (Immunex Corporation, 1991). Anti-G-CSF antibodies have not been reported (Crawford et al., 1991; Trillet-Lenoir et al., 1993; Amgen, 1994; Pettengell et al., 1992).

H. ASCO Recommendations for the Administration of CSFS

Guidelines for the administration of CSF therapy in patients with cancer were developed by a multidisciplinary panel who reviewed the clinical activity of CSF for a number of common clinical situations. These guidelines were generated to encourage use of CSF when reasonable benefit could be anticipated but to discourage their indiscriminate use when little benefit is expected. Table 5 provides a list of the situations where use of CSF is encouraged and discouraged in the management of patients with cancer.

Table 5. ASCO Guidelines for the Use of Hematopoietic Growth Factors

- Primary prophylaxis—chemotherapy regimens that produce febrile neutropenia in greater than 40% of treated patients
- Secondary prophylaxis—after a documented episode of febrile neutropenia in a prior cycle of chemotherapy when dose reduction is not appropriate
- After high-dose chemotherapy and stem cell infusion
- For mobilization of peripheral blood progenitor cells
- As therapy for febrile neutropenic patients with poor prognostic features, such as pneumonia, hypotension, fungal infection, or sepsis syndrome
- Patients with myelodysplastic syndrome with neutropenic infections
- After induction chemotherapy for acute myeloid leukemia in patients over the age of 55
- Pediatric patients should be handled in a fashion similar to adults
- Concurrent administration of CSF with chemotherapy and/or radiation therapy is to be avoided outside of clinical trials
- CSF should not be used to increase chemotherapy dose-intensity outside of clinical trials

IV. CLINICAL APPLICATIONS OF COLONY-STIMULATING FACTORS TO STIMULATE THROMBOPOIESIS

Thrombocytopenia remains a potential problem among cancer patients undergoing cytotoxic treatments, particularly those receiving induction chemotherapy for leukemia or those undergoing high-dose cytoreduction and stem cell transplantation. A number of cytokines with thrombopoietic activity have been evaluated and new agents are in the early phases of clinical testing (Table 6).

As these agents enter the final stages of development it is important to focus on the endpoints that will be used to decide their clinical merits. Death and serious bleeding are rare consequences of chemotherapy and it is unlikely that the use of agents that stimulate thrombopoiesis will be associated with a decrease in mortality. Reductions in the inconvenience and expense of prophylactic platelet transfusions will probably be the prime reason for their approval.

While much effort has been spent on development of IL-1, IL-3, and IL-6, relatively minimal activity in inducing thrombocytosis has been seen and constitutional toxicities have been problematic. Particularly troublesome has been the induction of fever, a side effect that may complicate myelosuppression by simulating infection and reducing circulatory platelet half-life. These actions might perversely increase the likelihood of patient admission for febrile neutropenia and worsen platelet transfusion requirements. IL-11 is relatively unique among the cytokines tested clinically in that it does not cause febrile reactions. A randomized placebo-controlled secondary prophylaxis trial in patients with grade 4 thrombocytopenia in a prior cycle of chemotherapy suggests that IL-11 can reduce the

Table 6. Developmental Status of Thrombopoietic CSF

CSF	*Fever*	*Dose Limiting Toxicities*	*Common Toxicities*	*Status*
IL-1	Yes	Hypotension, pulmonary edema, renal dysfunction	Chills, myalgia, headache,nausea, phlebitis	Halted
IL-3	Yes	Constitutional	Flu-like symptoms, headache, conjunctival inflammation, bone pain, injection site reaction	Phase III
IL-6	Yes	Constitutional Hepatic Atrial fibrillation	Chills, fatigue, myalgia, headache, bone pain, anorexia, nausea, injection site reaction, anemia	Phase III
IL-11	No	Constitutional Scope Atrial arrhythmias	Periperal edema, nasal congestion, fatigue, myalgia, injection site reaction	Phase III
PIXY-321	Yes	Undefined Anti-PIXYSZL Antibodies	Flu-like symptoms, injection site reaction	Phase III
SC-55494	Unknown	Unknown	Unknown	Phase I
TPO	Unknown	Unknown	Unknown	Phase I

subsequent need for platelet transfusion from 96% to 70% (Tepler et al., 1996). In this trial, 93 patients who had received one or more platelet transfusions for a nadir platelet count of less then 20,000/μL during the chemotherapy cycle immediately preceding study entry, were randomized to one of two doses of IL-11, 25 or 50 μg/kg/d or placebo. Twenty-four different chemotherapy regimens were administered without dose reduction. All but one of the 27 patients randomized to placebo again required platelet transfusion whereas eight of 27 patients treated with 50 μg/kg/d of IL-11 avoided platelet transfusion. Five of 28 patients randomized to 25 μg/kg/d of IL-11 did not require platelet transfusion. Only the higher dose of IL-11 produced a statistically significant reduction in the number of patients who required platelet transfusion. Subgroup analysis showed that patients who had received less prior chemotherapy were most likely to benefit from IL-11 administration. The difference in the median number of platelet transfusions required in patients treated with the higher dose of IL-11 approached statistical significance as compared with placebo (1 vs. 3). Two cardiovascular side effects potentially due to IL-11 were noted during this study; five patients experienced syncope or near-syncope at a median of 6.5 days after the start of treatment and six patients experienced symptomatic atrial arrhythmias at a median of eight days after treatment was started. Syncope is particularly disturbing as it may lead to head trauma at a time when the platelet count is low.

Two agents, PIXY321 and SC55494, are novel n that they are bioengineered molecules with potential advantages over native cytokines. PIXY321 is a fusion protein combining both IL-3 and GM-CSF into a single molecule in order to provide patients with the thrombopoietic action of IL-3 and the neutrophil stimulating activity induced by GM-CSF. SC55494 is a highly modified form of IL-3 designed to provide a better therapeutic ratio than the native cytokine; it has greater stimulatory activity for progenitor cells without a concomitant increase in pro-inflammatory activity. A randomized comparison of PIXY321 (375 $\mu g/m^2$ twice daily) and GM-CSF (250 $\mu g/m^2$ daily) shoed no thrombopoietic benefit for PIXY321 over GM-CSF in patients with breast cancer treated with five cycles of 5-fluorouracil, leucovorin, doxorubicin, and cyclophosphamide chemotherapy (O'Shaughnessy et al., 1996). There was no difference in platelet nadirs, duration of platelet nadirs less than 20,000/μL over all cycles of treatment, or platelet transfusion requirements for PIXY321 is compared to GM-CSF. PIXY321 was less well tolerated and the duration of neutropenia (ANC $< 1000/\mu L$) for all cycles was significantly longer for patients randomized to PIXY321, particularly during the final three cycles of treatment. This latter result is expected but may be due to induction of antibodies that neutralize the effects of PIXY321 (Miller et al., 1996). In this study, up to 92% of the patients treated with PIXY321 at doses of 750 $\mu g/m^2/d$ or greater developed neutralizing antibodies.

Although IL-11 has clinical promise as a platelet restorative agent, its toxicity, and incomplete efficacy suggest that alternatives will be developed. Thrombopoietin, which appears to have a primary role in thrombopoiesis, has shown significant activity in preclinical animal models and human clinical trials have been started.

V. CONCLUSIONS

Although many advances have been made in understanding the activity of hematopoietic growth factors, the ultimate goal of these agents must be to improve cancer therapy by extending life. Studies of the ability of these agents to improve dose-intensity of chemotherapy and to improve survival are needed and are underway.

Despite the capacity for these agents to reduce the myelotoxic effects of chemotherapy, marrow ablative regimens are still associated with significant periods of neutropenia, thrombocytopenia, and anemia. Novel uses of CSF for *ex vivo* expansion of hematopoietic progenitors may eliminate this current obligatory period of marrow aplasia and significantly reduce the cost and morbidity associated with these procedures. The results of these studies are eagerly awaited.

REFERENCES

Abels, R. (1993). Erythropoietin for anaemia in cancer patients. Eur. J. Cancer 19A, S2-S8.

Advani, R., Chao, N.J., Horning, S.J., Blume, K.G., Ahn, D.K., Lamborn, K.R., Fleming, N.C., Bonnem, E.M., & Greenberg, P.L. (1992). Granulocyte-macrophage colony-stimulating factor (GM-CSF) as an adjunct to autologous hemopoietic stem cell transplantation for lymphoma. Ann. Intern. Med. 116, 183-189.

Aglietta, M., Piacibello, W., Sanavio, F., Stacchini, A., Apra, F., Schena, M., Mossetti, C., Carnino, F., Caligaris-Cappio, F., & Gavosto, F. (1989). Kinetics of human hemopoietic cells after *in vivo* administration of granulocyte-macorphage colony-stimulating factor. J. Clin. Invest. 83, 551-557.

Aglietta, M., Monzeglio, C., Pasquino, P., Carnino, F., Stern, A.C., & Gavosto, F. (1993). Short-term administration of granulocyte-macrophage colony-stimulating factor deceases hematopoietic toxicity of cytostatic drugs. Cancer 72, 2970-2973.

Amgen, Inc. (1994). Neupogen (filgrastim). Package Insert.

Anaissie, E., Vartivarian, S., Bodey, G.P., Legrand, C., Kantarjian, H., BiSaid, D., Cabanillas, F., Karl, C., & Vadhan-Raj, S. (1994). Randomized comparison between antibotics alone and antibiotics plus granylocyte-macrophage colony stimulating factor (*E. coli*-derived) in cancer patients with fever and neutropenia. J. Clin. Oncol. In Press).

Antman, K.S., Griffin, J.D., Elias, A., Socinski, M.A., Ryan, L., Cannistra, S.A., Oette, D., Whitley, M., Frei, E., & Schnipper, L.E. (1988). Effect of recombinant human granulocyte-macrophage colony-stimulating factor on chemotherapy-induced myelosuppression. N. Engl. J. Med. 319, 594-598.

ASCO Ad Hoc Colony-Stimulating Factor Guidelines Expert Panel. (1994). American Society of Clinical Oncology recommendations for the use of hematopoietic colony-stimulating factor: Evidence-based clinical practice guidelines. J. Clin. Oncol. 12, 2471-2508.

ASCO Ad Hoc Colony-Stimulating Factor Guidelines Expert Panel. (1996). Update of recommendations for the use of hematopoietic colony-stimulating factors: Evidence-based clinical practice guidelines. J. Clin. Oncol. 14, 1957-1960.

Bajorin, D.F., Nichols, C.R., Schmoll, H.J., Kantoff, P.W., Bokemeyer, C., Demetri, G.D., Einhorn, L.H., & Bosl, G.J. (1995). Recombinant human granulocyte-macrophage colony-stimulating factor as an adjunct to conventional-dose ifosfamide-based chemotherapy for patients with advanced or relapsed germ cell tumors: A randomized trial. J. Clin. Oncol. 13, 79-86.

Bartley, T.D., Bogenberger, J., Hunt, P., Li, Y.S., Lu, H.S., Martin, F., Chang, M.S., Samal, B., Nichol, J.L., & Swift, S. (1994). Identification and cloning of a megakaryocyte growth and development factor that is a ligand for the cytokine receptor Mpl. Cell 77, 1117-1124.

Bensinger, W., Singer, J., Appelbaum, F., Lilleby, K., Longin, K., Rowley, S., Clarke, E., Clift, R., Hansen, J., & Shields, T. (1993). Autologous transplantation with peripheral blood mononuclear cells collected after administration of recombinant granulocyte stimulating factor. Blood 81, 3158-3163.

Beveridge, R.A., & Miller, J.A. (1993). Impact of colony-stimulating factors on the practice of oncology. Oncology 7, 43-48.

Beyer, J., Schwella, N., Zingsem, J., Strohscheer, I., Schwaner, I., Oettle, H., Serke, S., Huhn, D., & Siegert, W. (1995). Hematopoietic rescue after high-dose chemotherapy using autologous peripheral-blood progenitor cells or bone marrow: A randomized comparison. J. Clin. Oncol. 13. 1328-1335.

Biesma, B., de Vries, E.G., Willemse, P.H., Sluiter, W.J., Postmus, P.E., Limburg, P.C., Stern, A.C., & Vellenga, E. (1990). Efficacy and tolerability of recombinant human granulocyte-macrophage colony-stimulating factor in patients with chemotherapy-related leukopenia and fever. Eur. J. Cancer 26, 932-936.

Bishop, M.R., Anderson, J.R., Jackson, J.D., Bierman, P.J., Reed, E.C., Vose, J.M., Armitage, J.O., Warkentin, P.I., & Kessinger, A. (1994). High-dose therapy and peripheral blood progenitor cell transplantation: Effects of recombinant human granulocyte-macrophage colony stimulating factor on the autograft. Blood 83, 610-616.

Blaise, D., Vernant, J.P., Fiere, D., Gluckman, E., Reiffers, J., Harousseau, J.L., Sainty, D., Faucher, D., Marcel, C., & Maraninchi, D. (1992). A randomized, controlled, multicenter trial of recombinant human granulocyte colony stimulating factor (filgrastim) in patients treated by bone marrow transplantation (BMT) with total body irradiation (TBI) for acute lymphoblastic leukemia (ALL) or lymphoblastic lyumphoma (LL). Blood 80(10-Suppl. 1), 982a. (Abstract).

Blajchman, M.A., & Singal, D.P. (1989). The role of red blood cell antigens, histocompatibility antigens, and blood transfusions on renal allogrift survival. Transfus. Med. Rev. 3, 171-179.

Bodey, G.P., Buckley, M., Sathe, Y.S., & Freireich, E.J. (1966). Quantitative relationships between circulating leukocytes and infection in patients with acute leukemia. Ann. Intern. Med. 64, 328-340.

Boogaerts, M.A., Verhoef, G.E.G., & Scheers, W. (1990). The *in vivo* effects of GM-CSF on neutrophil function in patients with myelodysplastic syndrome. Blood 76(10-Suppl 1), 256a. (Abstract).

Boogaerts, M., Cavalli, F., Cortes-Funes, H., Gatell, J.M., Gianni, A.M., Khayat, D., Levy, Y., & Link, H. (1995). Granulocyte growth factors: achieving a consensus. Ann. Oncol. 6, 237-244.

Brandt, S.J., Peters, W.P., Atwater, S.J., Kurtzberg, J., Borowitz, M.J., Jones, R.B., Shpall, E., Bast, R.C., Gilbert, C.J., & Oette, D.H. (1988). Effects of recombinant human granulocyte-macrophage colony-stimulating factor on hematopoietic reconstitution after high dose chemotherapy and autologous marrow transplantation. N. Engl. J. Med. 318, 869-876.

Broxmeyer, H., Cooper, S., Williams, D., Hangoc, G., Gutterman, J., & Vadhan-Raj, S. (1988). Growth characteristics of marrow hematopoietic progenitor/precursor cells from patients on a phase I clinical trial with purified recombinant human granulocyte-macrophage colony-stimulating factor. Exp. Hematol. 16, 594-602.

Case, D.C., Jr., Carey, R.W., Fishkin, E.H., Henry, D.H., Jacobson, R.J., Jones, S.E., Keller, A.M., Kugler, J.W., & Nichols, C.R. (1993). Recombinant human erythropoietin therapy for anemic cancer patients on combinantion chemotherapy. J. Natl. Cancer Inst. 85, 801-806.

Chao, N.J., Schriber, J.R., Grimes, K., Long, G.D., Negrin, R.S., Raimondi, C.M., Horning, S.J., Brown, S.L., Miller, L., & Blume, K.G. (1993). Granulocyte colony-stimulating factor "mobilized" peripheral blood progenitor cells accelerate granulocyte and platelet recovery after high-dose chemotherapy. Blood 81, 2031-2035.

Chao, N.J., Schriber, J.R., Long, G.D., Negrin, R.S., Catolico, M., Brown, B.W., Miller, L.L., & Blume, K.G. (1994). A randomized study of erythropoietin and granulocyte colony-stimulating factor (G-CSF) versus placebo and G-CSF for patients with Hodgkin's and non-Hodgkin's lymphoma undergoing autologous bone marrow transplantation. Blood 83, 2823-2828.

Chung, M., Steinmetz, O.K., & Gordon, P.H. (1993). Perioperative blood transfusion and outcome after resection for colorectal carcinoma [see comments]. Br. J. Surg. 80, 427-432.

Crawford, J., Ozer, H., Stoller, R., Johnson, D., Lyman, G., Tabbara, I., Kris, M., Grous, J., Picozzi, V., & Rausch, G. (1991). Reduction by granulocyte colony-stimulating factor of fever and neutropenia induced by chemotherapy in patients with small-cell lung cancer (r-metHuG-CSF). N. Engl. J. Med. 325, 164-170.

Crawford, J., Glaspy, J., Vincent, M., Tomita, D., & Mazanet, R. (1994). Effects of filgrastim (R-methuG-CSF) on oral mucositis in patients with small cell lung cancer (SCLC) receiving chemotherapy (cyclophosphamide, doxorubicin and etoposide, CAE). Proc. Amer. Assoc. Cancer Res. 13, 442. (Abstract).

de Campos, E., Radford, J., Steward, W., Milroy, R., Dougal, M., Swindell, R., Testa, N., & Thatcher, N. (1995). Clinical and *in vitro* effects of recombinant human erythropoietin in patients receiving intensive chemotherapy for small-cell lung cancer. J. Clin. Oncol. 13, 1623-1631.

Demuynck, H., Pettengell, R., de Campos, E., Dexter, T.M., & Testa, N.G. (1992). The capacity of peripheral blood stem cells mobilised with chemotherapy plus G-CSF to repopulate irradiated marrow stroma *in vitro* is similar to that of bone marrow. Eur. J. Cancer 28, 381-386.

Doll, D.C., & Weiss, R.B. (1983). Chemotherapeutic agents and the erythron. Cancer Treat. Rev. 10, 185-200.

Donahue, R.E., Wang, E.A., Kaufman, R.J., Foutch, L., Leary, A.C., Witek Giannetti, J.S., Metzger, M., Hewick, R.M., Steinbrink, D.R., Shaw, G., Kamen, R., & Clark, S.C. (1986). Effects of N-linked carbohydrate on the *in vivo* properties of human GM-CSF. Cold Spring Harb. Symp. Quant. Biol. 51 Pt 1, 685-692.

Dorr, R.T. (1993). Clinical properties of yeast-derived versus *Escherichia coli*-derived granulocyte-macrophage colony-stimulating factor. Clin. Ther. 15, 19-29.

Dreger, P., Marquardt, P., Haferlach, T., Jacobs, S., Mulverstedt, T., Eckstein, V., Suttorp, M., Loffler, H., Muller-Ruchholtz, W., & Schmitz, N. (1993). Effective mobilisation of peripheral blood progenitor cells with "Dexa-BEAM" and G-CSF: Timing of harvesting and composition of the leukapheresis product. Br. J. Cancer 68, 950-957.

Dührsen, U., Villeval, J.L., Boyd, J., Kannourakis, G., Morstyn, G., & Metcalf, D. (1988). Effects of recombinant human granulocyte colony-stimulating factor on hematopoietic progenitor cells in cancer patients. Blood 72, 2074-2081.

Edmonson, J.H., Hartmann, L.C., Long, H.J., Colon-Otero, G., Fitch, T.R., Jeffries, J.A., Braich, T.A., & Maples, W.J. (1992). Granulocyte-macrophage colony-stimulating factor (GM-CSF): Preliminary observations on the influences of dose, schedules, and route of administration in patients receiving cyclophosphamide plus carboplatin. Cancer 70, 2529-2539.

Eschbach, J.W., Egrie, J.C., Downing, M.R., Browne, J.K., & Adamson, J.W. (1987). Correction of the anemia of end-stage renal disease with recombinant human erythropoietin. Results of a combined phase I and II clinical trial [see comments]. N. Engl. J. Med. 316, 73-78.

Eschbach, J.W., Abdulhadi, M.H., Browne, J.K., Delano, B.G., Downing, M.R., Egrie, J.C., Evans, R.W., Friedman, E.A., Graber, S.E., Haley, N.R., Korbet, S., Kranltz, S.B., Lundin, P., Nissenson, A.R., Ogden, D.A., Paganini, E.P., Rader, B., Rutsky, E.A., Stivelman, J., Stone, W.J., Teschan, P., Van Stone, J.C., Van Wyck, D.B., Zuckerman, K., & Adamson, J.W. (1989a). Recombinant human erythropoietin in anemic patients with end-stage renal disease. Results of a phase III multicenter clinical trial. Ann. Intern. Med. 111, 992-1000.

Eschbach, J.W., Kelly, M.R., Haley, N.R., Abels, R.I., & Adamson, J.W. (1989b). Treatment of the anemia of progessive renal failure with recombinant human erythropoietin. N. Engl. J. Med. 321, 158-163.

Estey, E.H., Kurzrock, R., Talpaz, M., McCredie, K.B., O'Brien, S., Kantarjian, H.M., Keating, M.J., Deisseroth, A.B., & Gutterman, J.U. (1991). Effects of low doses of recombinant human granulocyte-macrophage colony-stimulating factor (GM-CSF) in patients with myelodysplastic syndromes. Br. J. Haematol. 77, 291-295.

Falk, S., Seipelt, G., Ganser, A., Ottman, O.G., Hoelzer, D., Stutte, H.J., & Hubner, K. (1991). Bone marrow findings after treatment with recombinant human interleukin-3. Amer. J. Clin. Pathol. 95, 355-362.

Fukuda, M., Kojima, S., Matsumoto, K., & Matsuyama, T. (1992). Autotransplantation of peripheral blood stem cells mobilized by chemotherapy and recombinant human granulocyte colony-stimulating factor in childhood neutoblastoma and non-Hodgkin's lymphoma. Br. J. Haematol. 80, 327-331.

Fukuda, M., Nakano, A., Kinoshita, A., Watanabe, K., Itoh, N., Sakamoto, A., Fujino, R., Araki, J., Souda, H., & Hara, K. (1993). Optimal timing of G-CSF administration in patients receiving chemotherapy for non-small cell lung cancer (NSCLC). Proc. Amer. Soc. Clin. Oncol. 12, 447. (Abstract).

Gamucci, T., Thorel, M.F., Frasca, A.M., Giannarell, D., & Calabresi, F. (1993). Erythropoietin for the prevention of anaemia in neoplastic patients treated with cisplatin. Eur. J. Cancer 29A, Suppl. 2, S13-S14.

Gerhartz, H.H., Engelhard, M., Meusers, P., Brittinger, G., Wilmanns, W., Schlimok, G., Mueller, P., Huhn, D., Musch, R., Siegert, W., Gerhartz, D., Hartlapp, J.H., Thiel, E., Huber, C., Peschl, C., Spann, W., Emmerich, B., Schadek, C., Westerhausen M., Pees, H.-W., Radtke, H., Engert, A.,

Terhardt, E., Schick, H., Binder, T., Fuchs, R., Hasford, J., Brandmaier, R., Stern, A.C., Jones, T.C., Ehrlich, H. J., Stein, H., Parwaresch, M., Tiemann, M., & Lennert, K. (1993a). Randomized, double-blind, placebo-controlled, Phase III study of recombinant human granulocyte-macrophage colony-stimulating factor as adjunct to induction treatment of high-grade malignant non-Hodgkin's lymphomas. Blood 82, 2329-2339.

Gerhartz, H.H., Stern, A.C., Wolf-Hornung, B., Kazempour, M., Schmetzer, H., Gugerli, U., Jones, T.C., & Wilmanns, W. (1993b). Intervention treatment of established neutropenia with human recombinant granulocyte-macrophage colony-stimulating factor (rhGM-CSF) in patients undergoing cancer chemotherapy. Leuk. Res. 17, 175-185.

Gianni, A.M., Siena, S., Bregni, M., Tarella, C., Stern, A., Pileri, A., & Bonadonna, G. (1989). Granulocyte-macrophage colony-stimulating factor to harvest circulating haemopoietic stem cells for autotransplantation. Lancet 2(8663), 580-585.

Gisselbrecht, C., Prentice, H.G., Bacigalupo, A., Biron, P., Milpied, N., Rubie, H., Cunningham, D., Legros, M., Pico, J.L., & Linch, D.C. (1994). Placebo-controlled phase III trial of lenograstim in bone-marrow transplantation [published erratum appears in Lancet 1994 Mar 26; 343(8900):804]. Lancet 343, 696-700.

Gorin, N.C., Coiffier, B., Hayat, M., Fouillard, L., Kuentz, M., Flesch, M., Colombat, P., Boivin, P., Slavin, S., & Philip, T. (1992). Recombinant human granulocyte-macrophage colony-stimulating factor after high-dose chemotherapy and autologous bone marrow transplantation with unpurged and purged marrow in non-Hodgkin's lymphoma: A double-blind placebo-controlled trial. Blood 80, 1149-1157.

Greenberg, P., Taylor, K., Larson, R., Koeffler, P., Negrin, R., Saba, H., Ganser, A., Jakubowski, A., Gabrilove, J., Mufti, G., Cruz, J., Hammond, W., Broudy, V., Langley, G.R., Keating, A., Vardiman, ., Lamborn, K., & Brown, S. (1993). Phase III randomized multicenter trial of G-CSF vs. observation for myelodysplastic syndromes (MDS). Blood 82(10-Suppl. 1), 196a. (Abstract).

Gulati, S.C., & Bennett, C.L. (1992). Granulocyte-macrophage colony-stimulating factor (GM-CSF) as adjunct therapy in relapsed Hodgkin's disease. Ann. Intern. Med. 116, 177-182.

Haas, R., Ho, A.D., Bredthauer, U., Cayeux, S., Egerer, G., Knauf, W., & Hunstein, W. (1990). Successful autologous transplantajtion of blood stem cells mobilized with recombinant human granulocyte-macrophage colony-stimulating factor. Exp. Hematol. 18, 94-98.

Haas, R., Hohaus, S., Egerer, G., Ehrhardt, R., Witt, B., & Hunstein, W. (1992). Recombinant human granylocyte-macrophage colony-stimulating factor (rhGM-CSF) subsequent to chemotherapy improves collection of blood stem cells for autografting in patients not eligible for bone marrow harvest. Bone Marrow Transplant 9, 459-465.

Hamm, J., Schiller, J.H., Cuffie, C., Oken, M., Fisher, R.I., Shepherd, F., & Kaiser, G. (1994). Dose-ranging study of recombinant human ganulocyte-macrophage colony-stimulating factor in small-cell lung carcinoma. J. Clin. Oncol. 12, 2667-2676.

Ho, A., Del Valle, F., Haas, R., Engelhard, M., Hiddeman, W., Ruckle, H., Schlimok, G., Thiel, E., Andreesen, R., Fiedler, W., Frisch, J., Schulz, G., & Hustein, W. (1990). Sequential studies on the role of mitoxantyrone, high-dose cytarabine, and recombinant human granulocyte-macrophage colony stimulating factor in the treatment of refractory non-Hodgkin's lymphoma. Semin. Oncol. 17, 14-19.

Ho, A.D., Gluck, S., Germond, C., Sinoff, C., Dietz, G., Maruyama, M., & Corringham, R.E. (1993). Optimal timing for collections of blood progenitor cells following induction chemotherapy and granulocyte-macrophage colony-stimulating factor for autologous transplantation in advanced breast cancer. Leukemia 7, 1738-1746.

Immunex Corporation. (1991). Leukine (sargramostim). Package Insert.

Jagannath, S., Vesole, D.H., Glenn, L., Crowley, J., & Barlogie, B. (1992). Low-risk intensive therapy for multiple myeloma with combined autologous bone marrow and blood stem cell support. Blood 80, 1666-1672.

Janik, J., Miller, L., Smith, J., Kopp, W., Alvord, G., Gause, B., Curti, B., Urba, W.J., & Longo, D.L. (1993). Prechemotherapy granulocyte-macrophage colony-stimulating factor (GM-CSF) prevents topotecan-induced neutropenia. Proc. Amer. Soc. Clin. Oncol. 12, 437. (Abstract).

Janssen, W.E., Perkins, J., Heimenz, J.W., Fields, K.K., Zorsky, P.E., Bellester, O.F., Goldstein, S.C., Smilee, R., Kronish, L., & Elfenbein, G.J. (1994). Granulocyte recovery is not different following auto-transplant with G-CSF primed bone marrow or G-CSF mobilized peripheral blood "Stem cells." Blood 84 (10-Suppl. 1), 85a (Abstract).

Kaplan, L.D., Kahn, J.O., Crowe, F., Northfelt, D., Neville, P., Grossbergm, H., Abrams, D.I., Tracey, J., Mills, J., & Volberding, P.A. (1991). Clinical and virologic effects of recombinant human granulocyte-macrophage colony-stimulating factor in patients receiving chemotherapy for human immunodeficiency virus-associated non-Hodgkin's lymphoma: Results of a randomized trial. J. Clin. Oncol. 9, 929-940.

Kaplan, S.S., Basford, R.E., Wing, E.S., and Shadduck, R.K. (1989). The effects of recombinant GM-CSF on neutrophil activation in patients with refactory circinoma. Blood 73, 636-638.

Kessinger, A., & Armitage, J.O. (1991). The evolving role of autologous peripheral stem cell transplantation following high-dose therapy for malignancies. Blood 77, 211-213.

Kessinger, A., Reed, E.C., Vose, J.M., Bierman, P.J., & Armitage, J.O. (1992). Collection of transplantation of non-mobilised and rhGM-CSF primed autologous peripheral stem cells from previously treated cancer patients. Exp. Hematol. (Abstract).

Khwaja, A., Linch, D.C., Goldstone, A.H., Chopra, R., Marcus, R.E., Wimperis, J.Z., Russell, N.H., Haynes, A.P., Milligan, D.W., Leyland, M.J., Winifield, D.A., Hancock, B.W., Newland, A., Durrant, S.T., Devereux, S., Roitt, S., Collins, M., & Hudson, G.V. (1992). Recombinant human granulocyte-macrophage colony-stimulating factor after autologous bone marrow transplantation for malignant lymphoma: A British National Lymphoma Investigation double-blind, placebo-controlled trial. Br. J. Haematol. 82, 317-323.

Klaesson, S., Ringden, O., Ljungman, P., Lonnqvist, B., & Wennberg, L. (1994). Reduced transfusions requirements after allogeneic bone marrow transplantation: results of a randomised, double-blind study with high-dose erythropoietin. Bone Marrow Transplant. 13, 397-402.

Kritz, A., Crown, J.P., Motzer, R.J., Reich, L.M., Heller, G., Moore, M.P., Hamilton, N., Yao, T.J., Heelan, R.T., & Schneider, J.G. (1993). Beneficial impact of peripheral blood progenitor cells in patients with metastatic breast cancer treated with high-dose chemotherapy plus ganulocyte-macrophage colony-stimulating factor. A randomized trial. Cancer 71, 2515-2521.

Lane, T.A., Law, P., Maruyama, M., Young, D., Burgess, J., Mullen, M., Mealiffe, M., Terstappen, L.W., Hardwick, A., & Moubayed, M. (1995). Harvesting and enrichment of hematopoietic progenitor cells mobilized into the peripheral blood of normal donors by granulocyte-macrophage colony-stimulating factor (GM-CSF) or F-CSF: Potential role in allogeneic marrow transplantation. Blood 85, 275-282.

Lieschke, G., Maher, D., Cebon, J., O'Connor, M., Green, M., Sheridan, W., Boyd, A., Rallings, M., Bonnem, E., Metcalf, D., Burgess, A.W., McGrath, K., Fox, R.M., & Morstyn, G. (1989). Effects of bacterially synthesized recombinant human granulocyte-macrophage colony-stimulating factor in patients with advanced malignancy. Ann. Intern. Med. 110, 357-367.

Lieschke, G., Maher, D., O'Connor, M., Green, M., Sheridan, W., Rallings, M., Bonnem, E., Burgess, A.W., Mcgrath, K., Fos, R.M., & Morstyn, G. (1990). Phase I study of intravenously administered bacterially synthesized granulocyte-macrophage colony-stimulating factor and comparison with cubcutaneous administration. Cancer Res. 50, 606-614.

Linch, D.C., Scarffe, H., Proctor, S., Chopra, R., Taylor, P.R., Morgenstern, G., Cunningham, D., Burnett, A.K., Cawley, J.C., Franklin, I.M., et al. (1993). Randomised vehicle-controlled dose-finding study of glycosylated recombinant human ganulocyte colony-stimulating factor after bone marrow transplantation. Bone Marrow Transplant. 11, 307-311.

Link, H., Boogaerts, M.A., Carella, A.M., Ferrant, A., Gadner, H., Gorin, N.C., Harabacz, I., Harousseau, J.L., Herve, P., Holldack, J., Kolb, H.-J., Krieger, O., Labar, B., Linkesch, W., Mandelli, F., Maraninchi, D., Naparstek, E., Nicolay, U., Niederwieser, D., Reiffers, J., Rizzoli, V., Siegert, W. Vernant, J.-P., & de Witte, T. (1992). A controlled trial of recombinant human granulocyte-macrophage colony-stimulating factor after total body irradiation, high-dose chemotherapy, and autologous bone marrow transplantation for acute lymphoblastic leukemia or malignant lymphoma. Blood 80, 2188-2195.

Logothetis, G., Dexeus, F., Sella, A., Amato, R.J., Kilbourn, R.G., Finn, L., & Gutterman, J.U. (1990). Escalated therapy for refactory urothelial tumors: Methotrexate-vinblastine-doxorubicin-cisplatin plus unglycosylated recombinant human granulocyte-macrophage colony stimulating factor. J. Natl. Cancer Inst. 82, 667-672.

Lord, B.I., Bronchud, M.H., Owens, S., Chang, J., Howell, A., Souza, L., & Dexter, T.M. (1989). The kinetics of human granulopoiesis following treatment with granulocyte colony-stimulating factor *in vivo*. Proc. Natl. Acad. Sci. USA 86, 9499-9503.

Lord, B.I., Gurney, H., Chang, J., Thatcher, N., Crowther, D., & Dexter, T.M. (1992). Haemopoietic cell kinetics in humans treated with rGM-CSF. Int. J. Cancer 50, 26-31.

Ludwig, H., Fritz, E., Leitgeb, C., Pecherstorfer, M., Samonigg, H., & Schuster, J. (1994). Prediction of response to erythropoietin treatment in chronic anemia of cancer. Blood 84, 1056-1063.

Maher, D.W., Lieschke, G.J., Green, M., Bishop, J., Stuart-Harris, R., Wolf, M., Sheridan, W.P., Kefford, R.F., Cebon, J., Olver, I., McKendrick, J., Toner, G., Bradstock, K., Lieschke, M., Cruickshank, S., Tomita, D.K., Hoffman, E.W., Fox, R.M., & Morstyn, G. (1994). Filgrastim in patients with chemotherapy-induced febrile neutropenia. A double-blind, placebo-controlled trial. Ann. Intern. Med. 121, 492-501.

Mangan, K., Mullaney, M., Klumpp, T., Goldberg, S., & Macdonald, J. (1993). Mobilization of peripheral blood stem cells by subcutaneous injections of yeast-derived ganulocyte macrophage colony stimulating factor: A phase I-II study. Stem Cells 11, 445-454.

Mayer, P., Lam, C., Obenaus, H., Liehl, E., & Besemer, J. (1987). Recombinant human GM-CSF induces leukocytosis and activates peripheral blood polymorphonuclear neutrophils in nonhuman primates. Blood 70, 206-213.

Mayordomo, J.I., Rivera, F., Diaz-Puente, M.T., Lianes, P., Colomer, R., Lopez-Brea, M., Lopez, E., Paz-Ares, L., Hitt, R., Garcia-Ribas, I., Cubedo, R., Alonso, S., & Cortes-Funes, H. (1995). Improving treatment of chemotherapy-induced neutropenic fever by administration of colony-stimulating factors. J. Natl. Cancer Inst. 87, 803-808.

Miller, C.B., Jones, R.J., Piantadosi, S., Abeloff, M.D., & Spivak, J.L. (1990). Decreased erythropoietin response in patients with the anemia of cancer. N. Engl. J. Med. 322, 1689-1692.

Miller, L.L., Korn, E.L., Stevens, D.S., Janik, J.E., Gause, B.L., Kopp, W.C., & Longo, D.L. (1996). Abrogation of the hematological activities of the GM-CSF/IL-3 fusion protein (PIXY321) by anti-PIXY321 antibodies in cancer patients receiving high-dose carboplatin: Results of a phase I study. Thirty-Second annual meeting of the American Society of Clinical Oncology. J. Clin. Oncol. 14, 2012A, 1748. (Abstract).

Neidhart, J., Verma, S., Triozzi, P., Nemunaitis, J., Quick, D., Oette, D., Hayes, A., & Holcenberg, J. (1994). A randomized and placebo controlled trial of yeast-derived GM-CSF support of dose intensive cyclophosphamide, Etoposide and cisplatin (DICEP) in lymphoma and breast cancer. Proc. Amer. Soc. Clin. Oncol. 13, 69. (Abstract).

Nemunaitis, J., Rabinowe, S., & Singer, J. (1991). Recombinant granulocyte-macrophage colony-stimulating factor after autologous bone marrow transplantation for lymphoid cancer. N. Engl. J. Med. 324, 1773-1778.

Ohno, R., Tomonaga, M., & Kobayashi, T. (1990). Effect of granulocyte colony-stimulating factor after intensive induction therapy in relapsed or refactory acute leukemia. N. Engl. J. Med. 323, 871-877.

Ohno, R., Hiraoka, A., Tanimoto, M., Asou, N., Kuriyama, K., Kobayashi, T., Yoshida, M., Teshima, H., Saito, H., & Fujimoto, K. (1993). No increase of leukemia relapse in newly diagnosed patients with acute myeloid leukemia who received granulocyte colony-stimulating factor for life-threatening infection during remission induction and consolidated therapy (letter). Blood 81, 561-562.

Opelz, G., & Terasaki, P.I. (1978). Improvement of kidney-graft survival with increased numbers of blood transfusions. N. Engl. J. Med. 299, 799-803.

O'Shaughnessy, J.A., Tolcher, A., Riseberg, D., Venzon, D., Zujewski, J., Noone, M., Gossard, M., Danforth, D., Jacobson, J., Chang, V., Goldspiel, B., Keegan, P., Giusti, R., & Cowan, K.H., (1996). Prospective, randomized trial of 5-fluorouracil, leucovorin, doxorubicin, and cyclophosphamide chemotherapy in combination with the interleukin-3/granulocyte-macrophage colony-stimulating factor (GM-CSF) fusion protein (PIXY321) versus GM-CSF in patients with advanced breast cancer. Blood 87, 2205-2211.

Patrone, F., Ballestrero, A., Balleari, E., Bogliolo, F., Brema, F., Ferrando, F., Ghio, R., & Timitilli, S. (1992). Hsigh-dose cyclophosphamide followed by GM-CSF is a safe and effective procedure for the recruitment of trilineage circulating progenitor cells. Haematol. 77, 457-462.

Pettengell, R., Gurney, H., Radford, J.A., Deakin, D.P., James, R., Wilkinson, P.M., Kane, K., Bentley, J., & Crowther, D. (1992). Granulocyte colony-stimulating factor to prevent dose-limiting neutropenia in non-Hodgkin's lymphoma: A randomized controlled trial. Blood 80, 1430-1436.

Pettengell, R., Morgenstern, G.R., Woll, P.J., Chang, J., Rowlands, M., Young, R., Radford, J.A., Scarffe, J.H., Testa, N.G., & Crowther, D. (1993a). Peripheral blood progenitor cell transplantation in lymphoma and leuykemia using a single apheresis. Blood 82, 3770-3777.

Pettengell, R., Testa, N.G., Swindell, R., Crowther, D., & Dexter, T.M. (1993b). Transplantation protential of hematopoietic cells released into the circulation during routine chemotherapy for non-Hodgkin's lumphoma. Blood 82, 2239-2248.

Riikonen, P., Saarinen, U.M., Makipernaa, A., Hovi, L., Komulainen, A., Pihkala, J., & Jalanko, H. (1994). Recombinant human granulocyte-macrophage colony-stimulating factor in the treatment of febrile neutropenia: A double blind pacebo-controlled study in children. Pediatr. Infect. Dis. J. 13, 197-202.

Rose, C., Wattel, E., Bastion, Y., Berger, E., Bauters, F., Coiffier, B., & Fenaux, F. (1994). Treatment with very low-dose GM-CSF in myelodysplastic syndromes with neutropenia. A report on 28 cases. Leukemia 8, 1458-1462.

Rowe, J.M., Anderson, J., Mazza, J.J., Paietta, E., Bennett, J.M., & Wiernik, P.H. (1993). Phase III randomized placebo-controlled study of yeast derived GM-CSF in adult patients (55-70 years) with AML. Blood 82(10-Suppl 1), 329a. (Abstract).

Sanz, G.F., Sanz, M.A., Vallespi, T., Canizo, M.C., Torrabadella, M., & Garcia, S. (1989). Two regression models and a scoring system for predicting survival and planning treatment in myelodysplastic syndromes: A multivariate analysis of prognostic factors in 370 patients. Blood 74, 395-408.

Schmitz, N., Dreger, P., Zander, A., Peters, S., Ehninger, G., Wandt, H., Kolb, H.J., & Hecht, T. (1992). Recombinant human granulocyte colony stimulating factor (filgrastim) after autologous bone marrow transplantation for lymphoma: An open label randomized trial in Germany. Blood 80(10-Suppl. 1), 292a. (Abstract).

Schuster, M.W., Larson, R.A., Thompson, J.A., Coiffer, B., Bennett, J.M., & Israel, R.J. (1990). Granulocyte-macrophage colony-stimulating factor (GM-CSF) for myelodysplastic syndrome (MDS): Results of a multi-center, randomized, controlled trial. Blood 76(10-Suppl. 1), 318a. (Abstract).

Schwartzberg, L.S., Birch, R., Hazelton, B., Tauer, K.W., Lee, P.,Jr., Altemose, R., George, C., Blanco, R., Wittlin, F., & Cohen, J. (1992). Peripheral blood stem cell mobilization by chemotherapy with and without recombinant human granulocyte colony-stimulating factor. J. Hematother. 1, 317-327.

Schwartzberg, L., West, W., Birch, R., Heffernan, M., Tauer, K., Kalman, L., Middleman, E., Pendergrass, K., & Leff, R. (1993). Randomized prospective trial plus or minus pretreatment with GM-CSF prior to high-dose cyclophosphamide, etoposide, and cisplatin + G-CSF. Proc. Amer. Soc. Clin. Oncol. 12, 452. (Abstract).

Sheridan, W.P., Begley, C.G., Juttner, C.A., Szer, J., To, L.B., Maher, D., McGrath, K.M., Morstyn, G., & Fox, R.M. (1992). Effect of peripheral-blood progenitor cells mobilised by filgrastim (G-CSF) on platelet recovery after high-dose chemotherapy. Lancet 339, 640-644.

Shpall, E.J., Jones, R.B., Bearman, S.I., Franklin, W.A., Archer, P.G., Curiel, T., Bitter, M., Claman, H.N., Stemmer, S.M., & Purdy, M. (1994). Transplantation of enriched CD34-positive autologous marrow into breast cancer patients following high-dose chemotherapy: Influence of CD34-positive peripheral-blood progenitors and growth factors on engraftment. J. Clin. Oncol. 12, 28-36.

Socinski, M.A., Cannistra, S.A., Elias, A., Antman, K.H., Schnipper, L., & Griffin, J.D. (1986). Granulocyte-macrophage colony-stimulating factor expands the circulating hematopoietic progenitor cell compartment in man. Lancet 1(8596), 1194-1198.

Stahel, R.A., Jost, L.M., Cerny, T., Pichert, G., Honegger, H., Tobler, A., Jacky, E., Fey, M., & Platzer, E. (1994). Randomized study of recombinant human ganulocyte colony-stimulating factor after high-dose chemotherapy and autologous bone marrow transplantation for high-risk lymphoid malignancies. J. Clin. Oncol. 12, 1931-1938.

Stehle, B., Weiss, C., Ho, A.D., & Hunstein, W. (1990). Serum levels of tumor necrosis factor alpha in patients treated with granulocyte-macrophage colony-stimulating factor [letter; comment]. Blood 75, 1895-1896.

Steward, W., Scarffe, J., Austin, R., Bonnem, E., Thatcher, N., Morganstern, G., & Crowther, D. (1989). Recombinant human granulocyte-macrophage colony-stimulating factor given as daily short infusions—a phase I dose-toxicity study. Br. J. Cancer 59, 142-145.

Stone, R., George, S., Berg, D., Paciucci, P., & Schiffer, C. (1994). GM-CSF "V" placebo during remissin inductin for patients ≥ 60 years old with *de novo* acute myeloid leukemia. Proc. Amer. Soc. Clin. Oncol. 13, 304. (Abstract).

Tarella, C., Ferrero, D., Bregni, M., Siena, S., Gallo, E., Pileri, A., & Gianni, A.M. (1991). Peripheral blood expansion of early progenitor cells after high-dose cycleophosphamide and rhGM-CSF. Eur. J. Cancer 27, 22-27.

Tepler, I., Elias, L., Smith, J.W., II, Hussein, M., Rosen, G., Chang, A., Moore, J.O., Gordon, M.S., Kuca, B., Beach, K.J., Loewy, J.W., Garnick, M.B., & Kaye, J.A. (1996). A randomized placebo-controlled trial of reocmbinant human interleukin-11 in cancer patients with severe thrombocytopenia due to chemotherapy. Blood 87, 3607-3614.

Trillet-Lenoir, V., Green, J., Manegold, C., Von Pawel, J., Gatzemeier, U., Lebeau, B., Depierre, A., Johnson, P., Decoster, G., Tomita, D., & Ewen, C. (1993). Recombinant granulocyte colony-stimulating factor reduces the infectious complications of cytotoxic chemotherapy. Eur. J. Cancer 29A, 319-324.

Vadhan-Raj, S., Broxmeyer, H.E., Hittelman, W.N., Papadopoulos, N.E., Chawla, S.P., Fenoglio, C., Cooper, S., Buescher, E.S., Frenck, R.W., Holian, A., Perkins, R.C., Scheule, R.K., Gutterman, J.U., Salem, P., & Benjamin, R.S. (1992). Abrogating chemotherapy-induced myelosuppression by recombinant granulocyte-macrophage colony-stimulating factor in patients with sarcoma: Protection at the progenitor cell level. J. Clin. Oncol. 10, 1266-1277.

Vamvakas, E., & Moore, S.B. (1993). Perioperative blood transfusion and colorectal cancer recurrence: A qualitative statistical overview and meta-analysis [see comments]. Transfusion 33, 754-765.

Van Hoef, M.E., Baumann, I., Lange, C., Luft, T., de Wynter, E.A., Ranson, M., Morgenstern, G.R., Yvers, A., Dexter, T.M., & Testa, N.G. (1994). Dose-escalating induction chemotherapy supported by lenograstim preceding high-dose consolidation chemotherapy for advanced breast cancer. Selection of the most acceptable regimen to induce maximal tumor response and investigation of the optimal time to collect peripheral blood progenitor cells for haematological rescue after high-dose consolidation chemotherapy. Ann. Oncol. 5, 217-224.

Vose, J.M., & Armitage, J.O. (1995). Clinical applications of hematopietic growth factors. J. Clin. Oncol. 13, 1023-1035.

Vredenburgh, J.J., Ross, M., Meisenberg, B., Hussein, A., Kurtzberg, J., Gilbert, C., Petros, W., & Peters, W.P. (1992). A short course of G-CSF for priming of peripheral blood progenitor cells (PBPC). Blood 80(10-Suppl. 1), 291a. (Abstract).

Wexler, L.H., McClure, L., Steinberg, S., Jarosinski, P., Pizzo, P.A., & Horowitz, M.E. (1994). Lack of utility of rh-GM-CSF (*E. coli*, non-glycosylated, Schering-Plough/Sandoz) in reducing the myelosuppresson of vincristine, doxorubicin, cyclophosphamide (VAdriaC) and ifosfamide and etoposide (IE) in pediatric patients. Proc. Amer. Soc. Clin. Oncol. 13, 463. (Abstract).

Willemze, R., Van Der Lely, N., Zwierzina, H., Suciu, S., Solbu, G., Gerhartz, H., Labar, B., Visani, G., Peetermans, M.E., & Jacobs, A. (1992). A randomized phase-I/II multicenter study of recombinant human granulocyte-macrophage colony-stimulating factor (GM-CSF) therapy for patients with myelodysplastic syndromes and relatively low risk of acute leukemia. EORTC Leukemia Cooperative Group [published erratum appears in Ann Hematol 1992 June; 64(6):312]. Ann. Hematol. 64, 173-180.

Yuo, A., Kitagawa, S., Okabe, T., Urabe, A., Komatsu, Y., Itoh, S., & Takaku, F. (1987). Recombinant human granulocyte colony-stimulating factor repairs the abnormalities of neutrophils in patients with myelodysplastic syndromes and chronic myelogenous leukemia. Blood 70, 404-411.

INSULIN-LIKE GROWTH FACTOR BINDING PROTEINS

David R. Clemmons

ABSTRACT

The insulin-like growth factor binding proteins (IGFBPs) bind both IGF-I and IGF-II with high affinity. They control IGF half-lives, rates of efflux from the vascular space, distribution within extracellular fluids, and their equilibrium with cell surface receptors.

Growth Factors and Cytokines in Health and Disease
Volume 3A pages 191-222.

ISBN: 0-7623-0118-X

The six proteins that have been characterized to date have many similarities, but also significant differences. These differences include the presence of an R&D sequence which can bind to the α5β1 integrin receptor, adherence to cell surfaces, and adherence to extracellular matrix (ECM). Each protein is cleaved by specific proteases. Cleavage generally results in a major reduction in affinity for the IGFs and may be necessary for the IGFs to bind optimally to receptors. The identity of the proteases and the factors that control their abundance are undefined at present. IGFBP genes have been characterized and the individual promoter elements that control transcription are currently being determined. These proteins represent an important potential source of regulation of IGF activity in tissues.

I. INTRODUCTION

The insulin-like growth factors (IGFs) are present in extracellular fluids bound to high affinity carrier proteins. Six forms of IGF binding proteins (IGFBPs) have been cloned and their complete sequences have been obtained (Rechler and Brown, 1992). The structure of all six genes has been determined and chromosomal locations have been assigned. Different cell types have been shown to synthesize and secrete different combinations of binding proteins, but the physiologic significance of this observation remains undetermined. A major function of the binding proteins in the vascular compartment appears to be prolongation of the half-lives of the IGFs (Guler et al., 1989). Although stabilization of increased plasma IGF-I concentrations correlates with the anabolic response that follows IGF-I or growth hormone administration, the factors that control the rate of efflux of IGF-I and IGF-II from the vascular compartment to extracellular fluids and the importance of the IGFBPs in maintaining normal anabolism have not been determined. Several mechanisms, such as proteolytic cleavage of the binding proteins with subsequent release of the IGFs, have been proposed to control the rate of efflux (Zapf et al., 1981). After leaving the vascular compartment the IGFs are associated with binding proteins in all interstitial fluids that have been analyzed. Each of the binding proteins has an affinity constant that is higher than the type I IGF receptor, the refore, equilibrium favors IGF binding to IGFBPs in interstial fluid. Since no one has successfully identified a tripartite complex consisting of the IGF receptor, a form of binding protein and the growth factor, a mechanism is presumed to exist for release of the IGFs to the receptor. Several mechanisms have been shown to result in decreased affinity of IGFBPs that would allow release of the IGFs to receptors. These include proteolysis, polymerization, dephosphorylation, and association of specific forms of IGFBPs with extracellular matrix or cell surfaces. Exactly how many of these mechanisms are operative *in vivo* and which mechanism is predominant is unknown, however, clearly binding protein synthesis, proteolysis, and post-translational modifications that lower their affinities deserve further investigation. A major focus of this chapter will be to review the mechanisms that regulate IGFBP affinity and may thereby indirectly control IGF presentation to receptors. It is clear that understanding the factors that regulate the synthesis and secretion of these proteins as well as their proteolytic cleavage are also very important to understanding how they operate to control IGF presentation of receptors *in vivo*.

II. GENE AND PROTEIN STRUCTURES OF THE SIX FORMS OF IGFBPS

Complimentary DNA clones have been isolated and used to determine the complete sequence of each of the six forms of IGFBPs (Shimasaki et al., 1990). No other complete IGFBP gene beyond these six has been identified in the past four years, however, cDNA sequences of proteins that are homologous to specific regions of IGFBPs have been identified. An example is the retinoic acid inducible protein, Mac-25. A region of Mac-25 is homologous with the amino terminal sequence of IGFBP-1 (Swisshelm et al., 1995). Although forms of IGFBPs that have an eletrophorectic mobility which suggests that they are unique have been identified, to date these "unique bands" usually have been determined to be glycosylation variants or proteolytic cleavage products that retain some binding activity. Complete sequences of the six known proteins have either been determined directly or deduced from nucleotide sequencing and have been published (Rechler, 1993). All six contain leader sequences that are between 19 and 39 residues and are cleaved prior to secretion. The mature proteins vary between 201 and 289 residues. The amino acid sequences in the amino and carboxyl termini are highly conserved. There are 18 cysteines that are conserved in IGFBP 1-5, 12 are in the N-terminal region, and six in the carboxyl-terminal region. Human and rat IGFBP-6 lack two or four of the N-terminal cysteines respectively (Rechler, 1993). IGFBP-4 has two additional cysteines in the central core region of the molecule (Latour et al., 1990). Homology is particularly striking in the carboxyl-terminal region where conservation across species may be as high as 90% for certain IGFBPs. Disulfide bonding has been studied in IGFBP-1 and 3 and for IGFBP-1 it has been shown that at least 16 of the 18 cysteines are disulfide bonded. All are disulfide linked in IGFBP-3. IGFBP-1 and 2 contain an arg-gly-asp sequence near the carboxyl-terminus (Brown et al., 1989; Brewer et al., 1988). In the case of IGFBP-1 this sequence is required for binding to the $\alpha5\beta1$ integrin receptor (Jones et al., 1993). IGFBP-2 does not bind to this integrin and whether it binds to any other integrin receptor through this sequence has not been determined. All six IGFBPs contain the sequence CWCV in the carboxyl terminus. This is homologous to 10 repeat domains within human thyroglobulin but the significance of this homology is unknown. IGFBP-3 and 4 undergo N-linked glycosylation but no functional significance of this phenomenon has been discovered (Rechler, 1993). Likewise IGFBP-4, 5, and 6 are O-glycosylated but the physiologic significance of this is undetermined. The organizational structures of the human IGFBP-1, 2, and 3 genes as well as rat IGFBP-2 and mouse and human IGFBP-5 genes have been determined. Each gene contains four protein coding exons and the exon-intron boundaries occur at corresponding locations. Preliminary analyses of promoter regions of the four IGFBP genes (e.g., 1, 2, 3, and 5) have been reported (Suwanachkul et al., 1990; Brown and Rechler, 1990; Cubbage et al., 1989; Allander et al., 1994). IGFBP-1 and 3 contain TATA boxes that are recognized by the RNA polymerase II transcription initiation complex. The rat IGFBP-2 gene lacks TATA box (Brown and Rechler, 1990; Binkert et al., 1992).

IGFBP-1 has 234 amino acids with a 25 amino acid signal peptide and the mass estimate of 25,271 daltons. It is not glycosolated. The 3' untranslated region contains three AUUA sequences that are often associated with RNAs with short half-lives (Brewer et al., 1988; Julkunen et al., 1988). *In vivo* evidence exists to indicate that IGFBP-1 mRNA has a short half-life. The IGFBP-1 gene is a single copy 5.2 kb in length, and has been localized to chromosome 7, p12-13 (Brinkman et al., 1988). The five prime flanking region sequence has been reported for 1,205 base pairs upstream from transcription initiation site. A TATA sequence is present at position -28 and a CCAAT sequence at nucleotide -72. A consensus sequence for the transcription factor HNF1 is also present (Suwanichkul et al., 1990). Deletion and mutation analysis have indicated that the region containing the HNF-1 sequence is essential for basal promoter activity in HEP G2 cells and that region is protected in DNase footprinting assays (Suwanichkul et al., 1990). Gel retardation assay results have confirmed that DNA binding proteins in HEPG2 cell extracts bind to this region of the IGFBP-1 promoter and have the specificity for HNF1. IGFBP-1 gene transcription is also positively regulated by cyclic AMP (Powell et al., 1991) and negatively regulated by insulin (Unterman et al., 1991). The region between nucleotides -103 and -529 contains an insulin response element. This region contains two sequences that are similar to the putative insulin response element for the PEPCK gene (O'Brien and Granner, 1991). Expression of the gene is regulated by glucocorticoids with GREs located between the nucleotide -193 and -179 or -102 and -88 in the human (Suwanichkul et al., 1994).

IGFBP-2 cDNA clones were isolated from human fetal liver (Margot et al., 1989) and rat BRL3A cells (Jones et al., 1993). When the sequences of rat, human, and bovine IGFBP-2 are compared, they are 80% identical in exons 1 and 2, 98% in exon 3, and 89% in exon 4 (Bourner et al., 1992). Human and bovine IGFBP-2 contain an insert of 10-14 amino acids in exon 1. The mature protein contains 18 cysteines. The IGFBP-2 gene has been localized to chromosome 2q 33-34 (Argawal et al., 1991). The human gene is 32 kb due to a large intron containing 27 kb. The IGFBP-2 promoter is cell specific and is located between nucleotides -377 and -321. There are no known DNA binding protein binding sites in this region. Neither the rat nor the human promoters contain a TATA box. It has been proposed that GC rich regions can bind the transcription factor SP1 which binds to a TATA binding protein thereby indirectly regulating transcription (Binkert et al., 1992). The 5′ flanking region contains three GC rich regions and purified SP-1 protects these sites as do BRL-3A extracts. The protection activity can be reduced by mutations in other regions.

The IGFBP-3 gene also contains 18 conserved cysteines but the deduced amino acid sequences in the rat, human, porcine, and bovine IGFBP-3 are less homologous than IGFBP-2 with 64% identity in exon 1, 62% in exon 2, 95% in exon 3, and 80% in exon 4 (Rechler, 1993). The human IGFBP-3 gene also consists of four protein coding exons homologous to those in IGFBP-1 and 2. The gene is present in a single copy and spans 8.9 kb. It is located on chromosome 7 and is 20 kb from the IGFBP-1 gene in a tail-to-tail orientation (Ehrenborg et al., 1992). The promoter contains an upstream TATA box at nucleotide -30 and a GC element at a nucleotide -97. AP-2 and SP-1 binding sites overlap the GC element but no mutagenesis studies have been reported to determine if they are utilized.

IGFBP-4 contains 18 cysteines that are conserved and two additional cysteines that are not present in the other IGFBPs. Like IGFBP-1-3 it contains four exons. When rat and human amino acid sequences are compared there is 66% homology in exon 1, 74% in exon 2, 92% in exon 3, and 61% in exon 4 (Rechler, 1993). Three potential N linked glycosylation sites are present. IGFBP-4 is located on human chromosome 17q (Bajalica et al., 1992).

IGFBP-5 also contains four exons that are used. There are two potential AP-2 binding sites. One of these sites is used and accounts for c-AMP regulation (Allander et al., 1994). A TATA box is located 1,120 base pairs from ATG initiation codon. There are also cyclic AMP elements that are used and deletional analysis has indicated that the cyclic AMP promoter is used (Allander et al., 1994; Duan and Clemmons, 1995). IGFBP-5 is located on human chromosome 2q in a tail to tail orientation with the IGFBP-2 gene (Allander et al., 1994). The protein contains 18 cysteines and no N-glycosylation sites. It is O-glycosolated, , but has no RGD sequence. The IGFBP-5 sequence is highly conserved among the human, mouse, and rat with minimal amino acid substitutions. IGFBP-3 and 5 are closely related with 50% and 45% identity in the amino and carboxyl terminal sequences. Of particular importance is a heparin-binding motif in a carboxyl-terminal sequence between amino acids 201 and 218 (Arai et al., 1994). This sequence is also present in IGFBP-3 but not other forms of IGFBPs.

IGFBP-6 diverges in cysteine conservation with human IGFBP-6 lacking two of the 18 conserved cysteines and rat IGFBP-2 lacking four. There is an N-linked glycosylation site that is not used. Rat IGFBP-6 mRNA is 1.3 kb as determined by northern blotting. cDNA clones that contain complete protein coding sequence and 280 base pairs of the 3′ untranslated region have been analyzed (Shimasaki et al., 1991; Keifer et al., 1991). Human IGFBP-6 localized to chromosome 12q.

III. IGFBPS IN BLOOD

In the vascular compartment 85% of the IGF-I and II is bound to IGFBP-3 (Martin and Baxter, 1992). Three factors account for this, first is the abundance of IGFBP-3 in blood, second is its high affinity, and third is its ability to bind to a third protein termed acid labile subunit (ALS) which prevents its diffusion across capillaries. The ALS protein has no direct IGF-binding activity but when IGF-I or IGF-II is bound to IGFBP-3 a ternary complex will form (Baxter and Martin, 1989). Recent evidence has also shown that IGF-I and II bind to IGFBP-3 in serum independently of ALS and the IGFBP-3 itself can bind to ALS even in an unoccupied state (Barreca et al., 1995). The association of IGFBP-3 with ALS markedly prolongs the half-life of IGF-I and II thus making it more likely that these growth factors will be detected in the ternary complex form. When IGF-I and II associate with IGFBP-3 and ALS their half-lives are 16 hours (Guler et al., 1989), whereas when associated with free IGFBP-3 their half lives are reduced to approximately 90 minutes (Zapf et al., 1986). For these reasons the ternary complex forms a stable pool of IGF-I or II and IGFBP-3 in blood and this reservoir remains stable under normal physiologic conditions.

IGFBP-3 is growth hormone regulated (see below), therefore pathophysiologic conditions of growth hormone deficiency result in a decrease in the amount of this protein in serum (Blum et al., 1990; Zapf et al., 1989). When IGFBP-3 concentrations are decreased as in growth hormone deficiency, IGF-I is decreased and there is also a reduction in IGF-II. Presumably these changes are at least partially due to accelerated clearance of the IGFs although this has never been proven directly. ALS levels in blood are also hormonally regulated and increase after growth hormone administration (Dai and Baxter, 1994). In contrast, overexpression of IGF-I suppresses ALS and IGF-I infusions do not reconstitute the ternary complex in growth hormone deficient animals. N-glycosylation variants of IGFBP-3 are present in blood and there are often two to three bands detectable by immunoblotting with molecular weights ranging from 46 to 54,000 daltons even though the core protein has a molecular weight of 29,500 (Wood et al., 1988). The affinity of IGFBP-3 for IGF-II is threefold higher than for IGF-I.

Like other IGF-binding proteins IGFBP-3 binding to IGF-I is dependent upon the presence of four specific amino acids in the B domain of IGF-I, including positions 3, 4, 15, and 16 (Bayne et al., 1988). These residues are all divergent in insulin which accounts for its lack of binding activity. Unlike other forms of IGFBPs the A chain amino acids do not appear to be important for IGFBP-3 binding to IGF-I and substitutions for specific A chain residues not affect binding (Clemmons et al., 1990). However, substitution for the four B chain residues can result in a 500 fold reduction in the affinity of IGFBP-3 for IGF-I. In contrast, A chain mutations in IGFBP-1 in residues 49-51 result in a marked decrease in affinity for IGF-I (Clemmons et al., 1990). Similarly A chain mutations also reduce the affinities of IGFBP-2, 4, and 5 for IGF-I (Clemmons et al., 1992). Thus, it appears that structural determinants in both regions of the IGF-I molecule are necessary for normal affinity for all forms of IGFBPs except IGFBP-3. IGFBP-3 also contains a heparin-binding motif and it has been shown to bind to endothelial cell surfaces. Heparin-like compounds inhibit binding of IGFBP-3 to cell surfaces and this may be an important mechanism for regulating the amount of IGFBP-3 that comes in contact with endothelium under normal physiologic conditions (Baxter, 1990). Heparin binding is also associated with a threefold reduction in IGFBP-3 affinity for IGF-I (Arai et al., 1994). Whether this results in release of IGF-I after IGFBP-3 associates with endothelial cell surfaces is unknown.

A protease that is present in the serum of pregnant rats and humans cleaves IGFBP-3 into a 30-32 kDa fragment (Davenport et al., 1990; Guidice et al., 1990; Hossenlopp et al., 1990). This fragment has a marked reduction in affinity for IGF-I and II (Binoux et al., 1991). In the rat this protease is induced only in the last half of pregnancy, however, in the human it is present throughout pregnancy (Guidice et al., 1990; Hossenlopp et al., 1990). It appears to be a divalent cation dependent, serine protease. The serine protease activity is also induced in patients undergoing surgery (Davenport et al., 1992). Induction of this proteolytic activity is transient and it is not detectible in the serum 10 days after surgery. Cleavage of intact IGFBP-3 in these patients results in a marked reduction in serum IGF-I binding capacity. A protease that cleaves IGFBP-3 is also present in serum from children with severe protein/calorie malnutrition (Pucilowska et al., 1993). Whether these two proteases are identical to the protease activity in pregnancy serum is unknown. Likewise the exact molecular identity of these proteases is unclear. Although both proteolytic activities can be inhibited by serine protease inhibitors they are also divalent cation dependent. Fowlkes et al. (1994) have suggested that matrix metalloproteases can account for the IGFBP-3 protease activity in pregnancy serum. Whether IGFBP-3 proteolysis is necessary for IGF-I or II to be released from the ternary complex and transported out of the vascular compartment is unknown, but a low amount of proteolytic activity exists in normal serum.

IGFBP-1, 2, and 4 are also present in normal plasma. They are usually not completely saturated. Affinity cross-linking studies in normal pigs have suggested that IGFBP-1 and 2 are not saturated and therefore infused IGF-I is likely to bind first to these carriers (McCusker et al., 1988). Both IGFBP-1 and 2 have been shown to be transported out of the vascular compartment (Bar et al., 1990). These proteins cross intact capillary barriers and can function as transport proteins. Since their half lives are 90 minutes (Young et al., 1992) the IGF-I that is bound to these carriers has a considerably shorter half life than IGF-I that is bound to ALS and IGFBP-3. Correlative experiments have shown that when most of the IGF-I in plasma is bound to the high molecular ternary complex there is a positive anabolic response (Kupfer et al., 1993), but infusion of IGF-I analogues that bind receptor but do not bind to IGFBP-1 or 2, results in no stimulation of anabolism suggesting that binding to these proteins may be required for an in vivo anabolic response to IGF-I (Pell et al., 1996).

IV. PROTEIN STRUCTURE

All six reported IGFBP structures show clustering of their 18-20 cysteines at the amino and carboxyl terminal ends with a large cysteine free region in the center (Rechler, 1993). The exact disulfate bond linkage in any given IGFBP has not yet been elucidated, however, indirect studies indicate that all 18 cysteines are paired. IGFBP-4 contains two extra cysteines in the central core region of a molecule. The primary amino acid sequences are very homologous across the cysteine rich areas. For example the sequence in the carboxyl terminus of IGFBP-2 is highly conserved across species. Likewise among the six human or rat binding proteins there is significant sequence conservation in the amino and carboxyl-termini averaging approximately 70%. The central cysteine free regions of the molecules differ widely among the six forms although the degree of variability among species for a single form of binding protein does not show a consistent pattern. For example, rat and human IGFBP-1 vary markedly in this region, whereas mouse and human IGFBP-5 show minimal differences (James et al., 1996). Although not proven it is probable that most of the epitope determining regions of various antisera that have been raised against the IGFBPs are conferred by the antigenic differences of the sequences in the central cysteine free region.

Two proteins, IGFBP-1 and 2, contain arg-gly-asp sequences near their carboxyl terminus. In the case of IGFBP-1 this has been shown to mediate cell migration through binding to the $\alpha5\beta1$ (fibronectin) receptor (Jones et al., 1993). Interestingly, IGFBP-2 does not bind to $\alpha5\beta1$ or stimulate cell migration (Gockerman et al., 1995) suggesting that its folding pattern must be different and does not allow adequate access of its RGD sequence the $\alpha5\beta1$ receptor binding site. Proteolytic cleavage of IGFBPs occurs within the cysteine free region. Proteolytic fragments from several forms of IGFBPs have been shown to have IGF-II binding capacity. Therefore, it appears that there is a less stringent structural requirement for IGF-II binding to IGFBP-3 as compared to IGF-I (Zapf et al., 1981; Camacho-Hubner et al., 1992). Both C and N-terminal fragments of IGFBP-2 have been reported to retain IGF-II binding capacity although most of the data has been generated with N-terminal fragments (Rechler and Nissley). IGFBP-3 and 5 contain a heparin-binding motif near their carboxyl terminus in which 10 of 18 amino acids are basic. These sequences are identical within these two proteins and not present in the other IGF-binding proteins. The other forms of IGFBPs bind poorly, if at all, to heparin.

IGFBPs-3-6 are variably glycosylated. IGFBP-3 and 4 undergo N-linked glycosylation whereas IGFBP-5 and 6 have only O-linked glycosylation (Rechler, 1993). The functional significance of glycosylation is unknown. The cysteine free central region also contains several potential sites for phosphorylation. IGFBP-1 and 3 have been reported to be phosphorylated (Frost and Tseng, 1991, Jones et al., 1991; Hoeck and Mukku, 1994) and in the case of IGFBP-1, phosphorylation markedly increases its affinity for IGFI between six and 40-fold depending upon the degree of phosphorylation (Jones et al., 1991).

V. IGFBPS IN EXTRACELLULAR FLUIDS

Multiple cell types have been shown to synthesize IGFBPs. The general pattern of cellular secretion of IGFBPs that has been observed is one of differential combinations, that is a given cell type generally will secrete 2-4 binding proteins and usually one or two will predominate. Some tumor cell types have been shown secrete five or six forms of IGFBPs but this is unusual among normal diploid cells. Likewise some virally transformed cell lines have been shown to secrete only one form of binding protein (Martin et al., 1990). Again, the physiologic significance of this is unclear although some viral DNA elements contain sequences which can repress transcription of specific binding proteins and this could account for this phenomenon (Reeve et al., 1995). As yet the physiologic significance of the observation that certain cell types, such as skin fibroblasts, secrete predominately IGFBP-3 (Camacho-Hubner et al., 1992) whereas other cell types such lung epithelium secrete predominately IGFBP-2, has not been determined. Whether this is necessary for cell type specific regulation of IGF action is unclear. Likewise, certain binding proteins appear to be secreted in abundance during cell growth whereas others are secreted during differentiation. For example, stimulation of smooth muscle cell growth with IGF-I results in a induction of IGFBP-5 secretion (Duan et al., 1996). In contrast, skeletal muscle cells when they differentiate appear to make much more IGFBP-5 even though they have fused into myotubes and can no longer divide (Tollefsen et al., 1989).

VI. SYNTHESIS AND SECRETION OF IGFBPs BY CELLS AND TISSUES

A complete listing of the types of IGFBPs that are secreted by various cell types and the factors that regulate their secretion in cell culture is beyond the scope of this chapter. A partial listing has been published previously (Rechler, 1993). Some examples that may have important consequences for tissue specific responses will be discussed here. *In situ* hybridization of normal human tissues has shown that only decidua, liver and kidney contain substantial quantities of IGFBP-1 transcripts (Murphy et al., 1990). IGFBP-1 secretion has been studied in liver derived cell lines such as human HEP-G2 cells (Conover and Lee, 1990), fetal liver explants (Lewitt and Baxter, 1989), and H4-II-E rat hepatoma cells (Orlowski et al., 1990). Decidual cells (Thrailkill et al., 1990) or uterine explants (Rutanen et al., 1986) have also been used to study IGFBP-1 secretion. Agents that raise intracellular cyclic AMP levels or addition of slowly metabolizable forms of cyclic AMP increase IGFBP-1 synthesis (Suwanichkul et al., 1990). Insulin has been shown to significantly suppress both IGFBP-1 mRNA and peptide synthesis in some of these cell lines and dexamethasone has also been shown to increase its synthesis (Mohn et al., 1991; Orlowski et al., 1996). Insulin inhibits IGFBP-1 synthesis that has been stimulated by cyclic AMP (Suwanichkul et al., 1990). Glucose concentrations have been shown to regulate IGFBP-1 synthesis by human fetal liver explants and insulin further inhibits its secretion (Lewitt and Baxter, 1990; Snyder and Clemmons, 1990). The effect of insulin is believed to be mediated at the transcription level as is the effect of cyclic AMP (Powell et al., 1991). Human endometrial explants have also been shown to secrete IGFBP-1 (Rutanen et al., 1986). Whether the secretion is due to solely decidual cells has not been determined however, decidual cells are a major source of IGFBP-1 in these explants (Bell et al., 1991). Relaxin and medroxyprogesterone are important stimulants of IGFBP-1 production by placental explants (Thrailkill et al., 1990; Bell et al., 1991). Endometrial carcinoma cell lines have been studied and the HEC-1B cell line synthesizes IGFBP-1 (Camacho-Hubner et al., 1991). Likewise human granulosa cells express IGFBP-1 mRNA and synthesize IGFBP-1 protein (Jalkanen et al., 1989). Breast cancer cells *in vitro* have been shown to synthesize immunoreactive IGFBP-1 although this has only been shown with two cell lines, that is, HS578T and MDA-231 (Clemmons et al., 1990). Decidual cells in culture respond with striking increases in IGFBP-1 synthesis in response to dibutryl cyclic AMP or relaxin (Thrailkill et al., 1990). IGF-I and insulin in contrast inhibit this synthesis and the combination of insulin plus relaxin results in an inhibition of the relaxin effect.

IGFBP-2 has been shown to be synthesized by a much broader range of cell types. Several liver cell lines have been used to study IGFBP-2 secretion including BRL3A, HEP-G2, and TRL-1215 (Rechler, 1993). IGFBP-2 mRNA is detectable as a 1.6 kb transcript in human brain and liver (Tseng et al., 1989). The mRNA has also been detected in placenta as well as endometrium particularly during the secretory rather than the proliferative phase (Guidice et al., 1991). IGFBP-2 expression has been studied extensively in the rat during embryogenesis. It is detected at embryonic day 7 in all three germ layers but ectoderm derived cells appear to express it preferentially particularly surface ectoderm and neural ectoderm (Wood et al., 1990). After day 13 in the mouse its expression is restricted to neuroepithelial derived structures and the chorioid plexus. The brachial arches in the rat also continue to express this mRNA. Likewise notochord and mesodermal tissues in the splanchnic and mesonephric areas express IGFBP-2 (Chin and Bondy, 1992). Several rat-derived cell lines have been shown to synthesize IGFBP-2, specifically BRL-3A and TRL-1215, as well as rat hepatocytes, L6 myoblasts (McCusker and Clemmons, 1988), and human A673 rhabdomyosarcoma cells (McCusker et al., 1989). C2/C12 mouse myoblasts synthesize IGFBP-2 (Ernst et al., 1992). During differentiation C2 myoblasts express IGFBP-2 (Tollefson et al., 1989) as do BC3H1 (Tollefson et al., 1989; Orlowski et al., 1990) and L6 myoblasts (McCusker and Clemmons, 1988). Rat fetal calvarial cells secrete IGFBP-2 and its mRNA can be detected in these cultures (Schmid et al., 1989; Earnst and Rodan, 1990). Decidual cells and HEC-1A endometrial carcinoma cells synthesize this protein (Lamson et al., 1989; Guidice et al., 1991). Cultured granulosa cells have detectible IGFBP-2 mRNA and the protein is detectable in their medium (Mondschein et al., 1990). Peritubular cells in the testis synthesize IGFBP-2 (Smith et al., 1990). Rat astroglial cells synthesize IGFBP-2 as do neuronal cultures from 17 day rat fetuses and astrocytes have been shown to synthesize this mRNA (Lamson et al., 1989). Likewise, the neonatal choroid plexus is an abundant source of this protein (Tseng et al., 1989; Wood et al., 1990). Neuroepithelial-derived cells from the human synthesize IGFBP-2 as do neuroblastoma cells (Yee et al., 1990). IGFBP-2 has been shown to be synthesized by sheep thyroid cells in culture and its synthesis is regulated by cortisol in thyrotropin (Eggo et al., 1991). The kidney contains IGFBP-2 mRNA and MDBK cells (Cohick and Clemmons, 1991) are an abundant source of this protein as are bovine embryonic kidney cells (Upton et al., 1990). Endothelial cells have been shown to synthesize IGFBP-2 from bovine species as well as human (Moser et al., 1992). Normal breast epithelial cells as well as cell lines derived from breast carcinoma cells secrete IGFBP-2 and contain hybridizable transcripts (McGrath et al., 1991; DeLeon et al., 1989).

IGFBP-2 secretion is regulated *in vitro*. Dexamethasone and insulin decrease its mRNA abundance in cultured rat hepatocytes and insulin appears to be a potent inhibitor (Boni-Schnetzler et al., 1990). IGF-I induces a large increase in the synthesis of IGFBP-2 in rat fetal calvaria cultures (Schmid et al., 1992) and in rat osteoblasts (Chen et al., 1991). IGFBP-2 synthesis in bone can be increased by estradiol and decreased by dexamethasone (Chen et al., 1991). Porcine vascular smooth muscle cells synthesize IGFBP-2 and its synthesis is regulated by insulin, PDGF, and dexamethasone (Cohick et al., 1993). IGF-I stimulates its synthesis (Cohick et al., 1993). FSH increases its synthesis in porcine granulosa cells and forskolin decreases its synthesis in MDBK cells (Mondschein et al., 1990; Cohick and Clemmons, 1991). Endometrial stromal cells are sensitive to a combination of estrogen and progesterone which increase its synthesis (Guidice et al., 1991). When bovine mammary epithelial cells are maintained in culture in a differentiated state they are sensitive to the stimulatory effects of IGF-I on IGFBP-2 synthesis (McGrath et al., 1991). Human embryonic kidney 293 cells showed a large increase in synthesis of IGFBP-2 in response to IGF-I.

IGFBP-3 is detected as a single 2.5 kb mRNA transcript. There is minimal mRNA present in liver and abundant mRNA in placenta (Albiston and Herrington, 1991). The non-parenchymal cells of the liver appear to be a major source of the hepatic mRNA (Takenaka et al., 1991). Several adult tissues contain IGFBP-3m RNA in both rat and human, although the kidney is particularly rich as is stomach, heart, adrenal, ovary, uterus, lung, skin, intestine, spleen, testes, and muscle (Albiston and Herrington, 1991). In culture SK-HEP human carcinoma cells have been shown to synthesize IGFBP-3 as well as normal nonparenchymal cells derived from liver (Keifer et al., 1991). Osteoblasts from both the rat and human have been shown to synthesize IGFBP-3 mRNA and protein (Schmid et al., 1989; Earnst and Roden, 1990). Its abundance is increased with IGF-I and decreased by cortisol. Human porcine and rat granulosa cells synthesize IGFBP-3 (Guidice et al., 1991; Mondschein et al., 1990; Adashi et al., 1990). Rat C6 glial cells (Tseng et al., 1989) synthesize the protein as do B-104 neuroblastoma cells and T98G human glioblastoma cells (McCusker et al., 1990). Normal neonatal astroglial cells synthesize small amounts of this protein. Sheep thyroid follicular cells synthesize IGFBP-3 mRNA and protein (Bachrach et al., 1989). MDBK cells synthesize this protein but only when stimulated with forskolin. Human fibroblasts are an abundant source of IGFBP-3 mRNA and its synthesis can be simulated by cyclic AMP (Camacho-Hubner et al., 1992). IGFBP-3 synthesis by human fibroblasts is increased with IGF-I and TGFB treatment and it is decreased by dexamethasone (Martin and Baxter, 1991; Conover et al., 1989). There appears to be an increase in IGFBP-3 synthesis in senescent human fibroblasts (Goldstein et al., 1991). Swiss 3T3 fibroblasts synthesize IGFBP-3 as do bovine fibroblasts (Conover et al., 1990; Corps and Brown, 1991). Endothelial cells in culture are an important source of IGFBP-3 mRNA and protein (Moser et al., 1992). Breast cancer cell lines synthesize IGFBP-3 but only if they are estrogen receptor negative (Clemmons et al., 1990). Bovine IGFBP-3 is increased in fibroblasts after IGF-I or insulin treatment and in the mouse, IGFBP-3 is increased after IGF-I or PDGF. Calvarial cells increase IGFBP-3 synthesis in response to estradiol, growth hormone or IGF-I (Schmid et al., 1989; Earnst and Rodan, 1990). Rat sertoli cells increase synthesis in response to FSH or cyclic AMP (Smith et al., 1990). Bovine sheep thyroid cells produce IGFBP-3 mRNA and its synthesis is inhibited by EGF or phorbol esters (Eggo et al., 1991). Retinoic acid increases IGFBP-3 synthesis by MCF-7 breast carcinoma cells (Fontana et al., 1991). MDBK synthesis of IGFBP-3 mRNA and protein is increased by forskolin exposure. Normal bovine mammary epithelial cells respond to EGF, insulin, or IGF-I with release of this protein (McGrath et al., 1991).

IGFBP-4 is detected in the rat as a 2.6 kb transcript that is present in multiple tissues in adult animals including in the adrenal, testes, spleen, heart, lung, kidney, stomach, hypothalamus, and brain (Shimaski et al., 1990). In the human it is also present as a 2.4 kb transcript and is abundant in the liver and brain (Keifer et al., 1991). The mRNA is induced during pregnancy in the rat in the maternal liver. Rat hepatoma cells (H4-II-E) express this mRNA and human HEP-G2 cells also express it (Martin and Baxter, 1992). L6 myoblasts express this mRNA and its expression is increased with differentiation (McCusker and Clemmons, 1988). Mouse and rat bone cells (Ceda et al., 1991; Scharla et al., 1991) secrete the protein and it has been identified in TE-89 human osteosarcoma (Mohan et al., 1989), endometrial uterine, stromal (Guidice et al., 1991), and decidual cells (Thrailkill et al., 1990). Granulosa cells in culture from the human also secrete IGFBP-4 (Adashi et al., 1991). B-104 neuroblastoma cells secrete abundant amounts of IGFBP-4 and T98G glioblastoma cells also secrete this protein (McCusker et al., 1990; Cheung et al., 1991). Sheep thyroid cells secrete IGFBP-4 (Eggo et al., 1991). Human fibroblasts secrete IGFBP-4 and its secretion is increased with dexamethasone exposure (Camacho-Hubner et al., 1992). Bovine and mouse fibroblasts also secrete this protein (Corps and Brown, 1991; Conover and Powell, 1991). IGFBP-4 is an abundant protein secreted by endothelial cells in culture (Moser et al., 1992). Most breast carcinoma cell lines in culture, regardless of their estrogen receptor status secrete IGFBP-4 (Clemmons et al., 1990). TE85 osteosarcoma cells respond to forskolin with a large increase in IGFBP-4 (Latour et al., 1990; Mohan and Baylink, 1991). Likewise, IGFBP-4 mRNA and protein is increased by PTH exposure (Mohan and Baylink, 1991). Human decidual cells increase IGFBP-4 in response to insulin exposure (Thrailkill et al., 1990) and in mouse 3T3-E1 osteoblasts IGFBP-4 is increased 14-fold after 1,25 dihydoxy vitamin D exposure (Scharla et al., 1991). B-104 neuroblastoma cells increase their secretion of IGFBP-4 in response to insulin and IGF-II (Ceda et al., 1991). Endothelial cells respond to forskolin and cyclic AMP with increased synthesis. Forskolin increases IGFBP-4 synthesis in human fetal fibroblasts (Camacho-Hubner et al., 1992) and insulin is also a stimulant.

IGFBP-5 is detected as a single 6 kb transcript in multiple human and rat tissues (Schimasaki et al., 1991; Kiefer et al., 1991). It is most abundant in the kidney but high levels are present in the lung, heart, stomach, adrenal, large and small intestine. Very little is present in the brain, liver, spleen, and testis. During embryogenesis it is expressed in notochord, neural tube, and muscle progenitors as well as the olfactory bulb. Very little IGFBP-5 is detected in adult serum and liver is not a primary source (Rechler, 1993). Although it is an abundant transcript in cultured human fibroblasts, IGF-I has minimal effects on its transcription (Camacho-Hubner et al., 1992). In contrast, forskolin causes major increases in IGFBP-5 mRNA expression in these cultures (Camacho-Hubner et al., 1992). Basal expression in human fibroblasts is regulated by AP-2 (Duan and Clemmons, 1995). Mouse C2 cells express very large concentrations of IGFBP-5 mRNA and this is increased substantially when they undergo differentiation (James et al., 1993). Rat and human osteoblasts in culture synthesize IGFBP-5 mRNA and its synthesis can be increased by IGF-I or estrogen (Andress and Birnbaum, 1991; Rechler and Brown, 1992). T98G glioblastoma cells synthesize IGFBP-5 (McCusker et al., 1990). Thyroid cells in culture synthesize IGFBP-5 and its synthesis is inhibited by cortisol or thyrotropin (Wang et al., 1990). IGFBP-5 that is synthesized by fibroblasts adheres preferentially to extracellular matrix (Camacho-Hubner et al., 1992). Endothelial cells synthesize IGFBP-5 as do certain breast carcinoma cell lines and retinoic acid can stimulate its synthesis by these cells (LeRoith et al., 1995). Although IGF-I increases the abundance of IGFBP-5 in the conditioned medium of fibroblast cultures the mechanism accounting for this change appears to be inhibition of proteolysis rather than stimulation of synthesis (Camacho-Hubner et al., 1992). In contrast, osteoblasts respond to IGF-I with an increase in synthesis of IGFBP-5 (Conover and Keifer, 1993). IGFBP-5 is produced by L6 cells during differentiation (and Clemmons, 1988). C2 mouse myoblasts increase IGFBP-5 mRNA 50 fold during differentiation (James et al., 1993). FRTL5 cells synthesize IGFBP-5 mRNA and this is stimulated by IGF-I, IGF-II, and insulin (Backeljaw et al., 1993).

IGFBP-6 is expressed as a 1.3 kb transcript in multiple tissues in adult male rats (Keifer et al., 1991). It is highest in the lung. The levels are also high in the testes, small and large intestine, adrenal, kidney, stomach, spleen, heart, brain, and liver. There is minimal protein detectable in adult serum. Rat SK HEP cells express IGFBP-6. Human osteosarcoma U2 cells secrete this protein (Andress and Birnbaum, 1991). IGFBP-6 mRNA is synthesized by PC-12 pheochromocytoma cells and SV-40 transformed human fibroblasts (Martin et al., 1990). IGFBP-6 mRNA is expressed in estrogen receptor negative breast carcinoma cell lines (Adamo et al., 1992).

VII. REGULATION IN PLASMA

Several metabolic variables regulate the abundance of IGFBPs in serum. Generally IGFBP-3 is the most abundant protein although the source of this IGFBP-3 has not been determined. The next most abundant protein is IGFBP-2 and it is derived from liver. During fetal life IGFBP-2 appears to be the predominant form of IGFBP that is present, whereas at term the fetal concentrations of this protein decline and the concentrations of IGFBP-3 increase (Rechler and Nissley, 1996). The physiologic significance of having predominantly IGFBP-2 versus. IGFBP-3 in serum is unknown but this should allow more rapid equilibration of the IGFs with extravascular fluids. IGFBP-1 is also high in the fetal circulation and declines substantially after birth (Donovan et al., 1989). Plasma IGFBP-1 concentrations are increased dramatically after fasting and increase during the third trimester of pregnancy (Busby et al., 1988). Diabetes and hypopituitarism have also been shown to increase IGFBP-1 concentrations as has prolonged exercise (Suikkari et al., 1988). Refeeding after fasting and insulin infusions are potent suppressants and the half-life appears to be one to two hours in the human (Busby et al., 1988). Direct infusion studies in the rat confirms the half-life of approximately 90 minutes (Young et al., 1992). Insulin levels and substrate availability regulate plasma IGFBP-1 (Lewitt and Baxter, 1990). Although insulin suppresses release from the liver, glucose also appears to have a direct effect since fructose administration does not result in an increase in plasma insulin concentrations but it can suppress plasma IGFBP-1 (Snyder and Clemmons, 1990). Likewise insulin clamping shows that IGFBP-1 levels continue to decline as a function of insulin concentrations even when glucose is clamped (Brismar et al., 1988). Hypophysectomy results in a substantial increase in hepatic IGFBP-1 mRNA and this is not reversed totally with growth hormone administration (Ooi et al., 1992). Fasting results in a major increase in IGFBP-1 mRNA in the liver and plasma (Rechler, 1993). Likewise, experimental induction of diabetes in rats results in a major increase in hepatic IGFBP-1 mRNA and increases its serum level (Unterman et al., 1990). Kidney IGFBP-1 mRNA is also increased (Luo and Murphy, 1991). Glucocorticoids increase hepatic IGFBP-1 mRNA and plasma concentrations (Luo and Murphy, 1989). During liver regeneration there is a major increase in the expression of IGFBP-1 mRNA (Mohn et al., 1991).

Hypophysectomy results in a substantial increase in IGFBP-2 mRNA in the liver and plasma levels of IGFBP-2 are increased (Orlowski et al., 1990l Clemmons et al., 1991). Humans who are growth hormone deficient have high concentrations of IGFBP-2 which is lowered with growth hormone treatment (Clemmons et al., 1991). IGFBP-2 is increased substantially in blood in patients with noninsulin-dependent tumor cell hypoglycemia (Zapf et al., 1990) and some investigators have suggested this gene may be a synthetic product of the tumor or that its synthesis is increased in tumors that secrete IGF-II presumably as a result of an autocrine effect. Fasting also induces a significant increase in IGFBP-2 mRNA in liver and serum (Clemmons et al., 1991). Unlike IGFBP-1 the fasting-induced increase occurs over two to four days rather than in 6-8 hours (Clemmons et al., 1991). Similarly, plasma IGFBP-2 is suppressed with a refeeding but the time course of reduction is much slower than that for IGFBP-1. Infusion of IGF-I into humans results in a major increase in IGFBP-2 (Kupfer et al., 1993; Zapf et al., 1990) and this is an excellent marker of a response to infused IGF-I. IGFBP-2 levels have been shown to be increased in patients with multiple types of tumors but significance of this change is not clear.

Growth hormone deficiency results in major reduction in serum IGFBP-3 concentrations which can only be partially restored with IGF-I or growth hormone infusions (Kupfer et al., 1993; Zapf et al., 1990; Baxter, 1988). Growth hormone deficient humans have low IGFBP-3 which increases with growth hormone therapy and measurement of IGFBP-3 has been proposed as a test of GH deficiency in children (Blum et al., 1990; Baxter, 1988; Smith et al., 1993). GH deficient transgenic mice that express IGF-I have normal serum IGFBP-2 levels whereas IGFBP-3 is not increased to normal (Camacho-Hubner et al., 1991). IGFBP-3 levels are elevated in acromegaly but are only approximately twice normal and they are not increased in proportion to IGF-I (Grinspoon et al., 1995). Fasting results in a major decrease in IGFBP-3 and severe nutritional deprivation or surgical stress increase IGFBP-3 proteolysis (Davenport et al., 1992; Pucilowska et al., 1993). These changes can be reversed with refeeding or stress elimination. Plasma IGFBP-3 levels show a substantial increase, for example, eightfold, between birth and puberty then fall two to three fold from puberty to age 60 (Zapf et al., 1989). The plasma levels are slightly higher in women than men (Zapf et al., 1989).

VIII. MODULATION OF IGF ACTION

The IGFBPs have been shown to both positively and negatively regulate IGF action. The only exception to this rule is IGFBP-4 which has been shown to be consistently inhibitory and has never been shown to have positive modulatory effects (Jones and Clemmons, 1995). Studies with IGFBP-6 are minimal and, therefore, definitive conclusions regarding its biologic role cannot be drawn at this time. In general, for IGFBPs 1-5 the higher the affinity of the form of IGFBP for IGF-I or II, the more likely it is to be inhibitory. The initial reports describing IGFBP bioactivity used large quantities of high affinity binding proteins and therefore, uniformly reported inhibitory effects on IGF-I function (Rechler and Nissley, 1996; Jones and Clemmons, 1995). However, later studies that used lower affinity forms and lower ratios of IGFBP to IGF-I obtained different results (Elgin et al., 1987). An important variable in determining the affinity of IGFBPs for IGF-I and II is post-translational modification. Specifically phosphorylation (Jones et al., 1991), proteolysis (Binoux et al., 1991) and polymerization (Busby et al., 1989) have all been shown to result in major changes in the affinities of specific forms of IGFBPs. In addition, binding of IGFBP-5 to other proteins, such as proteoglycans, can indirectly determine the affinity for the IGFs (Jones et al., 1993a).

Phosphorylation of IGFBP-1 occurs on serines 101, 119, and 169 and serine 101 is the predominant phosphorylation site (Jones et al., 1993b). One of the serine threonine kinases that is responsible for this appears to be a casein kinase II-like enzyme (Ankrapp et al., 1995). Whether a phosphatase exists has not been determined. Phosphorylation increases the affinity of IGFBP-1 for IGF-I by eightfold. When IGFBP-1 is in the dephosphorylated state it can potentiate the effect of IGF-I on human fibroblast and porcine aortic smooth muscle cells DNA synthesis by several-fold. In contrast, the heavily phosphorylated form of the protein is inhibitory (Busby et al., 1988). Highly phosphorylated IGFBP-1 has also been shown to inhibit AIB uptake into JEG-3 choriocarcinoma cells (Ritvos et al., 1988), FRTL-5 cells (Frauman et al., 1989), chick embryo leaflets, and fibroblasts (Burch et al., 1990). That phosphorylation of IGFBP-1 serves to regulate growth has been illustrated by Frost and Tseng (1991) who demonstrated that during the proliferative phase of the menstrual cycle IGFBP-1 was secreted in primarily a dephosphorylated form, whereas in the secretory phase when proliferation ceased the predominant form appears to be phosphorylated. Generally inhibition of IGF action requires a molar excess of IGFBP-1 as compared to IGF-I. Usually this must be a least a 4:1 molar ratio to achieve significant inhibition. Using forms of IGFBP-1 that had unknown amounts of phosphorylation, Koistinen et al. (1990) have also demonstrated that IGFBP-1 plus IGF-I results in stimulation of DNA synthesis in human dermal fibroblasts. Kratz et al. (1992) demonstrated that IGFBP-1 potentiated the effect of IGF-I on proliferation of human keratinocytes by approximately threefold. Other biologic actions of IGF-I function have also been shown to be altered by IGFBP-1. Specifically, IGF-I is a potent stimulator of cell migration and IGFBP-1 addition can inhibit this response (Gockerman et al., 1995). However this effect of IGFBP-1 is complex since in certain cell types, such as CHO cells, that are non-IGF-I responsive IGFBP-1 binds to the fibronectin ($\alpha 5\beta 1$) receptor and directly stimulates migration through this IGF-independent mechanism (Jones et al., 1993). Mutants of IGFBP-1 that do not contain an intact RGD sequence do not stimulate migration, therefore, presence of this sequence appears to be required for receptor binding and stimulation of the biologic response (Jones et al., 1993). Infusion of phosphorylated forms of IGFBP-1 has been shown to inhibit IGF-I stimulated lowering of blood glucose (Lewitt et al., 1991). Direct infusion of IGFBP-1 into rat hearts showed that it was translocated across intact capillary endothelium and the amount that was translocated was increased with insulin (Bar et al., 1990). The addition of dephosphorylated IGFBP-1 to wounds, however, results in potentiation of IGF-I-stimulated wound healing (Jyung et al., 1984). Transgenic animals that overexpress IGFBP-1 (presumably phosphorylated forms) show either slight (Rajkumin et al., 1995) or no somatic growth retardation (Dai et al., 1994). Brain growth is decreased if IGFBP-1 is overexpressed (D'Ercole et al., 1994). In one study high levels in tissues such as liver were associated with mild glucose intolerance (Rajkumin et al., 1995). Since the phosphorylated state of the forms that were expressed are unknown the *in vivo* application of the data to normal physiology are difficult to interpret.

The combination of IGF-I and IGFBP-1 is also a potent growth stimulant *in vivo*. Kratz and Gilund (1994) demonstrated that this combination when adding an equimolar concentration stimulates linear incision in wound healing in rats. When IGF-I or IGFBP-1 was given alone there was no significant stimulation. The studies of Jyuan et al. (1994) showed that IGF-I had no direct effect on wound healing, however, the combination of IGF-I plus IGFBP-1 caused a 40% increase in wound breaking strength in rats. This required the dephosphorylated form of IGFBP-1 and addition to the phosphorphylated form showed no significant effect indicating that a high affinity form of IGFBP-1 could not act to potentiate the effect of IGF-I on this parameter. These investigators also noted increased wound connective tissue thickness and granulation layer within the wound as well as hypercellularity. These findings suggest that more cells were synthetically active and producing more extracellular matrix and that more were migrating into the wound tissue. A recent study in a compromised animal model reinforces these conclusions. Tsuboi et al. (1995) used an epithelial wound ulcer model in rabbits. IGF-I or IGFBP-1 were applied in various combinations (Tsuboi et al., 1995). An equimolar addition of IGF-I and IGFBP-1 resulted in a 35% stimulation of the rate of re-epithelialization and an increase in capillary number but no change in granulation tissue formation (Tsuboi et al., 1995). This study was repeated in diabetic mice (the db/db animal model) and again coadministration of IGF-I and IGFBP-1 resulted in potent stimulation of re-epithelization without a significant change in granulation tissue formation. 50µg of IGF-I with 100µg of IGFBP-1 was the most effective combination when IGFBP-1 was applied to wounds using this compromised animal model. There was no effect on re-epithelization but an increase in granulation tissue and in capillary number formation was present.

IGFBP-2 has also been shown to be a potent modulator of IGF action. Purified IGFBP-2 when added in excess has been shown to inhibit IGF-I stimulated ^{3}H-thymidine incorporation in the chick embryo fibroblasts and in rat astroglial cells (Rechler and Nissley, 1996; Knauer and Smith, 1980). Likewise DNA synthesis by human lung carcinoma cell lines or MDBK cells is inhibited by IGFBP-2 (Reeve et al., 1993; Ross et al., 1989). Des 1-3 IGF-I which does not bind to IGFBP-2 is much more potent than IGF-II or IGF-I in stimulating DNA synthesis in MDBK cells suggesting that IGFBP-2 binding of IGF-I or IGF-II inhibits their effects (Ross et al., 1989). Likewise, the ability of IGF-I or IGF-II to inhibit protein breakdown of these cells is due to the ability to bind to the receptors, not IGFBPs (Ross et al., 1989). Schwander et al. (1989) also showed that IGFBP-2 inhibited IGF-I actions in cultured fibroblasts. Bar has reported that IGFBP-2 is a weak potentiator of IGF-II enhanced AIB uptake in endothelial cells (Bar et al., 1989) and IGFBP-2 has been reported to be a weak enhancer of IGF action in porcine aortic smooth muscle cells but it is much less potent than in IGFBP-1 (Bourner et al., 1992). No IGFBP-2 transgenic animals have been prepared but targeted gene disruption experiments have shown that IGFBP-2 is required for the attainment of normal spleen size *in utero* (John Pinter, personal communication).

A major source of IGFBP-3 appears to be endothelial cells (Moser et al., 1992). These cells contain substantial amounts of protein by immunohistochemical staining and in culture secrete substantial amounts into their conditioned medium. The function of this large quantity of vascular IGFBP-3 is unknown, however, it has been shown to associate with endothelial cell surfaces and is dissociable from endothelial cell surfaces with administration of heparin (Booth et al., 1995). IGFBP-3 has been reported to enhance or to inhibit IGF-I action in fibroblasts (De Mellow and Baxter, 1988) depending upon whether it is associated with cell surfaces. When breast carcinoma cells are transfected with IGFBP-3 it appears to be inhibitory (Oh et al., 1993). This effect appears to occur even in the absence of IGF-I. In contrast, transfection of another breast cancer cell line resulted in potentiation of IGF-I action. Transfection of mice with a tissue specific promoter results in sustained high levels of IGFBP-3. This results in inhibition of the mammary cell involution that occurs following lactation. Administration of the IGFBP-3/IGF-I complex to experimental animals has been shown to result in marked potentiation of osteoblast growth and collagen synthesis. Likewise, in vivo administration of this complex to growing rats resulted in potentiation of growth in these animals. Clark et al. (1993) showed that combination of IGFBP-3 plus IGF-I given an equimolar amount was a potent stimulant of hypophysectomized rat growth and more potent than IGF-I alone. However, this required a specific post translationally modified form of IGFBP-3 since the bacterially-derived material was not effective.

When differentiated functions have been examined IGFBP-3 has been shown to be inhibitory. Schmid et al. (1991) showed inhibition of collagen synthesis by cultured rat osteoblasts and Bisack et al. (1990) showed inhibition of IGF-I stimulated c-AMP generation in rat granulosa cells. Walton et al. (1989) showed inhibition of insulin-like actions in porcine fat cells. IGFBP-3 may have independent inhibitory effects, however, since Cohen et al. (1993) showed that Balb/c 3T3 cells that had been transfected with IGFBP-3 grew less well in response to serum. The reduction in growth was not believed to be the result of the IGF binding function of IGFBP-3 since when this protein was overexpressed it suppressed ^{3}H-thymidine incorporation in the absence of added IGF-I. Blat et al. (1989) have shown that addition of IGFBP-3 and IGF-I with or without serum to 3T3 cells results in inhibition of IGF-I stimulate DNA synthesis and that IGFBP-3 can also inhibit FGF stimulated DNA synthesis. DeMellow and Baxter (1988) showed that if cells were preincubated with IGFBP-3 plus IGF-I inhibited the effect on DNA synthesis whereas if they are preincubated with IGFBP-3 which was then removed then IGF-I subsequently added, there was a twofold increase in the cellular response to IGF-I. Likewise, Conover (1991) has shown that if cells are incubated with IGFBP-3 for extended time periods, for example, 72 h, to allow proteolytic processing and then IGF-I is added to the cultures there is a slight potentiation of AIB uptake. Ernst and Rodan (1990) observed that rat osteoblast cultures secrete more IGFBP-3 in response to IGF-I. Blum et al. (1989) showed that extended incubation of IGFBP-3 with baby hamster kidney cells resulted in potentiation of IGF-I action by 90% compared to IGF-I alone. These results suggest that the mechanism by which IGFBP-3 alters cellular responsiveness to IGF-I is complex.

Some of these discrepancies in cellular responsiveness to IGFBP-3 may be explained by proteolysis. IGFBP-3 is proteolytically cleaved into a 30Kd fragment which binds IGF-II avidly (Davenport et al., 1990; Guidice et al., 1990; Hossenlopp et al., 1990). This protease is abundant in human pregnancy serum and is induced in mid-gestation in the rat. It is also induced by stress such as postoperative states or in starving children (Davenport et al., 1992; Pucilowska et al., 1993). Whether this protease activity is the same as that present in human pregnancy serum is unknown. Proteases that cleave IGFBP-3 have also been noted in conditioned medium and when the matrix metalloproteases are disassociated from endogenous tissue inhibitors of metalloprotease they cleave IGFBP-3 into various fragments. Other investigators have shown that serine protease inhibitions can block IGFBP-3 proteolytic activity (Davenport et al., 1990; Hossenlopp et al., 1990). Therefore, multiple proteases may cleave this form of IGFBP thus lowering its affinity for IGF-I or II, and these lower affinity forms may then act to potentiate IGF action as compared to the high affinity, intact protein. When Schmid, et al. tested the effect of purified 30 kDa fragment on IGF-I ^{3} stimulated H-thymidine incorporation by rat osteoblast lines, they noted that it had significant intrinsic potentiating activity and could potentiate the effect of IGF-I. Other investigators have suggested that this effect is mediated by inhibiting IGF receptor down-regulation. One common observation that occurs in conjunction with IGFBP-3 potentiating effects is the appearance of IGFBP-3 adherence to cell surfaces. Specifically, when IGFBP-3 localizes on cell surfaces it has approximately a 10 fold reduction in affinity for IGF-I (McCusker et al., 1990). Many of the papers that have shown a potentiating effect have shown abundant amounts of IGFBP-3 associated with the cell surface. When present the cell-associated IGFBP-3 often represents greater than 90% of the cell surface binding sites and clearly it has the potential to control the release of IGFs to receptors. Whether proteolysis is required for cell surface associated potentiation of IGF action to occur is unknown.

In contrast to IGFBPs 1-3, IGFBP-4 consistently has been shown to inhibit IGF actions. Phosphorylation has been reported but there is no cell surface or extracellular matrix association of IGFBP-4. Likewise, polymerization has not been shown to occur in a physiologically relevant system. Finally, although proteolytic fragments are present they appear to have low or no binding capacity for the IGFs (Parker et al., 1995). Two groups of investigators, while purifying an unknown growth inhibitor and monitoring purification using a biologic test systems, purified IGFBP-4. One was based on its ability to inhibit FSH-induced granulosa cell steroidogenesis (Ui et al., 1989). The second was based on its ability to inhibit growth of a clonal colon carcinoma cell line, HD-29 cells (Coulouscou and Shoyab, 1991). IGFBP-4 was shown to inhibit the PTH-stimulated increase in osteoblast IGF-I synthesis (Latour et al., 1990). This protein also has been shown to inhibit IGF-I stimulated smooth muscle cell division and osteosarcoma cell growth (Cohick et al., 1993; Mohan et al., 1989). Generally a molar excess of IGFBP-4 has been required. Like IGFBP-3, IGFBP-4 is cleaved into proteolytic fragments (Chernausek et al., 1995), however, unlike IGFBP-3 these fragments have no biologic activity. The IGFBP-4 protease has the interesting property of having its activity markedly accelerated when IGF-I and II bind to IGFBP-4 (Ling et al., 1993). The IGFs appear to have to bind to IGFBP-4 for proteolysis to be enhanced and binding of the protease itself to IGFBP-4 is insufficient to activate it. It appears to be a serine protease and cleaves IGFBP-4 at dibasic residues (Chernausek et al., 1995). When IGFBP-4 is cleaved it apparently loses biologic activity and can no longer inhibit IGF-I-stimulated DNA synthesis (Chernausek et al., 1995). The form of the protease that is secreted by smooth muscle cells is a serine protease that is calcium dependent (Parker et al., 1995).

The initial biologic studies that were conducted with IGFBP-5 showed inhibition of steroidogenesis by granulosa cells (Ling et al., 1993) and of glycogen synthesis by human osteosarcoma cells (Kiefer et al., 1992). However, the important property of IGFBP-5 binding to extracellular matrix was unknown when these studies were conducted. Jones et al. (1993a) showed that IGFBP-5 adheres to fibroblast ECM and following adherence there is an eightfold reduction in the affinity for IGF-I and II. Following this affinity shift it has been shown that IGFBP-5 can then potentiate the effect of IGF-I on fibroblast growth by twofold (Jones et al., 1993a). In contrast, when IGFBP-5 fibroblast is in conditioned medium it is present as a proteolytically cleaved fragment which does not potentiate cell growth response to IGF-I. The 23 kDa fragment that is generated as a result of proteolysis does not bind to extracellular matrix (Jones et al., 1993a). This protease is a serine protease (Nam et al., 1994). It is cation-dependent and it is not activated by IGF-I binding to IGFBP-5 (Nam et al., 1994). IGFBP-5 binding to the extracellular matrix appears to require association with proteoglycans since high salt inhibits binding and heparin has been shown to disassociate IGFBP-5 from the ECM. The heparin binding domain of IGFBP-3 and IGFBP-5 is located near the carboxyl-terminus. A specific disaccharide structure (Arai et l., 1994) is required for this binding to occur. Specifically, heparin or heparan sulfate must be O-sulfated in the 2 or 3 position of iduronic acid ring (Arai et al., 1994). N-linked sulfates in these positions are much less effective. Chondroitin sulfate A and C which do not contain O-sulfated sugars in the 2 and 3 position of the iduronic acid ring are ineffective in inhibiting the binding of IGFBP-5 to heparin sepharose. The specific residues that are required within IGFBP-5 are between R201-R218 but the exact basic residues within the sequence that are required for glycosaminoglycan binding have not been defined (Arai et al., 1994). That proteoglycans are the primary components to the extracellular matrix that bind to IGFBP-5 is known not only from salt inhibition studies but also from specific peptide inhibition studies and from heparinase digestion of ECM proteins, such as tenascin which results in decreased IGFBP-5 binding. Andress et al. (1993) have demonstrated that a 23 kDa fragment of IGFBP-5 has stimulatory activity for osteoblasts that is independent of IGF binding and therefore they have postulated that proteolysis may enhance IGFBP-5 activity. The cellular binding site for this fragment is unknown. Bautista et al. (1991) showed that coincubation of IGFBP-5 and IGF-I results in potentiation of IGF-II action.

IX. SUMMARY

The insulin like growth factor binding proteins bind the IGFs with high affinity and occur ubiquitously in physiologic fluids. Therefore, their capacity to direct the distribution of the IGFs among tissues and to control the amount of IGF that can bind to receptors is an important determinant of the capacity of the IGFs to control tissue and organ growth. This versatile yet complex system is subject to regulation by several variables including other hormones and growth factors, proteolytic enzymes, protein kinases, and proteins of physiologic relevance such as extracellular matrix proteins that can bind directly to the binding proteins themselves. Whether the binding proteins have discrete receptors and initiate biologic reactions that are important for growth that are independent of IGF binding is a matter of intense study. A unifying hypothesis which will help to understand how such proteins can function coordinately to regulate this system will be important for the development of new therapeutic approaches that utilize these proteins to modulate IGF action.

ACKNOWLEDGMENTS

The author greatfully acknowledges the contribution of Ms. Leigh Elliott who helped in preparing the manuscript. This work was supported in part by grants from the National Institutes of Health, HL26309 and AG02331.

REFERENCES

Adamo, M.L., Shao, Z.M., Lanau, F., Chen, J.C., Clemmons, D.R., Roberts, C.T., LeRoith, D., & Fontana, J.A. (1992). Insulin like growth factor I (IGF-I) and retinoic acid modulation of IGF binding proteins (IGFBP's). IGFBP-2, 3 and 4 gene expression and protein secretion in a breast cancer cell line. Endocrinology 131, 1858-1866.

Adashi, E.Y., Resnick, C.E., Hernandez, E.R., Hurwitz, A., & Rosenfeld, R.G. (1990). Follicle-stimulating hormone inhibits the constitutive release of insulin-like growth factor binding proteins by cultured rat ovarian granulosa cells. Endocrinology 126, 1305-1307.

Adashi, E.Y., Resnick, C.E., Hurwitz, A., Ricciarelli, E., Hernandez, E.R., & Rosenfeld, R.G. (1991). Ovarian granulosa cell-derived insulin-like growth factor binding proteins: Modulatory role of follicle-stimulating formone. Endocrinology 128, 754-760.

Albiston, A.L., & Herrington, A.C. (1991). Tissue distribution and regulation of insulin-like growth factor IGF binding protein-3 messenger ribonucleic acid (mRNA) in the rat: Comparison with IGF-I mRNA expression. Endocrinology 130, 497-502.

Allander, S.V., Norsson, C., Ehronborg, E., Suwannichkul, A., Weber, G., Morris, S.L., Bijalica, S., Kiefer, M.C., Luthman, H., & Powell, D.R. (1994). Characterization of the chromosomal gene and promoter for human insulin like growth factor binding protein-5. J. Biol. Chem. 269, 10891-10898.

Andress, D.L., & Bimbaum, R.S. (1991). A novel human insulin-like growth factor binding protein secreted by osteoblast cells. Biochem. Biophys. Res. Commun. 176, 213-218.

Andress, D.L., Loop, S.M., Zapf, J., & Kiefer, M.C. (1993), Carboxy truncated insulin like-growth factor binding protein-5 stimulates mitogenesis in osteoblast like cells. Biochem. Biophys. Res. Commun. 195, 25-30.

Ankrapp, D.P., Jones, J.I., & Clemmons, D.R. (1996). Characterization of insulin like growth factor binding protein-1 kinases from human hepatoma cells. J. Cell. Biochem. 60, 387-399.

Arai, T., Parker, A.J., Busby, W.H., & Clemmons, D.R. (1994). Heparin, heparan sulfate, and dermatan sulfate regulate formation of the insulin-like growth factor-I and insulin-like growth factor-binding protein complexes. J. Biol. Chem. 269, 20388-20393.

Argawal, N., Hsieh, C.L., Sills, D., Swaroop, M., Desai, B., Franke, V., & Swaroop, A. (1991). Sequence analysis expression and chromosomal localization of a gene isolated from a substraction library from human retinoic DNA library, that encodes insulin like growth factor binding protein-2. Exp. Cell. Res. 52, 549-561.

Bachrach, L.K., Liu, F.R., Burrow, G.N., & Eggo, M.C. (1989). Characterization of insulin-like growth factor-binding proteins from sheep thyroid cells. Endocrinology 125, 2831-2838.

Backeljaw, P.F., Clemmons, D.R., & D'Ercole, A.J. (1993). Synthesis and regulation of insulin like growth factor binding protein-5 (IGFBP-5) in FRTL-5 cells. Endocrinology 132, 1677-1681.

Bajalica, S., Allander, S.V., Ehrenborg, E., Brodum-Neilson, K., Luthman, H., & Larsson, C. (1992). Localization of human insulin-like growth factor binding protein-4 gene to region 17 12.21. Hum. Genet. 89, 234-237.

Bar, R.S., Boes, M., Clemmons, D.R., Busby, W.H., Sandra, A., Dake, B.L., & Booth, B.A. (1990). Insulin differentially alters transcapillary movement of intravascular IGFBP-1, IGFBP-2 and endothelial cell IGF binding proteins in rat heart. Endocrinology 127, 497-499.

Bar, R.S., Booth, B.A., Bowes, M., & Drake, B.L. (1989). Insulin-like growth factor binding proteins from cultured endothelial cells: Purification, characterization, and intrinsic biologic activities. Endocrinology 125, 1910-1920.

Bar,. R.S., Clemmons, D.R., Boes, M., Busby, W.H., Booth, B.A., Dake, B.L., & Sandra, A. (1990). Transcapillary permeability and subendothelial distribution of endothelial and amniotic fluid IGF-binding proteins in rat heart. Endocrinology 127, 1078-1086.

Barreca, A., Ponzani, P., Arvigo, M., Giordano, G., & Minuto, F. (1995). Effect of the acid labile subunit on the binding of insulin like growth factor (IGF) binding protein-3 and ^{125}I-IGF-I. J. Clin. Endocrinol. Metab. 80, 138-1324.

Bautista, C.M., Baylink, D.J., & Mohan, S. (1991). Isolation of a novel insulin-like growth factor (IGF) binding protein from human bone: A potential candidate for fixing IGF-II in human bone. Biochem. Biophys. Res. Commun. 176, 756-763.

Baxter, R.C. (1988). Characterization of the acid labile subunit of the growth hormone dependent insulin like growth factor binding protein complex. J. Clin. Endocrinol. Metab. 67, 265-272.

Baxter, R.C. (1990). Glycosaminoglycans inhibit formation of the 140kDa insulin like growth factor binding protein complex. Biochem. J. 271, 773-777.

Baxter, R.C., & Martin, J.L. (1989). Structure of the Mr 140,000 growth hormone-dependent insulin-like growth factor binding protein complex: Determination by reconstitution and affinity labeling. Proc. Natl. Acad. Sci. USA. 86, 6898-6902.

Bayne, M.L., Applebaum, J., Chicchi, G.G., Hayes, N.S., Green, B.G., & Cascieri, M.A. (1988). Structural analogues of human insulin-like growth factor-I with reduced affinity for serum binding protein and the type II insulin-like growth factor receptor. J. Biol. Chem. 263, 6233-6239.

Bell, S.C., Jackson, J.A., Ashmone, J., Zhu, H.H., & Tseng, L. (1991). Regulation of insulin like growth factor binding protein-1 synthesis and secretion by progestin and relaxin in long term cultures of human endometrial stromal cells. J. Clin. Endocrinol. Metab. 72, 1014-1024.

Bicsak, T.A., Shimonaka, M., Malkowski, M., & Ling, N. (1990). Insulin-like growth factor-binding protein (IGF-BP) inhibition of granulosa cell function: Effect of cyclic adenosine 3',5'-monophosphate, deoxyribonucleic acid synthesis, and comparison with the effect of an IGF-I antibody. Endocrinology 126, 2184-2189.

Binkert, C., Margot, J.B., Handwehr, J., Heinrich, G., & Schwander, J. (1992). Structure of the human insulin like growth factor binding protein-2 gene. Mol. Endocrinol. 6, 826-836.

Binoux, M., Hossenlopp, P., Lassarre, C. & Segovia, B. (1991). Degradation of IGF binding protein-3 by proteases: physiological implications. In: Modern Concepts of Insuin-like Growth Factors, pp. 329-336. ed. by Spencer, E.M.

Binoux, M., Roghani, M., Hossenlopp, P., Hariodin, S., & Gourmelen, M. (1991). Molecular forms of human IGF binding proteins: Physiological implications. Acta Endocrinol 124 (Supp. 12), 41-47.

Blat, C., Bohlen, P., Villaudy, J., Chatelain, P.G., Golde, A., & Harel, L. (1989). Isolation and amino-terminal sequence of a novel cellular growth inhibitor (Inhibitory Diffusible Factor 45) secreted by 3T3 Fibroblasts. J. Biol. Chem. 264, 6021-6024.

Blum, W.F., Jenne, E.W., Reppin, F., Kietzmann, K., Ranke, M.B., & Bierich, J.R. (1989). Insulin-like growth factor I (IGF-I) binding protein complex is a better mitogen than free IGF-I. Endocrinology 125, 766-772.

Blum, W.F., Ranke, M.B., Kietzmann, K., Gauggel, E., Zeisel, H.J., & Bierich, J.R. (1990). A specific radioimmunoassay for the growth hormone (GH)-dependent somatomedin binding protein: Its use for diagnosis of GH deficiency. J. Clin. Endocrinol. Metab. 70, 1292-1298.

Boni-Schnetzler, M., Schmid, C., Mary, J-L., Zimmerli, B., Meier, P.J., Zapf, J., Schwander, J., & Froesch, E.R. (1990). Insulin regulates the expression of the insulin-like growth factor binding protein 2 mRNA in rat hepatocytes. Mol. Endocrinol. 1320-1326.

Booth, B.A., Boes, M., Andress, D.L., Dake, B.L., Keifer, M.C., Maack, C., Linkhard, T., Bar, R., Caldwell, F.E., & Weiler, J. (1995). IGFBP-3 and IGFBP-5 association with endothelial cells: role of c-terminal heparin binding domain. Growth Regulation 5, 1-17.

Bourner, M.J., Busby, W.H., Siegel, N.R., Krivi, G.G., McCusker, R.H., & Clemmons, D.R. (1992). Cloning and sequence determination of bovine insulin-like growth factor binding protein-2 (IGFBP-2): comparison of its structural and functional properties with IGFBP-1. J. Cell. Biochem. 48, 215-226.

Brewer, M.T., Stetler, G.L., Squires, C.H., Thompson, R.C., Busby, W.H., & Clemmons, D.R. (1988). Cloning, characterization and expression of a human insulin-like growth factor binding protein. Biochem. Biophys. Res. Commun. 152, 1289-1297.

Brinkman, A., Groffen, C.A.H., Kortleve, D.J., & Drop, S.L.S. (1988). Organization of the gene encoding the insulin-like growth factor binding protein IGF-BP-1. Biochem. Biophys. Res. Commun. 157, 898-907.

Brismar, K., Gutniak, M., Povoa, G., Werner, S., & Hall, K. (1988). Insulin regulates the 35 kDa IGF binding protein in patients with diabetes mellitus. J. Endocrinol. Invest. 11, 599-602.

Brown, A.L., Chiariotti, L., Orlowski, C.C., Mehlman, T., Burgess, W.H., Ackerman, E.J., Bruni, C.B., & Rechler, M.M. (1989). Nucleotide sequence and expression of a cDNA clone encoding a fetal rat binding protein for insulin-like growth factors. J. Biol. Chem. 264, 5148-5154.

Brown, A.L., Rechler, & M.M. (1990). Cloning of the rat insulin like growth factor binding protein-2 gene and identification of a promoter lacking a TATA box. Mol. Endocrinol. 4, 2039-2051.

Burch, W.W., Correa, J., Shaveley, J.E., & Powell, D.R. (1990). The 25 k Dalton insulin-like growth factor (IGF) binding protein inhibits both basal and IGF mediated growth in chick embryonic pelvic cartilage *in vitro*. J. Clin. Endocrinol. Metab. 70, 173.

Busby, W.H., Hossenlopp, P., Binoux, M., & Clemmons, D.R. (1989) Purified preparations of the amniotic fluid IGF binding protein contain multimeric forms that are biologically active. Endocrinology 125, 773-777.

Busby, W.H., Klapper, D.G., & Clemmons, D.R. (1988). Purification of a 31000 dalton insulin-like growth factor binding protein from human amniotic fluid. J. Biol. Chem. 263, 14203-14210.

Busby, W.H., Snyder, D.K., & Clemmons, D.R. (1988). Radioimmunoassay of a 26,000 dalton plasma insulin like growth factor binding protein: Control by nutritional variables. J. Clin. Endocrinol. Metab. 67, 1225-1230.

Camacho-Hubner, C., Busby, W.H., McCusker, R.H., Wright, G., & Clemmons, D.R. (1992). Identification of the forms of insulin-like growth factor binding proteins produced by human fibroblasts and the mechanisms that regulate their secretion. J. Biol. Chem. 267, 11949-11956.

Camacho-Hubner, C., McCusker, R.H., & Clemmons, D.R. (1991). Secretion and biologic actions of the insulin-like growth factor binding proteins in two human tumor derived cell lines *in vitro*. J. Cell. Physiol. 148, 281-289.

Camacho-Hubner, C., Clemmons, D.R., & D'Ercole, A.J. (1991). Regulation of insulin-like growth factor binding proteins in transgenic mice with altered expression of growth hormone and IGF-I. Endocrinology 129, 1201-1206.

Ceda, G.P., Fielder, P.J., Henzel, W.J., Louie, A., Donovan, S.M., Hoffman, A.R., & Rosenfeld, R.G. (1991). Differential effects of insulin-like growth factor (IGF)-I an IGF-II on the expression of IGF binding proteins (IGFBPs) in a rat neuroblastoma cell line: Isolation and characterization of two forms of IGFBP-4. Endocrinology 128, 2815-2824.

Chen, T.L., Liu, F., Bates, R.L., & Hintz, R.L. (1991). Further characterization of insulin-like-growth factor binding proteins in rat osteoblast-like cell cultures: Modulation by 1794Symbol"b-estradiol and human growth hormone. Endocrinology 128, 2489-2496.

Chernausek, S.D., Smith, C.E., Duffin, K.L., Busby, W.H., Wright, G., & Clemmons, D.R. (1995). Proteolytic cleavage of insulin like growth factor binding protein-4 (IGFBP-4). J. Biol. Chem. 270, 11377-11382.

Cheung, P.T., Smith, E.P., Shimasaki, S., Ling, N., & Chernausek, S. (1991). Characterization of insulin-like growth factor binding protein (IGFBP-4) produced by B104 rat neuronal cell line: Chemical and biologic properties and differential synthesis by sublines. Endocrinology 129, 1006-1015.

Chin, E., & Bondy, C.A. (1992). Insulin like growth factor system gene expression in human kidney. J. Clin. Endocrinol. Metab. 75, 962-968.

Clark, R.G., Mortensen, D., Reigsupreder, D., Mohler, M., Etchevery, T., & Mukkev, V. (1993). Recominant human insulin like growth factor binding protein 3: Effects on the glycine and growth promotions activities of rh IGF-I in the rat. Growth Regulation 3:46-50.

Clemmons, D.R., Busby, W.H., & Snyder, D.K. (1991). Variables controlling the secretion of insulin-like growth factor binding protein-2 in normal human subjects. J. Clin. Endocrinol. Metab. 73, 727-733.

Clemmons, D.R., Camacho-Hubner, C., Coronado, E.B., & Osborne, C.K. (1990). Insulin-like growth factor binding protein secretion by breast carcinoma cell lines: Correlation with estrogen receptor status. Endocrinology 127, 2679-2686.

Clemmons, D.R., Cascieri, M.A., Camacho-Hubner, C., McCusker, R.H., & Bayne, M.L. (1990). Discrete alterations of the IGF-I molecule that alter its affinity for IGF binding proteins result in changes in bioactivity. J. Biol. Chem. 265, 12210-12216.

Clemmons, D.R., Dehoff, M.H., Busby, W.H., Bayne, M.L., & Cascieri, M.A. (1992). Competition for binding to insulin-like growth factor (IGF) binding protein-2, 3, 4, and 5 by the IGFs and IGF analogs. Endocrinology 131, 890-895.

Cohen, P., Hanson, G., Okajama, T., & Rosenfeld, R.G. (1993). Transfection of human IGFBP-3 gene into balb c fibroblasts a model for the cellular funtions of IGFBPs. Growth Regulation 3, 23-26.

Cohick, W.S., & Clemmons, D.R. (1991). Regulation of insulin-like growth factor binding protein synthesis and secretion in a bovine epithelial cell line. Endocrinology 129, 1347-1354.

Cohick, W.S., Gockerman, A., & Clemmons, D.R. (1993). Vascular smooth muscle cells synthesize two forms of insulin-like growth factor binding proteins which are regulated differently by the insulin like growth factors. J. Cell. Physiol. 157, 52-60.

Conover, C.A. (1991). Glycosylation of insulin like growth factor binding protein-3 (IGFBP-3 is not required for potentiation of IGF-I action: Evidence for processing of cell bound IGFBP-3. Endocrinology 129, 3259-3268.

Conover, C.A., & Keifer, M.C. (1993). Regulation of biological effect of endogenous insulin like growth factor binding protein 5 in human osteoblast cells. J. Clin. Endocrinol. Metab. 76, 1153-1159.

Conover, C.A., & Lee, P.D.K. (1990). Insulin regulation of insulin-like growth factor-binding protein production in cultured HepG2 cells. J. Clin. Endocrinol. Metab. 70, 1062-1067.

Conover, C.A., & Powell, D.R. (1991). Insulin-like growth factor (IGF) binding protein-3 blocks IGF-I-induced receptor down-regulation and cell densitization in cultured bovine fibroblasts. Endocrinology 129, 710-716.

Conover, C.A., Liu, F., Powell, D.R., Rosenfeld, R.G., & Hintz, R.L. (1989). Insulin like growth factor binding proteins from cultured human fibroblasts: Characterization and Hormonal Regulation. J. Clin. Invest. 83, 852-859.

Conover, C.A., Ronk, M., Lombana, F., & Powell, D.R. (1990). Structural and biological characterization of bovine insulin-like growth factor binding protein-3. Endocrinology 127, 2795-2803.

Corps, A.N., & Brown, K.D. (1991). Mitogens regulate the production of insulin-like growth factor-binding protein by Swiss 3T3 cells. Endocrinology 128, 1057-1064.

Coulouscou, J.M., & Shoyab, M. (1991). Purification of a colon cancer cell growth inhibitor and its identification as insulin-like growth factor binding protein-4. Cancer Res. 51, 2813-2819.

Cubbage, M.L., Suwanichkul, A., & Powell, D.R. (1989). Structure of the human chromosomal gene for the 25 kilodalton insulin-like growth factor binding protein. Mol. Endocrinol. 3, 846-851.

D'Ercole, A.J. & Clemmons, D.R. (1994). Brain growth retardation due to the expression of human insulin-like growth factor binding protein-1 (IGFBP-1) in transgenic mice an *in vivo* model for the analysis of IGF function in the brain. Devel. Brain. Res. 82, 213-222.

Dai, A., Xing, Y., Boney, C.M., Clemmons, D.R., & D'Ercole, A.J. (1994). Human insulin-like growth factor binding protein-1 (hIGFBP-1) transgenic mice: Characterization and insights into the regulation of IGFBP-1 expression. Endocrinology 135, 1316-1327.

Dai, J., & Baxter, R.C. (1994). Regulation in vivo of acid labile subunit of the rat serum insulin like growth factor binding protein complex. Endocrinology 135, 2335-2341.

Davenport, M.L., D'Ercole, A.J., & Underwood, L.E. (1990). Effects of maternal fasting on fetal growth serum insulin like growth factors (IGF's) and tissue IGF messenger ribonucleaic acids. Endocrinology 126, 2062-2067.

Davenport, M.L., Isley, W.I., Pucilowska, J., Penberton, B., Lyman, B., Underwood, L.E., & Clemmons, D.R. (1992). Insulin-like growth factor binding protein-3 proteolysis is induced following elective surgery. J. Clin. Endocrinol. Metab. 130, 2505-2512.

De Mellow, J.S.M., & Baxter, R.C. (1988). Growth hormone dependent insulin-like growth factor binding protein both inhibits and potentiates IGF-I stimulated DNA synthesis in skin fibroblasts. Biochem. Biophys. Res. Commun. 156, 199-204.

DeLeon,, D.D., Wilson, D.M., Bakker, B., Lamsom, G., Hintz, R.L., & Rosenfeld, R.G. (1989). Characterization of insulin-like growth factor binding proteins from human breast cancer cells. Mol. Endocrinol. 3, 567-574.

Donovan, S.M., Oh, Y., Pham, H., & Rosenfeld, R.G. (1989). Ontogeny of serum insulin-like growth factor binding proteins in the rat. Endocrinology 125, 2621-2627.

Duan, C., Hawes, S., Prevette, T., & Clemmons, D.R. (1996). Insulin like growth factor-I (IGF-I) stimulates IGF binding protein-5 synthesis through transcriptional activation of the gene in aortic smooth muscle cells. J. Biol. Chem. 271, 4280-4288.

Duan, C.M., & Clemmons, D.R. (1995). Transcription factor AP-2 regulates human insulin like growth factor binding protein-5 gene expression. J. Biol. Chem. 270, 24844-24851.

Earnst, M., & Rodan, G.A. (1990). Increased activity of IGF-I in osteoblastic cells in the presence of GH: Positive correlation with the presence of GH induced IGFBP-3. Endocrinology 127, 807-814.

Eggo, M.C., Bachrach, L.K., Brown, A.L., & Burrow, G.N. (1991). Thyrotropin inhibits while insulin, epidermal growth factor and tetradecanoyl phorbol acetate stimulate insulin-like growth factor binding protein secretion from sheep thyroid cells. Growth Factors 4, 221-230.

Ehrenborg, E., Larsson, C., Stern, I., Janson, M., Powell, D.R., & Luthman, H. (1992). Contiguous localization of the genes encoding human insulin-like growth factor binding proteins 1 (IGFBP1) and 3 (IGFBP3) on chromosome 7. Genomics 12, 497-502.

Elgin, R.G., Busby, W.H., & Clemmons, D.R. (1987). An insulin-like growth factor binding protein enhances the biologic response to IGF-I. Proc. Natl. Acad. Sci. USA 84, 3254-3258.

Ernst, C.W., McCusker, R.H., & White, M.E. (1992). Gene expression and secretion of insulin like growth factor binding protein during myoblast differentiation. Endocrinology 130, 607-615.

Fontana, J.A., Burrows-Mezu, A., Clemmons, D.R., & LeRoith, D. (1991). Retinoid modulation of insulin-like growth factor binding proteins and inhibition of breast carcinoma proliferation. Endocrinology 128, 1115-1122.

Fowlkes, J., Enghild, J.J., Suzukik, N., & Nagase, H. (1994). Matrix metalloproteases degrade insulin like growth factor binding protein-3 in dermal fibroblast cultures. J. Biol. Chem. 269, 25742-25746.

Frauman, A.G., Tsuzaki, S., & Moses, A.C. (1989). The binding characteristics and biological effects in FRTL5 cells of placental protein-12, and insulin-like growth factor-binding protein purified from human amniotic fluid. Endocrinology 124(5), 2289-2296.

Frost, J.P., & Tseng, L.T. (1991). Insulin-like growth factor binding protein-1 is phosphorylated by cultured human endometrial cells and multiple protein kinases *in vitro*. J. Biol. Chem. 266, 18082-18088.

Gockerman, A., Jones, J.I., Prevette, T., & Clemmons, D.R. (1995). Insulin like growth factor binding protein-2 inhibits smooth muscle cell migration response to IGF-I. Endocrinology 136, 4168-4173.

Goldstein, S., Moerman, E.J., Jones, R.A., & Baxter, R.C. (1991). Insulin-like growth factor binding protein 3 accumulates to high levels in culture medium of senescent and quiescent human fibroblasts. Proc. Natl. Acad. Sci. USA 88, 9680-9684.

Grinspoon, S., Clemmons, D.R., Swearingen, B., & Klibanski, A. (1995). Diagnostic use of IGFBP-3 levels in acromegaly. J. Clin. Endocrinol. Metab. 80, 927-932.

Guidice, L.C., Farrell, E.M., Pham, H., Lamson, F.G., & Rosenfeld, R.G. (1990). Insulin like growth factor binding proteins in maternal serum throughout gestation and in the puerperium. J. Clin. Endocrinol. Metab. 71, 806-816.

Guidice, L.C., Milkowski, D.A., Fielder, P.J., & Irwin, J.C. (1991). Characterization of steroid dependence of insulin like growth factor binding protein 2 synthesis and mRNA expression in human endometrial stromal cells. Hum. Reprod. 6, 632-640.

Guidice, L.C., Milkowski, D.A., Larsson, G., Rosenfeld, R.G., & Irwin, J.C. (1991). Insulin like growth factor binding proteins in human endometrium: Steroid dependent mRNA expression and proteins synthesis. J. Clin. Endocrinol. Metab. 72, 779-787.

Guler, H-P, Zapf, J., Schmid, C., & Froesch, E.R. (1989). Insulin-like growth factors I and II in healthy man. Estimations of half-lives and production rates. Acta Endocrinol. 121, 753-758.

Hoeck, W.G., & Mukku, V.R. (1994). Identification of the major sites of phosphorylation in IGF binding protein-3. J. Cell. Biochem. 56, 262-273.

Hossenlopp, P., Segovia, B., Lassare, C., Roghani, M., Bredon, M., & Binoux, M. (1990). Enzymatic evidence of degradation of insulin-like growth factor binding protein in 150 k complex during pregnancy. J. Clin. Endocrinol. Metab. 71, 797-805.

Jalkanen, J., Suikkari, A-M., Koistinen, R., Butzow, R., Ritvos, O., Seppala, M., & Ranta, T. (1989). Regulation of insulin-like growth factor-binding protein-1 production in human granulosa-luteal cells. J. Clin. Endocrinol. Metab. 69, 1174-1179.

James, P.L., Jones, S.B., Busby, W.H., Clemmons, D.R., & Rotwein, P. (1996). IGF binding protein-5 is expressed in myoblast differentiation and is highly conserved. J. Biol. Chem. 268, 22305-22312.

James, P.L., Jones, S.B., Busby, W.H., Clemmons, D.R., & Rotwein, P. (1993). IGF binding protein-5 is expressed in myoblast differentiation and is highly conserved. J. Biol. Chem. 268, 22305-22312.

Jones, J.I., & Clemmons, D.R. (1995). Insulin like growth factor and their binding proteins: biologic actions. Endocrine. Rev. 16, 3-34.

Jones, J.I., Busby, W.H., Wright, G., Smith, C.E., Kimack, N.M., & Clemmons, D.R. (1993b). Identification of the sites of phosphorylation in insulin-like growth factor binding protein-1: Regulation of its affinity by phosphorylation of serine 101. J. Biol. Chem. 268, 1125-1131.

Jones, J.I., D'Ercole, A.J., Camacho-Hubner, C., & Clemmons, D.R. (1991). Phosphorylation of insulin-like growth factor binding protein in cell culture and *in vivo*: Effects on affinity for IGF-I. Proc. Natl. Acad. Sci. 88, 7481-7485.

Jones, J.I., Gockerman, A., Busby, W.H., Camacho-Hubner, C., & Clemmons, D.R. (1993a). Extracellular matrix contains insulin-like growth factor binding protein-5: Potentiation of the effects of IGF-I. J. Cell. Biol. 121, 679-687.

Jones, J.I., Gockerman, A., Busby, W.H., Jr., Wright, G., & Clemmons, D.R. (1993b). Insulin-like growth factor binding protein 1 stimulates cell migration and binds to the α5β1 integrin by means of its Arg-Gly-Asp sequence. Proc. Natl. Acad. Sci. USA 90, 10553-10557.

Julkunen, M., Koistinen, R., Aalto-Setala, K., Seppala, M., Janne, O.A., & Kontula, K. (1988). Primary structure of human insulin-like growth factor-binding protein/placental protein 12 and tissue-specific expression of its mRNA. FEBS Lett 236, 295-302.

Jyung, R.W., Mustoe, T., Busby, W.H., & Clemmons, D.R. (1994). Increased wound breaking strength induced by insulin-like growth factor-1 in combination with IGF binding protein-1. Surg 115, 223-239.

Keifer, M.D., Masiarz, F.R., Bauer, D., & Zapf, J. (1991). Identification and molecular cloning of two new 30 kDa insulin-like growth factor binding proteins isolated from adult human serum. J. Biol. Chem. 266, 9043-9049.

Kiefer, M.C., Ioh, R.S., Bauer, D.M., & Zapf, J. (1991). Molecular cloning of a new human insulin-like growth factor binding protein. Biochem. Biophys. Res. Comm. 176, 219-225.

Kiefer, M.C., Schmid, C., Waldvogel, M., Schlapfer, I., Futo, E., Masiarz, F.R., Green, K., Barr, P.J., & Zapf, J. (1992). Characterization of recombinant human insulin-like growth factor binding proteins 4, 5 and 6 produced in yeast. J. Biol. Chem. 267, 12692-12699.

Knauer, D.J., & Smith, G.L. (1980). Inhibition of biologic activity of multiplication stimulating activity by binding to its carrier protein. Proc. Natl. Acad. Sci. USA 77, 7252-7254.

Koistinen, R., Itkinen, P., Selenius, P., & Seppala, M. (1990). Insulin like growth factor binding protein-1 inhibits binding of IGF-I on fetal skin fibroblasts but stimulates their DNA synthesis. Biochem. Biophys. Res. Commun 173, 408-415.

Kratz, G., & Gilund, M. (1994). Insulin like growth factor 1 and 2 and their role in epithelization of wounds: investigation in insulin like growth factor binding protein type I. Scan. J. Plastic. Reconstruct. Surg. 28, 107-112.

Kratz, G., Lake, M., Ljungstrom, K., Forsberg, G., Haegerstrand, A., & Gidlund, M. (1992). Effect of recombinant IGF binding protein-1 on primary cultures of human keratinocytes and fibroblasts: selective enhancement of IGF-I but not IGF-II induced cell proliferation. Exp. Cell. Res. 202, 381-385.

Kupfer, S.R., Underwood, L.E., Baxter, R.C., & Clemmons, D.R. (1993). Enhancement of the anabolic effects of growth hormone and insulin like growth factor-I by the use of both agents simultaneously. J. Clin. Invest. 91, 391-397.

Lamson, G., Oh, Y., Pham, H., Giudice, L.C., & Rosenfeld, R.G. (1989). Expression of two insulin-like growth factor-binding proteins in a human endometrial cancer cell line: Structural, immunological and genetic characterization. J. Clin. Endocrinol. Metab. 69, 852-859.

Lamson, G., Pham, H., Oh, Y., Ocrant, I., Schwander, J., & Rosenfeld, R.G. (1989). Expression of the BRL-3A insulin-like growth factor binding protein (rBP-30) in the rat central nervous system. Endocrinology 123, 1100-1102.

Latour, D., Mohan, S., Linkhart, T.A., Baylink, D.J., & Strong, D. (1990) Inhibiting insulin-like growth factor binding protein cloning complete sequence and physiologic regulation. Mol. Endocrinol. 4, 1806-1814.

LeRoith, D., Baserga, R., Helman, L., & Roberts, C.T. (1995). Insulin like growth factors and cancer. Ann. Intern. Med. 122, 54-59.

Lewitt, M.S., & Baxter, R.C. (1989). Regulation of growth hormone independent insulin-like growth factor binding protein (BP-2) in cultured human fetal liver explants. J. Clin. Endocrinol. Metab. 69, 246-252.

Lewitt, M.S., & Baxter, R.C. (1990). Inhibitors of glucose uptake stimulate the production of insulin-like growth factor-binding protein 1 (IGFBP-1) by human fetal liver. Endocrinology 126, 1527-1533.

Lewitt, M.S., Denyer, G.S., Cooney, G.J., & Baxter, R.C. (1991). Insulin-like growth factor binding protein-1 modulates blood glucose levels. Endocrinology 129, 2254-2256.

Ling, N.C., Liu, X.J., Malkowski, M., Guo, Y.L., Erickson, G.F., & Schwandi, S. (1993). Structural and functional stimulation of insulin like growth factor binding proteins in the ovary. Growth Regulation 3, 70-74.

Luo, J., & Murphy, L.J. (1991). Differential expression of insulin-like growth factor-I and insulin-like growth factor binding protein-1 in the diabetic rat. Mol. Cell. Biochem. 103, 41-50.

Luo, J.,. & Murphy, L.J. (1989). Dexamethasone inhibits growth hormone induction of insulin-like growth factor-I (IGF-I) messenger ribonucleic acid (mRNA) in hypophysectomized rats and reduces IGF-I mRNA abundance in the intact rat. Endocrinology 125, 165-171.

Margot, J.B., Binkert, C., Mary, J-L., Landwehr, J., Heinrich, G., & Schwander, J. (1989). A low molecular weight insulin-like growth factor binding protein from rat: cDNA cloning and tissue distribution of its messenger RNA. Mol. Endocrinol. 3, 1053-1060.

Martin, J.L., & Baxter, R.C. (1991). Transforming growth factor-beta stimulates production of insulin-like growth factor binding protein-3 by human skin fibroblasts. Endocrinology 128, 1425-1433.

Martin, J.L., & Baxter, R.C. (1992). Insulin like growth factor binding protein-3: Biochemistry and physiology. Growth Regulation 2, 88-99.

Martin, J.L., Willetts, K.E., & Baxter, R.C. (1990). Purification and properties of a novel insulin-like growth factor-II binding protein from transformed human fibroblasts. J. Biol. Chem. 26, 4124-4130.

McCusker, R.H., & Clemmons, D.R. (1988). Insulin-like growth factor binding proteins (IGFBPs) secretion by muscle cells: Effect of cellular differentiation and proliferation. J. Cell. Physiol. 137, 505-512.

McCusker, R.H., Camacho-Hubner, C., & Clemmons, D.R. (1989). Identification of the types of insulin like growth factor binding proteins that are secreted by muscle cells *in vitro*. J. Biol. Chem. 264, 7795-7800.

McCusker, R.H., Camacho-Hubner, C., Bayne, M.L., Cascieri, M.A., & Clemmons, D.R. (1990). Insulin-like growth factor (IGF) binding to human fibroblast and glioblastoma cells: The modulating effect of cell released IGF binding proteins (IGFBPs). J. Cell. Physiol. 144, 244-253.

McCusker, R.H., Campion, D.R., & Clemmons, D.R. (1988). The ontogeny and regulation of a 31,000 Mr insulin-like growth factor/somatomedin (IGF) binding protein in fetal porcine plasma and sera. Endocrinology 122, 2071-2079.

McGrath, M.F., Collier, R.J., Clemmons, D.R., Busby, W.H., Sweeny, C.A., & Krivi, G.C. (1991). The direct effect of insulin-like growth factors (IGFs) on normal bovine mammary cell proliferation and production of IGF binding proteins: An *in vitro* study. Endocrinology 129, 671-678.

Mohan, S., & Baylink, D.J. (1991). Evidence that inhibition of TE85 human bone cell proliferation by agents which stimulate cAMP production may in part be mediated by changes in the IGF-II regulatory system. Growth Regulation 1, 110-118.

Mohan, S., Bautista, C.M., Wergedal, J., & Baylink, D.J. (1989). Isolation of inhibitory insulin-like growth factor (IGF) binding protein from bone cell conditioned medium: A potential local regulator of IGF action. Proc. Natl. Acad. Sci. USA 86, 8338-8342.

Mohn, K.L., Melby, A.E., Tewari, D.S., Laz, T.M., & Taub, R. (1991). The gene encoding rat insulin-like growth factor binding protein 1 is rapidly and highly induced in regenerating liver. Mol. Cell. Biol. 11, 1393-1401.

Mondschein, J.S., Smith, S.A., & Harmond, J.M. (1990). Production of insulin like growth factor binding proteins (IGFBPs) by porcine granulosa cells: Identification of IGFBP-2 and 3 and regulation by hormones and growth factors. Endocrinology 127, 2298-2306.

Moser, D.R., Lowe, W.L., Dake, B.L., Booth, B.A., Boes, M., & Clemmons, D.R. (1992). Endothelial cells produce IGFBP-2 IGFBP-3 and IGFBP-4. Mol. Endocrinol. 6, 1805-1814.

Murphy, L.J., Seneviratne, C., Bellejo, G., Croze, F., & Kennedy, T.G. (1990). Identification and characterization of a rat decidual insulin-like growth factor-binding protein complementary DNA. Mol. Endocrinol. 4, 329-336.

Nam, T.J., Busby, W.H., & Clemmons, D.R. (1994). Human fibroblasts secrete a serine protease that cleaves insulin like growth factor binding protein-5. Endocrinology 135, 1385-1391.

O'Brien, R.M., & Granner, D.K. (1991). Regulation of gene expression by insulin. Biochem. J. 278, 609-619.

Oh, Y., Muller, H.L., Lamson, G., & Rosenfeld, R.G. (1993). Insulin like growth factor (IGF) independent actions of IGF binding protein-3 in HS578 in human breast carcinoma cells. J. Biol. Chem. 268, 14964-14971.

Ooi, G.T., Tseng, L.Y., Tran, M.Q., & Rechler, M.M. (1992). Insulin rapidly decreases insulin-like growth factor-binding protein-1 gene transcription in streptozotocin-diabetic rats. Mol. Endocrinol. 6, 2219-2228.

Orlowski, C.C., Brown, A.L., Ooi, G.T., Yang, Y.W-H., Tseng, L.Y-H., & Rechler, M.M. (1990). Tissue, developmental and metabolic regulation of messenger ribonucleic acid encoding a rat insulin-like growth factor-binding protein. Endocrinology 126, 644-652.

Orlowski, C.C., Ooi, G., & Rechler, M.M. (1996). Dexamethasone stimulates transcription of insulin like growth factor binding protein-1 (IGFBP-1) gene in the H4-II-E rat hepatoma cells. Mol. Endocrinol. 4, 1592-1599.

Orlowski, C.C., Ooi, G.T., Brown, D.R., Young, Y., Tseng, L., & Rechler, M.M. (1990). Insulin rapidly inhibits insulin like growth factor binding protein-1 expression in H4-II-E rat hepatoma cells. Mol. Endocrinol. 5, 1180-1187.

Parker, A., Gockerman, A., Busby, W.H., & Clemmons, D.R. (1995). Properties of an insulin like growth factor binding protein 4 protease that is secreted by smooth muscle cells. Endocrinology 135, 2470-2476.

Pell, J.M., Flicksmith, H.C., Dye, S., & Hill, R.A. (1996). Further characterization of an IGF-I enhancing antibody on IGF-I induced hypoglycemia and interaction with the analogue LR3 IGF-I.

Presented to the 3rd International Symposium on IGF Binding Proteins, October 6-8: Tuebingen, Germany.

Powell, D.R., Suwanichkul, A., Cubbage, M.L., DePaolis, L.A., Snuggs, M.B., & Lee, P.D.K. (1991) Insulin inhibits transcription of the human gene for insulin-like growth factor-binding protein-1. J. Biol. Chem. 266, 18868-18876.

Pucilowska, J.B., Davenport, M.L., Kabir, I., Clemmons, D.R., Thissen, J.B., Butler, T., & Underwood, L.E. (1993). The effect of dietary protein supplementation on insulin like growth factors (IGF's) and IGF binding proteins in children with shigellosis. J. Clin. Endocrinol. Metab. 77, 1516-1521.

Rajkumin, K., Barron, D., Lewitt, M., & Murphy, L.J. (1995). Growth retardation and hyperglycemia in insulin like growth factor binding protein-1 transgenic mice. Endocrinology 136, 4029-4034.

Rechler, M.M. & Nissley, S.P. (1996). Insulin like growth factors. In: Peptide Growth Factors and Their Receptors (Swan, M., & Roberts, A.B., Eds), pp. 263-287. Springer Verlag, Berlin.

Rechler, M.M. (1993). Insulin like growth factor binding proteins. Vitamins and Hormones 47, 1-114.

Rechler, M.M., & Brown, A.L. (1992). Insulin like growth factor binding proteins: Gene structure and expression. Growth Regulation 2, 55-68.

Reeve, J.G., Schwanda, J., & Bleecham, N.M. (1993). IGFBP-2: An important regulator of insulin like growth factor action in human lung tissues. Growth Regulation 3, 82-84.

Reeve, J.G., Xiong, J., Morgan, J., & Bleeher, N.M. (1995). Diminished expression of insulin like growth factor (IGF) binding protein-5 and activation of autocrine inhibited growth in simian virus-40 transformed human fibroblasts. J. Biol. Chem. 270, 135-142.

Ritvos, O., Ranta, T., Jalkanen, J., Suikkari, A.M., Voutilainen, R., Bohn, H,. & Rutanen, E.M. (1988). Insulin-like growth factor (IGF) binding protein from human decidua inhibits the binding and biological action of IGF-I in cultured choriocarcinoma cells. Endocrinology 122, 2150-2157.

Ross, M., Francis, G.L., Szabo, L., Wallace, J.C., & Ballard, F.J. (1989). Insulin-like growth factor (IGF)-binding proteins inhibit the biological activities of IGF-I and IGF-2 but not des 2D(1-3)-IGF-I. Biochem. J. 258, 267-272.

Rutanen, E-M., Menabawey, M., Isaka, K., Bohn, H., Chard, T., & Grudzinskas, J.G. (1986). Synthesis of placental protein 12 by decidua from early pregnancy. J. Clin. Endocrinol. Metab. 63, 675-679.

Scharla, S.H., Strong, D.D., Mohan, S., Baylink, D.J., & Linkhart, T.A. (1991). 1,25 Dihydroxyvitamin D3 differentially regulates the production of insulin-like growth factor I (IGF-I) and IGF-binding protein-4 in mouse osteoblasts. Endocrinology 129, 3139-3146.

Schmid, C., Ernst, M., Zapf, J., & Froesch, E.R. (1989). Release of insulin-like growth factor carrier proteins by osteoblasts: stimulation by estradiol and growth hormone. Biochem. Biophys. Res. Commun. 160, 788-794.

Schmid, C., Rutishauser, J., Schlapfer, I., Froesch, E.R., & Zapf, J. (1991). Intact but not truncated insulin-like growth factor binding protein-3 blocks IGF-I induced stimulation of osteoblasts: Control of IGF signalling to bone cells by IGFBP-3 specific proteolysis. Biochem. Biophys. Res. Commun. 179, 579-585.

Schmid, C., Schlapfer, I., Waldvogel, M., Meier, P.J., Schwander, J., Boni-Schnetzler, M., Zapf, J., & Froesch, E.R. (1992). Differential regulation of insulin-like growth factor binding protein (IGFBP)-2 mRNA in liver and bone cells by insulin and retinoic acid *in vitro*. FEBS Lett 303, 205-209.

Schwander, J., Mary, J-L., Landwehr, J., Boni, M., Margot, J. & Binkert, C. (1989). The regulation of the mRNA of an insulin-like growth factor binding protein (IBP-2) in the rat. In: Insulin-like Growth Factor Binding Proteins. Excerpta Medica International Congress Series 881, (Drop, S.L.S., & Hintz, R.L., Eds.). pp. 125-131. Amsterdam: Elsevier Science Publishers B.V.

Shimasaki, S., Gao, L., Shimonaka, M., & Ling, N. (1991). Isolation and molecular cloning of insulin-like growth factor binding protein-6. Mol. Endocrinol. 5, 938-941.

Shimasaki, S., Shimonaka, M., Zhang, H.P., & Ling, N. (1991). Identification of five different insulin-like growth factor binding proteins (IGFBPs) from adult rat serum and molecular cloning of a novel IGFBP in rat and human. J. Biol. Chem. 266, 10646-10653.

Shimasaki,S., Uchlyama, F., Motoyuki, S., & Ling, N. (1990) Molecular cloning of the complementary DNAs encoding a novel insulin-like growth factor binding protein from rat and human. Mol. Endocrinol. 4, 1451-1458.

Smith, E.P., Dickson, B.A., & Chemausek, S.D. (1990). Insulin like growth factor binding protein-3 secretion from cultured rat sertoli cells: Dual regulation by follicle stimulating hormone and IGF-I. Endocrinology 127, 27441-27451.

Smith, W.J., Nam, T.J., Underwood, L.E., Busby, W.J., Celniker, A., & Clemmons, D.R. (1993). Use of insulin-like growth factor binding protein 2 (IGFBP-2), IGFBP-3 and IGF-I for assessing growth hormone status in short children. J. Clin. Endocrinol. Metab. 77, 1294-1299.

Snyder, D.K., & Clemmons, D.R. (1990). Insulin dependent regulation of insulin like growth factor binding protein-1. J. Clin. Endocrinol. Metab. 71, 1632-1636.

Suikkari, A.M., Koivisto, V.A., Rutanen, E.M., Yki-Jarvinen, H., Karonen, S.L., & Seppala, M. (1988). Insulin regulates the serum levels of low molecular weight insulin-like growth factor binding protein. J. Clin. Endocrinol. Metab. 66, 266-272.

Suwanichkul, A., Cubbage, M.L., & Powell, D.R. (1990). The promoter of the human gene for insulin like growth factor binding protein-1: Basal promoter activity in Hep G2 cells depends upon liver factor B1. J. Biol. Chem. 265, 21185-21193.

Swanichkul, A., Allander, S.V., Morris, S.L., & Powell, D.R. (1994). Glucocorticoids and insulin regulate expression of the human gene for insulin like growth factor binding protein-1 through proximal promoter elements. Proc. Natl. Acad. Sci. 269, 30835-30841.

Swisshelm, K., Ryan, K., Tsuchiy, K., & Sager, R. (1995). Enhanced expression of an insulin like growth factor binding protein (Mac-25) in seenescent human using epithelial cells and induced expression with retinoic acid. Proc. Natl. Acad. Sci. 92, 4472-4476.

Takenaka, A., Miura, Y., Mori, M., Hirosawa, M., Kato, H., & Noguchi, T. (1991). Distribution of messenger RNAs of insulin-like growth factor (IGF)-binding proteins-1 and-3 between parenchymal and nonparenchymal cells in rat liver. Agric. Biol. Chem. 55, 1191-1193.

Thrailkill, K.M., Clemmons, D.R., Busby, W.H., & Handwerger, S.R. (1990). Differential regulation of insulin-like growth factor binding protein secretion from human decidual cells by IGF-I, insulin and relaxin. J. Clin. Invest. 86, 878-883.

Tollefsen, S.E., Lajara, R., McCusker, R.H., Clemmons, D.R., & Rotwein, P. (1989). Insulin-like growth factors (IGF) in muscle development. J. Biol. Chem. 264, 13810-13817.

Tseng, L.Y-H., Brown, A.L., Yang, Y.W-H., Romanus, J.A., Orlowski, C.C., Taylor, T., & Rechler, M.M. (1989). The fetal rat binding protein for insulin-like growth factors is expressed in the choroid plexus and cerebrospinal fluid of adult rats. Mol. Endocrinol. 3, 1559-1568.

Tsuboi, R., Shi, C.M., Soto, C., Cox, C.N., & Ogawa, H. (1995). Coadministration of insulin like growth factor I (IGF-I) with IGF binding protein-1 stimulates wound healing in animal models. J. Invest. Dermatol. 104, 199-203.

Ui, M., Shimonaka, M., Shimasaki, S., & Ling, N. (1989). An insulin-like growth factor-binding protein in ovarian follicular fluid blocks follicle-stimulating hormone-stimulated steroid production by ovarian granulosa cells. Endocrinology 125(2), 912-916.

Unterman, T.G., Oehler, D.T., Murphy, L.J., & Lacson, R.G. (1991). Multihormonal regulation of insulin like growth factor binding protein-1 in rat H4 IIE hepatoma cells: the dominant role of insulin. Endocrinology 128, 2693-2701.

Unterman, T.G., Patel, K., Kumar, M.V., Rajamohan, G., Oehler, D.T., & Becker, R.E. (1990). Regulation of low molecular weight insulin-like growth factor binding proteins in experimental diabetes mellitus. Endocrinology 126, 2614-2626.

Upton, F.Z., Szabo, L., Walker, J.C., & Ballard, F.J. (1990). Characterization and cloning of bovine insulin-like growth factor binding protein. J. Mol. Endocrinol. 5, 77-84.

Walton, P.E., Gopinath, R., & Etherton, T.D. (1989). Porcine insulin-like growth factor (IGF) binding protein blocks IGF-I action on porcine adipose tissue. Proc. Soc. Exper. Biol. Med. 190, 315-319.

Wang, J.F., Decks, G.F., Buckingham, K.D., & Hill, D.J. (1990). Characterization of insulin-like growth factor-binding proteins secreted by isolated sheep thyroid epithelial cells. Endocrinology 125, 439-448.

Wood, T.L., Brown, A.L., Rechler, M.M., & Pintar, J.E. (1990). The expression pattern of an insulin like growth factor (IGF) binding protein gene is distinct from IGF-II in mid gestation embryo. Mol. Endocrinol. 4, 1257-1263.

Wood, W.I., Cachianes, G., Henzel, W.J., Winslow, G.A., Spencer, S.A., Hellmiss, R., Martin, J.L., & Baxter, R.C. (1988). Cloning and expression of the GH dependent insulin like growth factor binding protein. Mol. Endocrinol. 2, 1176-1185.

Yee, D., Favoni, R.E., Lebovic, G.S., Lombana, F., Powell, D.R., & Reynolds, C.P. (1990). Insulin-like growth factor I expression by tumors of neuroectodermal origin with the t(11;22) chromosomal translocation. A potential autocrine growth factor. J. Clin. Invest. 86, 1806-1814.

Young, S.C.J., Miles, M.V., & Clemmons, D.R. (1992). Determination of the pharmacokinetic profiles of IGFBP-1 and IGFBP-2 in rats. Endocrinology 131, 1867-1873.

Zapf, J., Froesch, E.R., & Humbel, R.E. (1981). The insulin-like growth factors (IGF) of human serum: chemical and biological characterization and aspects of their possible physiologic role. Current topics in Cellular Regulation 19, 257-39.

Zapf, J., Hauri, C., Waldvogel, M., & Froesch, E.R. (1986). Acute metabolic effects and half-lives of intravenously administered insulinlike growth factors I and II in normal and hypophysectomized rats. J. Clin. Invest. 77, 1768-1775.

Zapf, J., Hauri, C., Waldvogel, M., Futo, E., Hasler, K., Binz, K., Guler, H-P, Schmid, C., & Froesch, E.R. (1989). Recombinant human insulin-like growth factor I induces its own specific carrier protein in hypophysectomized and diabetic rats. Proc. Natl. Acad. Sci. USA 86, 3813-3817.

Zapf, J., Schmid, C., Guler, H.P., Waldvogel, M., Hauri, C., Futo, E., Hossenlopp, P., Binoux, M., & Froesch, E.R. (1990). Regulation of binding proteins for insulin-like growth factors in humans: Increased expression of IGF binding proteins during treatment of healthy adults and in patients with extrapancreatic tumor. J. Clin. Invest. 86, 952-961.

GROWTH FACTORS AND CYTOKINES IN THE REPRODUCTIVE TRACT OF WOMEN:

PHYSIOLOGY AND PATHOPHYSIOLOGY

Linda C. Giudice, Yasmin A. Chandrasekher, Nicholas A. Cataldo, Thierry van Dessel, Walid Saleh, O.W. Stephanie Yap, and Gary A. Ulaner

Growth Factors and Cytokines in Health and Disease
Volume 3A pages 223-291.

ISBN: 0-7623-0118-X

ABSTRACT

The remarkably coordinated changes that occur in the female reproductive system during the menstrual cycle are unparalleled by any other integrated organ system in the body. The predictable and precise cyclic changes in this system set the stage for ovulation of generally a single oocyte from the ovary, its fertilization, nourishment, and transport in the oviduct, and implantation in the endometrium (Figure 1). In the absence of implantation, the ovarian corpus luteum undergoes regression, and the endometrium is shed as the menses. These events occur prior to a repeat performance in each organ in a subsequent cycle, with the single-minded goal of successfully establishing a pregnancy. In women, dysfunctions in ovarian follicle development, oocyte maturation, the process of ovulation, ovum pickup by and transport of gametes and an embryo through the Fallopian tubes, and nonreceptivity of the endometrium to implantation all have a common endpoint of unsuccessfully establishing a pregnancy. Seekers of contraception have exploited several of these possibilities, however, to couples desiring fertility, abnormalities in one or more of these processes (as well as in the male partner) can pose formidable barriers to achieving their goals. In addition to an impact on fertility, dysfunction in the female reproductive system can lead to other sequelae. Anovulation, resulting in no progesterone to oppose the effects of estradiol, predisposes women to dysfunctional uterine bleeding, endometrial hyperplasia, and/or endometrial cancer. Blocked Fallopian tubes can result in severe lower abdominal pain, and poorly developed endometrium can result in abnormal uterine bleeding and repetitive miscarriage. Endometriosis is a benign gynecologic condition in which endometrial tissue is found outside of its normal location, mainly on the ovaries and in the cul de sac. It responds to changes in circulating steroids and is associated with severe dysmenorrhea (pain with menses) and infertility. Another member of the system, the uterine myometrium (Figure 1), is most commonly thought of as an active participant in the process of labor and as a structural bystander in the nonpregnant state, aside from its potential contributions to dysmenorrhea. However, benign smooth muscle tumors of this compartment of the female reproductive tract, called leiomyomas or "fibroids," often impair fertility and have other side-effects that are detrimental to women's health, including discomfort, excessive uterine bleeding and repetitive miscarriage. These unwelcome intruders comprise a leading cause of hysterectomy in women of reproductive age. Clearly, mechanisms regulating the processes at work for normal reproductive function are critical to continuation of the species and for the well-being of women. These processes involve auto- para- and/or juxtacrine interactions of growth modulators. This chapter reviews what is currently known about growth factors and cytokines in normal ovarian follicle development as well as in abnormal development (in polycystic ovarian syndrome [PCOS]), in

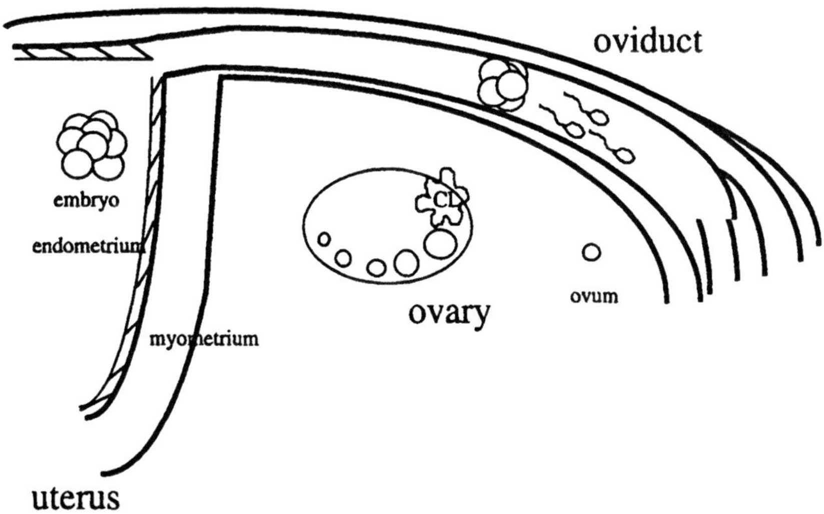

Figure 1. Schematic representation of the female reproductive tract showing a portion of the uterus with myometrium and endometrium, an ovary with follicles of varying stages, an ovulated oocyte, and an oviduct with sperm traveling to meet the ovulated oocyte. Fertilization normally occurs in the ampullary portion of the oviduct. A cleaving embryo is shown traversing its way through the Fallopian tube, and an embryo is shown in the uterine cavity prior to implantation into the maternal endometrium. Most of the events shown occur *in vivo* at distinct times of the menstrual cycle and not simultaneously as depicted here (see text). CL=corpus luteum.

oviductal function, in endometrial development, implantation, and endometrial shedding, in endometriosis, and in the growth of uterine leiomyomas. Our information is at best incomplete, which prompts us to propose models for the potential roles of growth factors and related peptides in the pathogenesis of these processes. Some systems have been more extensively studied than others due to tissue availability. Most systems have also been extensively studied in animal models, although some do not have natural homologues in most animals (e.g., PCOS, endometriosis, and uterine fibroids). Where appropriate we have made reference to animal models, although the primary focus of the chapter is on growth modulators in the human female reproductive system in health and disease.

I. OVARY

A. Intraovarian Peptides and Normal Ovarian Physiology

Normal Ovarian Function

The production of estradiol by the ovary is essential for the pubertal development of secondary sexual characteristics and growth of steroid-responsive tissues such as the breast and the uterus. When regular monthly cycles are established, there is a repetitive process of cyclic estradiol production, and after ovulation the production of progesterone. These processes are highly dependent on a normally functioning hypothalamic-pituitary axis. Follicular development within the ovary (Figure 2a) is independent of gonadotropin action up until the early antral stage, and then growth and acquisition of steroidogenic capability are dependent upon the presence of pituitary-derived FSH and LH (for review see Gougeon, 1996; Richards, 1994). In the late luteal phase, just before the onset of the menses, under FSH action, follicles are recruited for the next cycle (Figure 2b). As FSH levels increase in the early follicular phase, the follicles grow to about 4-6 mm, and then as FSH levels fall, in the mid-follicular phase, one follicle gains dominance and the remainder of the cohort undergoes atresia (for reviews see Hsueh et al., 1994; Richards, 1994). It has been suggested that the production of sufficient amounts of estradiol is essential for further follicle development and prevention of atresia (McNatty et al., 1979), and only in the follicle that has gained dominance is a substantial increase in granulosa aromatase gene expression and estradiol production observed (for review see Adashi, 1991). In a recent study using transvaginal sonography, it was demonstrated that a rise in serum estradiol levels in women coincides with the appearance of the dominant follicle (van Santbrink et al., 1994). Enhancement of FSH action by local growth modulators is believed to be crucial for the major increase in aromatase activity in the follicle gaining dominance and in the growth of the follicles per se. Mechanisms underlying follicular selection remain unknown, although atresia of the remaining cohort is believed to be effected by apoptotic mechanisms which may also be under the control of growth factors and related peptides (Hsueh et al., 1994). The induction of ovulation by luteinizing hormone (LH Figure 2) is also likely to involve intraovarian growth modulators, as does functioning of the corpus luteum and its regression during the luteal phase (Giudice et al., 1992a). This section reviews growth factors and cytokines found in the human ovary and their potential roles in normal ovarian follicular development and abnormal follicular development, specifically in polycystic ovarian syndrome (PCOS).

Insulin-like Growth Factor System

The IGF system is comprised of two IGF peptides (IGF-I and IGF-II), the Type I and Type II IGF receptors, six IGF binding proteins (IGFBPs), and a family of IGFBP proteases (Jones and Clemmons, 1995). IGF-I and IGF-II are small mitogenic and differentiative peptides (Rotwein, 1991). IGF-I (somatomedin-C) is a mediator of growth hormone (GH) in some systems, and most of the growth promoting and metabolic activities of the IGFs in childhood and in adults, respectively, are due to IGF-I. In contrast, IGF-II is abundant in the fetus and in tumors, although it is also expressed in selective organs, including human ovarian granulosa

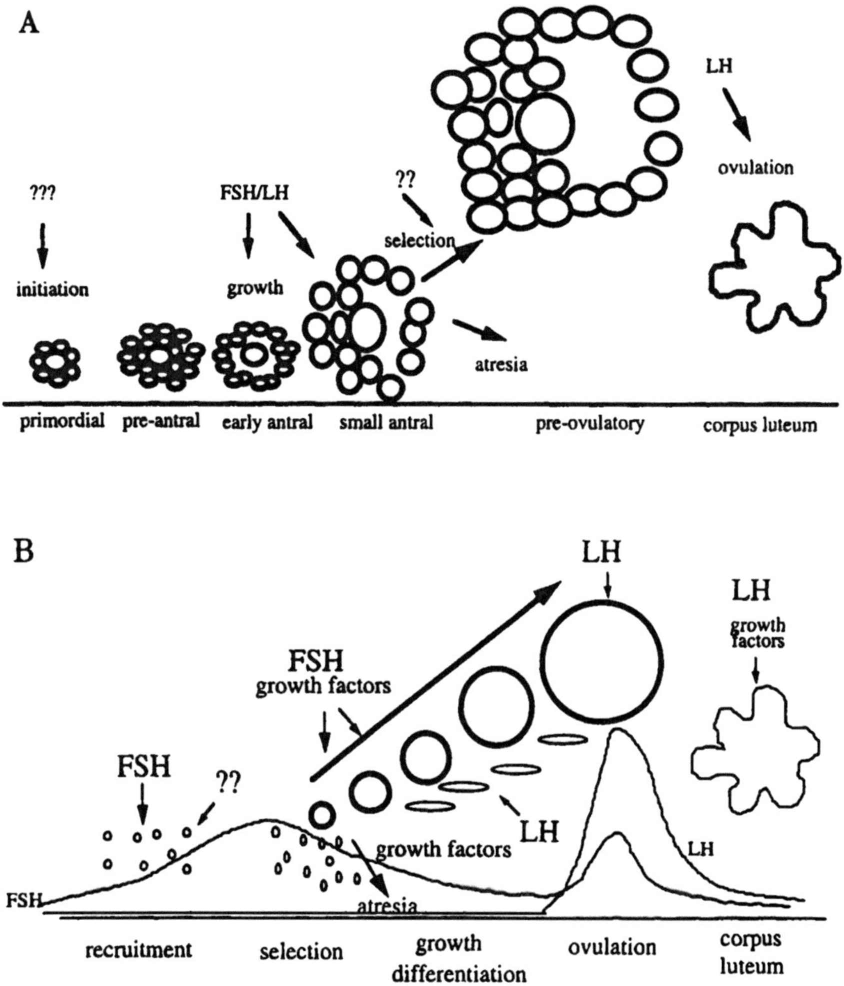

Figure 2. Schematic representations of: (A). gonadotropin-independent and -dependent stages of ovarian follicular development, and (B). complex interactions between gonadotropins and growth factors and the processes of ovarian follicle recruitment, selection, growth and differentiation, and ovulation, as well as corpus luteum function (see text).

(see below). The Type I IGF receptor, a classical growth factor receptor with tyrosine kinase activity, primarily mediates the actions of the IGFs at their target cells. The Type II IGF receptor is identical to the mannose-6-phosphate cation-independent receptor which shuttles lysosomal enzymes intracellularly and is believed to play a role in IGF-II turnover. It may have signal transduction properties by G-proteins, although this is not well established (Nissley and Lopaczynski, 1991). The IGFs circulate bound to IGFBPs which prolong their half-lives and also modulate (mostly inhibit) IGF action at target cells (Rosenfeld et al., 1990; Jones and Clemmons, 1995). The IGFBPs have high affinities for the IGFs, approximately two orders of magnitude higher than the IGF receptors, and posttranslational modifications of the IGFBPs, including phosphorylation, glycosylation, and proteolysis, are often required to increase IGF availability to their receptors (Lamson et al., 1993). IGFBPs may also have IGF-independent actions (Clemmons, 1993).

The pioneering work of Hsueh and colleagues (Davoren and Hsueh, 1986) using the hypophysectomized (hypox) rat model demonstrated that administration of GH resulted in increased ovarian immunoreactive IGF-I and provided the first evidence for a potential role for IGFs in the ovary. Subsequently, Murphy and colleagues found that IGF-I mRNA was expressed in third highest abundance in the rat ovary (Murphy et al., 1987). Adashi and colleagues, using the hypox/diethylstilbesterol (DES)-treated rat model (reviewed in Adashi 1991), demonstrated IGF production, action, and reception, and Hammond and colleagues have extensively investigated the role of this growth factor system in the porcine model (Hammond et al., 1993, review). Recent studies demonstrate important differences between human and animal ovarian IGF systems. For example, in the rat, IGF-I mRNA is expressed exclusively in granulosa, and IGF-II is expressed exclusively by theca, whereas in the human, IGF-II is the *major* IGF peptide in the ovary. A summary of the expression of the IGFs, their receptors, and their binding proteins in human ovary is presented in Table 1. Human preovulatory granulosa express IGF-II, but not IGF-I mRNA (Geisthovel et al., 1989a; Hernandez et al., 1992; El-Roeiy et al., 1993, 1994; Zhou and Bondy, 1993), and whether IGF-I mRNA is expressed at all in human ovary is controversial (El- Roeiy et al., 1993, 1994; Zhou and Bondy 1993). It is unlikely that GH effects on human ovary are mediated via IGF-I since human granulosa, which contain GH receptors (Carlsson et al., 1992), do not express IGF-I (Geisthovel et al., 1989a; Hernandez et al., 1992; El-Roeiy et al., 1993, 1994; Zhou and Bondy 1993) and human theca do not have GH receptors (Katz et al., 1993). Gonadotropins regulate the expression of IGF-II mRNA as well as secretion of IGF-II peptide by cultured human granulosa cells (Ramasharma and Li, 1987; Voutilainen and Miller, 1987). IGF receptor mRNAs are expressed in granulosa and theca in the human ovary (Hernandez et al., 1992; El Roiey et al., 1994; Zhou and Bondy, 1993), and specific binding sites have been demonstrated in granulosa cells (Gates et al., 1987; Balboni et al., 1987). *In vitro* the effects of primarily IGF-I have been investigated on human granulosa and theca, although the endogenous ligand *in vivo* is probably IGF-II. Most of the effects of the IGFs in human ovary are likely effected through the Type I receptor, although some actions of IGF-II may also be mediated via the Type II IGF receptor (Cataldo et al., 1993). IGF actions at the level of the ovary include mitogenic effects and augmentation of steroidogenesis (Table 2). IGF-I stimulates the synthesis of DNA in human granulosa and granulosa-luteal cells (Olsson et al., 1990; Angervo et al., 1991; Yong

Table 1. Summary OF IGFs, IGFBPs, and IGF Receptors (Ir,IIr) in Human Ovary[†]

	IGF-I	*IGF-II*	*IGFBP-1*	*IGFBP-2*	*IGFBP-3*	*IGFBP-4*	*IGFBP-5*	*IGFBP-6*	*Ir*
G	-/-	-/1+	-/-	3+/4+	-/-	2+/	1+/	-/	2+/3
T	2+/-	1+/1+	-/-	3+/4+	2+/	4+/	2+/	-/	-/2+
S	-/-	-/	-/-		-/-	4+/4+	4+/4+	-/	-/
V			-/-			/4+			
G	-/-	2+/	-/-	/4+					3+/3
T	2+/-	2+	-/-	/4+					-/2+
S	-/-		-/-					-/	-/
V			-/-					-/	
G	-/-	4+/4+	4+/	-/4+	3+/-	-/4+	-/-4+cu	-/	4+/4
T	-/-	-	-/-	4+/4+	3+/	2+/2+	2+/	-/	-/2+
S	-/-		-/-	-/	-/-	4+/3+	4+/4+	-/	-/
V				/4+	/4+	/4+		-/	
G	-	4+/4+	4+/						-/*
T	-		-/-						/
S	-	-/-							-/
V					/4+				

from Zhou and Bondy (1993), Hernandez et al. (1992) and El-Roeiy et al. (1993). Since there are discrepancies between the groups using s are reported from El Roeiy et al. and Zhou and Bondy. Data are presented as strong hybridization (4+) to weak hybridization (1+). If no e not reported. G=granulosa; T=theca; S=stroma; V=vascular endothelium; cu=cumulus. *Type I receptor mRNA expression present in rnandez et al., 1992).

Table 2. IGF Effects in Human Ovary

IGF-I
stimulates granulosa and granulosa-luteal DNA synthesis
stimulates estradiol secretion by granulosa and granulosa-luteal cells
synergizes with gonadotropins in augmenting estradiol and progesterone production by granulosa and granulosa-luteal cells
stimulates cytochrome P450 aromatase activity and mRNA in granulosa and granulosa-luteal cells
enhances DNA synthesis in theca
synergizes with LH in thecal androstenedione production
augments oocyte maturation
IGF-II
stimulates basal progesterone and estradiol secretion by cultured granulosa-luteal cells
stimulates granulosa proliferation *in vitro*

et al., 1992), and basal estradiol secretion by granulosa and granulosa-luteal cells is stimulated by IGF-I (Wood et al., 1994). In human granulosa, IGF-I also synergizes with FSH and hCG in augmenting estradiol (Erickson et al., 1989; Mason et al., 1993a) and progesterone (Erickson et al., 1991) production. In addition, basal progesterone and estradiol secretion by cultured human granulosa is significantly stimulated by IGF-II (Kamada et al., 1992; Kubota et al., 1993; Mason et al., 1994). IGF-II also stimulates human granulosa proliferation *in vitro* (Di Blasio et al., 1994). In granulosa cells from unstimulated preovulatory follicles and gonadotropin-stimulated follicles, IGF-I alone and synergistically with gonadotropins stimulates cytochrome P450 aromatase activity and mRNA (Steinkampf et al., 1988; Erickson et al., 1989, 1990; Christman et al., 1991; Bergh et al., 1991). There are also reports of IGF action in human theca cells and oocytes. In cultured human thecal cells, IGF-I enhances DNA and androgen synthesis (Hillier et al., 1991), and the spontaneous maturation of immature human oocytes can be augmented by IGF-I (Gómez et al., 1993).

In humans, follicular fluid (FF) from patients stimulated with gonadotropins contains IGF-I and IGF-II peptides (Ramasharma et al., 1986; Geisthovel et al., 1989b; Rabinovici et al., 1990a; Giudice et al., 1990a; van Dessel et al., 1995; Yap et al., 1995). In FF from normally cycling women, IGF-I levels were found to be higher in dominant versus cohort follicles (Eden et al., 1988). However, in a recent study using different methodology, IGF-I levels were not found to be statistically different in FF from estrogen- versus androgen-dominant follicles and were not correlated with follicular size (van Dessel et al., 1995). The latter study suggests that the origin of IGF-I in FF may be by transudation from serum (and, less likely, by local production in the theca). For IGF-II, FF levels are higher in estrogen-dominant compared to androgen-dominant follicles and correlate positively with follicle size and cycle day (van Dessel et al., 1995). These data suggest that FF IGF-II levels are derived primarily by local production in the granulosa (and perhaps the theca) with some contribution from the circulation as well. Recent studies further demonstrate no cyclic dependence of serum levels of IGF-I and IGF-II in normally cycling women, underscoring the importance of the local intra-ovarian IGF system (van Dessel et al., 1995). That circulating IGF-I is not requisite for normal ovarian follicular development in humans is supported by the findings of patients with Laron-type dwarfism. In these patients, where the GH receptor is absent or defective (Laron et al., 1966; Godowski et al., 1989), ovulation and fertility are not compromised and spontaneous conceptions have been reported (Laron, 1984; Lunenfeld et al., 1991; Menashe et al., 1991; Dor et al., 1992). Furthermore, normal follicular development in response to injectable gonadotropins, ovulation, and conception have been reported in a woman whose serum GH levels were markedly elevated, GH receptors deficient, and serum and FF IGF-I levels barely detectable (Lunenfeld et al., 1991; Dor et al., 1992). Serum IGF-II levels were about 25% of normal, however, FF IGF-II was not measured. These clinical observations suggest that circulating IGF-I is not essential for normal follicular development and that IGF-II is the major, bioavailable IGF in the human ovary. At the other extreme is the setting of elevated serum IGF-I levels. It is currently not known whether FF levels of IGF-I are elevated in acromegaly, in parallel with elevated serum levels commonly observed in this disorder. However, in normal women treated with pharmacologic doses of IGF-I, enhancement of the preovulatory LH surge has been reported (Duffy et al., 1993), suggesting a role for this peptide in the events surrounding ovulation.

Of six the IGFBPs which have been identified (Shimasaki and Ling, 1992), five have been detected in the human ovary (Zhou and Bondy, 1993; El-Roeiy et al., 1994). IGFBP-1, -2, and -3 have been detected in FF from gonadotropin-stimulated, luteinizing follicles (Seppala et al., 1984; Giudice et al., 1990a; Hartsthorne et al., 1990; Hamori et al., 1991a), whereas IGFBP-1 (Holly et al., 1990) and IGFBP-2, IGFBP-3, and IGFBP-4 (Cataldo and Giudice, 1992a) were detected in FF from cycling women. Luteinizing GC synthesize IGFBP-1 and IGFBP-2 *de novo* (Suikkari et al., 1989a; Cataldo et al., 1993) and secrete IGFBP-1, IGFBP-2, and IGFBP-3 (Jalkanen et al., 1989; Giudice et al., 1991; Hamori et al., 1991b; San Ramon and Magoffin, 1993). IGFBP-1 is expressed in granulosa cells of the dominant follicle following the LH surge, with expression of this IGFBP also present in the corpus luteum (El-Roeiy et al., 1994), and IGFBP-2, -3, -4, and -5 are expressed in human theca from small antral as well as dominant follicles (Table 1; Zhou and Bondy 1993; El-Roeiy et al., 1994). An important role for IGFBPs as modulators of IGF action in the ovary *in vivo* is likely, based on their action *in vitro*. In human granulosa, IGFBP-1 inhibits IGF-I stimulated ^{3}H-thymidine incorporation (Angervo et al., 1991). An inhibitory effect of IGFBPs on granulosa steroidogenesis has also been reported. IGFBP-1 and IGFBP-3 inhibit IGF-I-stimulated granulosa estradiol and progesterone production (Mason et al., 1992). A role for inhibitory IGFBPs in contributing to arrested follicular development has been postulated. The IGFBP profile in FF from normally cycling women is dependent on the functional status of the follicle (Table 3), with different profiles observed in estrogen-dominant (healthy, growing follicles) and androgen-dominant, atretic, or growth-arrested follicles (Cataldo and Giudice, 1992a). Androgen-dominant follicles (with low estrogen to androgen ratios) have high levels of IGFBP-2 and IGFBP-4 compared to healthy estrogen-dominant follicles. In support of these observations are additional reports of higher levels of IGFBP-2 in atretic, androgen-dominant follicles (San Roman and Magoffin, 1993), and *in situ* hybridization studies revealing abundant IGFBP-2 mRNA expressed in granulosa of atretic follicles (El-Roeiy et al., 1994). It is likely that elevated levels of inhibitory IGFBPs in androgen-dominant follicles in normally cycling women and in patients with PCOS (*vide infra*) decrease intrafollicular levels of bioavailable IGF peptides, thereby contributing to arrested follicular development in the ovary.

IGFBP secretion by granulosa is regulated (mostly inhibited) by gonadotropins, insulin-like peptides, and activin-A (Cataldo et al., 1993, 1994a; Mason et al., 1993a, 1993b; Chandrasekher et al., 1994). Control of IGFBP production may be one mechanism to regulate gonadotropin and IGF action in the ovary. Modulation of IGF action is further influenced by IGFBP proteases which affect the affinity of the IGFBPs for the IGFs (Table 3). IGFBP proteolysis was first suspected when western ligand blot analysis showed low levels of IGFBP-3 in human pregnancy compared to nonpregnancy serum and plasma (Hossenlopp et al., 1990; Giudice et al., 1990b), despite approximately twofold higher levels in pregnancy compared to nonpregnancy samples, determined by RIA (Baxter and Martin, 1986). Subsequent studies confirmed the presence of specific IGFBP proteases in a variety of systems (Lamson et al., 1993). In the ovary, proteases for IGFBP-4 and IGFBP-5 have recently been identified in the conditioned medium of rat granulosa (Erickson et al., 1994; Fielder et al., 1993). With porcine granulosa, regulation of IGFBP-3 by proteolysis occurs, and, furthermore, IGFs inhibit IGFBP-3 proteolysis in this model (Grimes and Hammond, 1994). In the human ovary, an IGFBP-4 protease is present in estrogen-dominant but not androgen-dominant FF (Chandrasekher et al., 1994). This protease activity may be one of several mechanisms in estrogen-dominant follicles which increase the levels of bioavailable IGF peptides for synergism with gonadotropins in steroidogenesis and follicle development.

The IGF system has also been implicated in follicular atresia (for review see Hsueh et al., 1994). Follicular atresia is associated with apoptosis and regular

Table 3. IGFBPs and IGFBP-4 Protease in Estrogen- and Androgen-Dominant Follicles in Human Ovary

IGFBP/protease	*FFe*	*FFa*
IGFBP-2,-4	⇓	⇑
IGFBP-4 protease	⇑	⇓

cleavage of nuclear DNA by endonucleases (Tilly et al., 1992a). *In vitro* this process is suppressed by IGF-I and gonadotropins and is enhanced by the presence of IGFBPs (Chun et al., 1994). The cellular mechanisms underlying this action by IGF-I are currently not known, and whether apoptotic mechanisms involving the IGF system operate in the human ovary remains to be investigated.

A model of gonadotropin, IGF, IGFBP, and IGFBP protease actions in human granulosa and theca is presented in Figure 3. In healthy, growing, estrogenic follicles, androgen from theca is used as substrate for gonadotropin- and IGF-stimulated aromatase activity for the production of estradiol. High bioavailable levels of IGFs are effected within the follicle by a combination of decreased production of IGFBPs (mainly IGFBP-2) and increased degradation by IGFBP-specific proteases (mainly IGFBP-4). Granulosa-derived IGFBPs are in turn regulated by IGFs, insulin, LH, and activin-A (*vide infra*). In androgen-dominant follicles, however, the amount of bioavailable IGFs is markedly decreased by high levels of inhibitory IGFBPs.

Other Growth Factor Systems and Cytokines

While the IGF system is one of the most extensively investigated growth factor systems in the human and nonhuman ovary, other growth factors systems have also

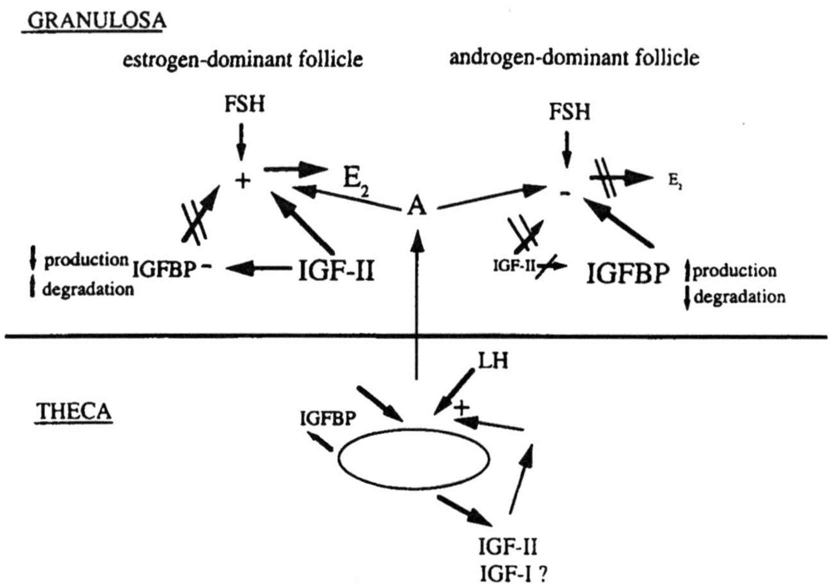

Figure 3. Model of gonadotropin, IGF, and IGFBPs in human granulosa and theca (see text).

received attention. The potential roles of members of the EGF FGF, PDGF, and TGF-β families are summarized in Table 4 and discussed in detail below. The data are consistent with EGF and FGF as mitogens to granulosa and a counter-regulatory relationship between TGF-β and EGF/TGF-α in induction of aromatase, FSH receptor mRNA, and FSH-induced LH receptors in granulosa. Considerable investigation has been pursued with regard to activin, inhibin, follistatin, and MIS, and this is also reviewed in detail below.

Epidermal growth factor family. The epidermal growth factor (EGF) "family" is comprised of EGF, transforming growth factor-α (TGF-α), heparin-binding EGF-like growth factor (HB-EGF), amphiregulin, betacellulin, and the pox virus growth factors (Carpenter, 1987; Massague and Pandiella, 1993; Watanabe et al., 1994). Members of the EGF family are synthesized as transmembrane glycoprotein precursors which are biologically active, and cleavage of the precursor represents switching between two active forms of the growth factor (Massague, 1990). The ability of membrane-bound pro-EGF to activate receptors on adjacent cells ("juxtacrine" action) may play a role in discrete cell-cell interactions such as between granulosa and thecal cells. The most intensively studied members of this family are EGF and TGF-α, due to the recent discoveries of the other members.

Table 4. Effects of EGF/TGF-α, FGF, PDGF, and TGF-β in the Ovary

EGF/TGF-α
stimulates mitosis of granulosa and granulosa-luteal cells
inhibits basal and FSH-induced aromatase activity in granulosa and granulosa-luteal cells
decreases FSH receptor mRNA expression in granulosa
decreases FSH-stimulated LH receptor acquisition by granulosa in culture
FGF
mitogenic to granulosa and granulosa-luteal cells
inhibits FSH-stimulated steroidogenesis in granulosa
decreases FSH-stimulated LH receptor acquisition by granulosa in culture
suppresses spontaneous onset of apoptosis
stimulates oocyte germinal vesicle breakdown
stimulates tPA
PDGF
enhances FSH-stimulated LH receptor acquisition by granulosa in culture
synergizes with FSH in progesterone production by granulosa
TGF-β
enhances FSH-stimulated LH receptor acquisition by granulosa in culture
enhances FSH-mediated cAMP and estradiol synthesis in granulosa
enhances FSH receptor mRNA expression in granulosa

EGF, a 6-kD peptide (Carpenter and Cohen, 1979), has diverse effects on granulosa cell function. It stimulates mitosis of human granulosa cells, especially those originating from small- and medium-sized follicles (Gospodarowicz and Bialecki, 1979) and granulosa-luteal cells (Tapanainen et al., 1987). EGF also affects steroidogenesis. It decreases human granulosa cell aromatase activity in vitro (Steinkampf et al., 1988) and FSH-induced aromatase mRNA expression in granulosa cells from normally cycling women (Mason et al., 1990). There is a more pronounced decrease in estradiol production by granulosa-luteal cells in response to EGF with increasing follicular size (Hurst et al., 1993). In granulosa-luteal cells, EGF stimulates the production of IGFBP-1, which inhibits IGF-I action in stimulating estradiol production (Angervo et al., 1992; Yap et *al.,* 1994). Likewise in cultured granulosa-luteal cells, EGF also stimulates progesterone production, an effect which is inhibitable by Mullerian-inhibiting substance (Kim et al., 1992). Immunoreactive EGF and EGF receptor have been detected in granulosa and theca internal cell layers beginning in the antral follicle and continuing through the corpus luteum, with both signals intensifying in the mid-luteal phase (Maruo et al., 1993), suggesting roles for these growth factors in corpus luteum function. In the preantral follicle immunoreactive EGF and EGF receptor have been observed only in the oocyte. In atretic follicles, EGF and EGF receptor immunostaining is negligible in granulosa, whereas theca interna show intense staining for both. EGF is also present in human FF (Westergaard and Andersen, 1989; Eden et al., 1988; Hofman et al., 1990; Volpe et al., 1991) and is likely derived from intraovarian sources as well as from serum.

TGF-α, a 5.5-kD peptide exhibits 30-40% amino acid homology with EGF and shares the same receptor with the latter (Derynck, 1986; Carpenter and Zendegui, 1986). Immunohistochemical studies have localized TGF-α in human ovarian tissue to oocytes in primary follicles, granulosa and theca cell layers, and small and large luteal cells (Chegini and Williams, 1992). In small follicles, granulosa cell layers immunostain at higher intensity than the surrounding theca, however, the intensity of the staining increases in both cell types with increasing follicle size. TGF-α is also reduced in the late luteal phase, compared to both early and mid luteal phases. A role for TGF-α and EGF in the periovulatory period has recently been suggested, based on studies on the human ovarian plasminogen activator system (Piquette et al., 1993), important in the process of ovulation. In addition, increased immunostaining for TGF-α in the mid-luteal phase cells suggests a potential luteotropic role for TGF-α in human luteal tissue (Chegini and Williams, 1992). The expression of TGF-α and its receptor in the same and neighboring cells during follicle maturation suggest auto and paracrine actions of this growth factor in the ovary.

Studies in other species reveal that TGF-α and EGF attenuate granulosa cell aromatase activity (Hsueh et al., 1984; Tsafriri et al., 1989; Adashi et al., 1987; Mondschein and Schomberg, 1981a;) and increase progesterone production (Hsueh et al., 1984; Jones et al., 1982). Both TGF-α and EGF also exert potent regulatory effects on granulosa and theca cell proliferation and differentiation (Knecht and Catt, 1983; May et al., 1988; Skinner and Coffey, 1988). Studies in the rat suggest a role for TGF-α and EGF in increasing the number of follicles developing to a periovulatory state by inhibiting apoptosis, and therefore, follicular atresia (Tilly et al., 1992b). TGF-α also appears to have a role in inhibiting gonadotropin-induced cytodifferentiation. TGF-α suppresses hCG-induced androstenedione and progesterone production by bovine thecal cells *in vitro* (Roberts and Skinner, 1991). TGF-α (and EGF) also attenuate FSH-stimulated aromatase activity *in vitro* in murine granulosa (Hsueh et al., 1981; Adashi and Resnick, 1986) and adult porcine theca cells (Caubo et al., 1989). A recent study demonstrates opposite regulation of FSH receptor mRNA and protein expression in the rat ovary by TGF-α and TGF-β (Dunkel et al., 1994). While TGF-β stimulates FSH receptor mRNA and protein expression in cultured murine granulosa cells, TGF-α decreases basal and TGF-β-stimulated FSH mRNA expression (Dunkel et al., 1994). Contrary to the above observations, a stimulatory effect of TGF-α on estrogen production by prepubertal porcine granulosa and thecal cells has been also observed (Gangrade et al., 1991). These contrasting observations may be due to differential regulation of the aromatase enzyme complex in adult versus prepubertal ovaries.

Evidence for the multiple modulatory roles for EGF and its related peptide, TGF-α, in other animal species and the presence of EGF, TGF-α, and EGF receptor in human ovarian tissue suggest that these peptides may be para- auto- or juxtacrine regulators of granulosa and theca cell proliferation, differentiation, and steroidogenesis. Given the cell type-specific coexpression of EGF and EGF receptor in various stages of follicular development an autocrine mode of action may exist to regulate oocyte maturation, follicular growth, and atresia in the human ovary. Messenger RNA transcripts encoding other members of the EGF "family"—amphiregulin and HB-EGF—along with transcripts for the EGF receptor, EGF, TGF-α have been detected in porcine corpora lutea (Kennedy et al., 1993). However, insufficient information is available regarding physiological roles for the newer members of the EGF family, and studies on their presence and potential actions in human ovary are wanting.

Fibroblast growth factors. The fibroblast growth factor (FGF) family consists of seven members: acidic FGF (aFGF), basic FGF (bFGF), int-2, hst-1/K-FGF, FGF-5, FGF-6, and KGF/FGF-7 (Baird and Klagsbrun, 1991; Basilico and Moscatelli, 1992). aFGF and bFGF, which exhibit 55% sequence homology, are the best known members of the family. There are two types of FGF receptors: high-affinity receptors with tyrosine kinase activity, thought to initiate signal transduction, and low-affinity heparin sulfate proteoglycans (Robinson, 1991). Alternative mRNA splicing gives rise to diverse forms of the high affinity receptor which may influence the biological effects of FGF (Robinson, 1991; Jaye et al., 1992). Much of the research on aFGF and bFGF in the ovary has been conducted in animal models, where act as autocrine and/or paracrine modulators of

FSH-mediated LH receptor synthesis, aromatase activity, and progesterone production (see Terranova, 1991, Giudice et al., 1992a for reviews). Bovine granulosa (Neufeld et al., 1987), corpora lutea (Stirling et al., 1991), and ovarian germinal epithelial cells (Gospodarowicz et al., 1989) produce bFGF, and in rat ovary expression of aFGF and bFGF mRNAs have been described (Koos and Olson, 1989; Koos and Seidel, 1989; Dissen et al., 1994). FGF receptors are expressed in rat granulosa and are inducible by gonadotropins and bFGF (Shikone et al., 1992). FGF may also have a role in modulating ovarian function during the periovulatory period. Rat granulosa cell tissue plasminogen activator (tPA) activity and tPA mRNA levels, important in ovulation, are stimulated by bFGF (LaPolt et al., 1990). Additionally, bFGF has been shown to induce rat oocyte germinal vesicle breakdown and stimulate prostaglandin E production (LaPolt et al., 1990). In the rat bFGF inhibits gonadotropin-mediated granulosa cell differentiation and steroidogenesis as well as FSH-mediated LH receptor acquisition in cultured granulosa cells (Gospodarowicz and Bialecki, 1978; Hsueh et al., 1984; Jones et al., 1982; Mondschein and Schomberg, 1981a; Baird and Hsueh, 1986; Hurwitz et al., 1990). Recent evidence points to another role for bFGF, possibly via an autocrine mechanism, in cultured rat granulosa cells—suppression of the spontaneous onset of apoptosis—suggesting a modulatory effect of bFGF on follicular atresia (Tilly et al., 1992b). With regard to the role of FGFs in the human ovary, the 18-kD bFGF peptide (Esch et al., 1985) is a known mitogen for human granulosa cells from small to medium-sized follicles (Gospodarowicz and Bialecki, 1979) and granulosa-luteal cells (Tapanainen et al., 1987; Wandji et al., 1992; Neufeld et al., 1987). Nevertheless, little is known about its role as an intraovarian modulator of follicular development and regression in humans.

Platelet-derived growth factor. Platelet-derived growth factor (PDGF) is a dimeric protein composed of two closely related A- and B- chain polypeptides (Heldin et al., 1985). These polypeptides can be combined as homodimers or heterodimers to give rise to three dimeric forms which bind to two distinct receptors, α and α (Ross et al., 1986). A potent mitogen for mesenchymally-derived cells (Heldin et al., 1985), PDGF enhances FSH-dependent LH receptor formation (Mondschein and Schomberg, 1981b) and synergizes with FSH in progesterone production in rat granulosa (Knecht and Catt, 1983; Terranova, 1991). PDGF and its two receptors are expressed in the stroma but not in the epithelium of normal tissue (Henriksen et al., 1993). In contrast, PDGF and its a receptor are localized to the epithelium and to a lesser extent in the stroma, while the β receptor is only detected in the stroma of neoplastic tissue. To date, its function in the human ovary remains to be defined.

Transforming growth factor-β. Transforming-growth factor-βs (TGF-βs), are homodimeric proteins that stimulate as well as inhibit cellular proliferation and differentiation (Sporn et al., 1985). In animal models, TGF-βs modulate steroidogenesis (Hernandez et al., 1990; Caubo et al., 1989; Magoffin et al., 1989), granulosa cell mitosis (Dorrington et al., 1988), oocyte cumulus mucification (Salustri et al., 1990; Magoffin et al., 1989), and oocyte germinal vesicle breakdown (Tsafriri et al., 1989; Magoffin et al., 1989; Table 4). In vitro studies on growth and differentiation of bovine thecal cells from various stages of antral follicle development suggest differential regulation of these processes by TGF-b and TGF-a (Roberts and Skinner, 1991). While TGF-b inhibits cell proliferation but stimulates steroidogenesis, TGF-a acts conversely. In humans, mRNAs encoding TGF-b1 and TGF-b2 are expressed in granulosaluteal and oocyte-cumulus cells (Mulheron et al., 1992), and by immunocytochemistry, TGF-b has been observed in granulosa and theca cell layers of developing follicles (Chegini and Williams, 1992). The intensity of TGF-b immunostaining increases with follicular size, suggesting not only a proliferative but also a differentiative role for TGF-b in human follicular development. Immunoreactive TGF-b is present in small and large luteal cells, however, the intensity of the signal is reduced in the late luteal phase (Chegini and Williams, 1992). The immunolocalization data in the human ovary suggests that TGF-bs may act as regulators of granulosa and theca cell differentiation after they have been stimulated to proliferate by other growth factors. TGF-bs appear to be a counter-regulatory growth factors to the effects of the EGF/TGF-a system in the ovary. However, a precise role for granulosa-derived TGF-b in human follicular development remains to be identified.

Activin. Activin is a homodimeric (ββ) member of the TGF-β superfamily originally isolated as a stimulator of pituitary FSH release, in contrast to the related, heterodimeric inhibin (αβ). Two closely related β subunits, βA and βB, encoded by separate genes, give rise to activins A (βAβA), AB (βAβB), and B (βBβB; Ying, 1988). Expression of activin βA and βB subunit mRNA by ovarian granulosa cells of several species has been shown by both northern and *in situ* hybridization analysis (Woodruff et al., 1987; Schwall et al., 1990; Roberts et al., 1993), and in the human ovary, activin protein subunits and dimers have been localized by immunocytochemistry to granulosa and luteal cells (Yamoto et al., 1992; Rabinovici et al., 1992; Roberts et al., 1993). In primate follicles, inhibin/activin α, βA, and βB mRNAs are differentially expressed: small antral follicles express predominantly βB, while large, healthy antral follicles express both α and βA (Schwall et al., 1990; Roberts et al., 1993). Dimeric activin-A production by human granulosa cells *in vitro* is stimulated by cyclic AMP (Rabinovici et al., 1992). Development of immunoassays for activin in serum, tissue fluids, and cell-conditioned medium (CM) has been confounded by the presence of activin-binding proteins as well as by cross-reactivity with inhibin. Recently developed double-antibody sandwich immunoassays for activins A and B have shown that levels in the serum of normal women are 0.35-6.5 and 0.4-10.4 ng/mL, respectively (Wong et al., 1993). Detailed in vitro studies of the regulation of activin secretion by granulosa cells await the validation of an immunoassay for activin in CM.

Two types (II and IIB) of specific, high-affinity receptors for activin have been identified by expression cloning (Mathews and Vale, 1991; Attisano et al., 1992), and both have been localized by *in situ* hybridization to granulosa cells and oocytes in the rat (Cameron et al., 1994). Similar analysis did not reveal expression of either receptor type in the human ovary (Roberts et al., 1994), although the potency of activin on human granulosa cells is suggestive of their presence (Rabinovici et al., 1992; Cataldo et al., 1994b). The effects of recombinant human activin-A (Schwall et al., 1988) on granulosa and theca cell function have been studied; a diagrammatic representation of the actions of activin (and inhibin) is presented in Figure 4. In numerous *in vitro* studies, activin-A modulates granulosa cell steroidogenesis in a

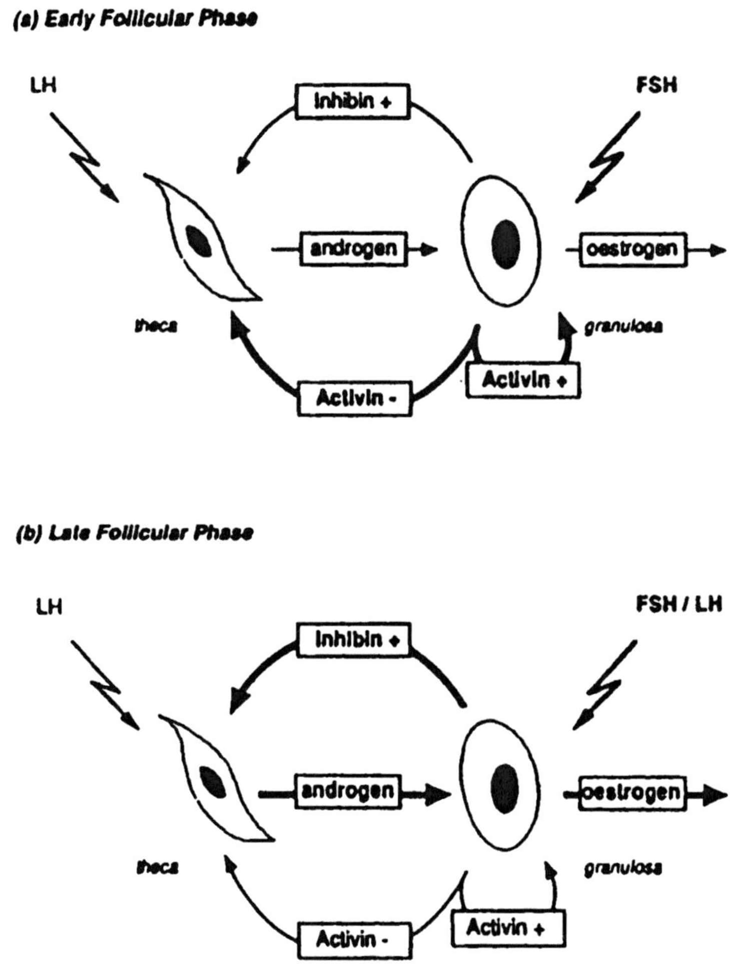

Figure 4. Schematic of regulatory functions of activin and inhibin in the control of follicular estradiol production (see text). Adapted, with permission, from Hillier and Miro (1993).

fashion dependent on the state of follicular differentiation. In granulosa cells from antral follicles prior to the ovulatory LH surge, activin-A stimulates aromatase activity, FSH receptor expression, and inhibin production (LaPolt et al., 1989; Findlay, 1993; Hillier and Miró, 1993). In luteinized granulosa and luteal cells, activin-A inhibits progesterone and estrogen production (Findlay, 1993; Rabinovici et al., 1992; Li et al., 1992; Brannian et al., 1992). Activin-A also promotes proliferation of luteinizing human and immature rat granulosa cells (Rabinovici et al., 1990b; Li et al., 1995). Activin A stimulates production of IGFBP-2 and IGFBP-4 by cultured human granulosa-luteal cells (Cataldo et al., 1994a), suggesting that it may decrease steroidogenesis and direct the human follicle toward atresia in part by antagonizing the actions of the IGFs through increased IGFBP production. Activin acts on theca cells to inhibit androgen production (Hsueh et al., 1987; Hillier and Miró, 1993) and on rat oocytes to induce resumption of meiotic maturation (Sadatsuki et al., 1993). The ovarian effects of activin have also been studied *in vivo*. In immature, DES-treated rats, intrabursal activin-A accelerates follicular atresia (Woodruff et al., 1990), although similar results were not obtained in cycling adult rats (Woodruff et al., 1993a). In rhesus monkeys, activin-A administered systemically leads to the demise of the dominant follicle, but allows new follicular recruitment after its withdrawal (Stouffer et al., 1993).

Inhibin. Inhibin is a 32 kDa heterodimer isolated as a gonadal inhibitor of FSH production (Ying, 1988), which can act as an antagonist to some but not all the actions of activin. Inhibin is produced by granulosa cells and corpora lutea in every species examined. Its production is positively regulated by gonadotropins (Tsonis et al., 1987; Hillier and Miró, 1993), and α-subunit expression increases as the dominant follicle matures (Schwall et al., 1990). Measurement of circulating levels of inhibin by immunoassays which employ a single antibody directed against the α subunit, including the widely used Monash heterologous assay (McLachlan et al., 1986), is confounded by the presence in serum of cross-reacting free a subunit forms (Schneyer et al., 1990). New double-antibody sandwich assays for inhibin (Woodruff et al., 1993b; Groome et al., 1994) have revealed undetectable levels in post-menopausal women and levels near 5 pg/mL in the early and mid-follicular phase, 40-50 pg/mL near ovulation, and 50-65 pg/mL in the mid-luteal phase. Much higher levels are observed during ovarian hyperstimulation with exogenous gonadotropins and in pregnancy, the latter likely the result of placental inhibin production (Groome et al., 1994; Lambert-Messerlian et al., 1994). The effects of recombinant human inhibin-A on the ovary have been studied. On the granulosa cell, inhibin has no effect on steroidogenesis and does not modulate the effect of activin-A (Rabinovici et al., 1992; Hillier and Miró, 1993; Cataldo et al., 1994a). In the theca, however, inhibin stimulates androgen production and antagonizes the effect of activin (Hsueh et al., 1987; Hillier and Miró, 1993). Inhibin also inhibits the meiotic maturation of oocytes (Oh et al., 1989). While inhibin binding to granulosa cells has been observed (Woodruff et al., 1990), a receptor for inhibin distinct from activin receptors has not been identified.

A 57 kDa precursor of the inhibin α subunit found in human follicular fluid (Robertson et al., 1990) is suspected of being an FSH receptor antagonist (Schneyer et al., 1991). The biologic significance of this observation remains uncertain, but a similar FSH receptor inhibitor has been reported in the serum of women with premature ovarian failure (Sluss and Schneyer, 1992).

Follistatin is a high-affinity activin-binding protein with multiple isoforms of mol wt 32-39 kDa originally isolated from follicular fluid as an inhibitor of pituitary FSH release distinct from inhibin (Ying, 1988; Nakamura et al., 1990). By ligand blotting analysis, follistatin has one binding site for the activin β subunit and thus can bind either one activin or one inhibin molecule. Each activin molecule can bind two molecules of follistatin (Shimonaka et al., 1991). The association of activin with follistatin is of similar or greater affinity than that of activin with its receptors; activin associated with follistatin is essentially irreversibly bound (Schneyer et al., 1994). As predicted from this observation, follistatin antagonizes activin in every system examined, including ovarian granulosa and theca cells. In cultured granulosa cells, follistatin reverses the effects of activin on steroidogenesis, FSH receptor induction, and inhibin production (Findlay, 1993). Full antagonism requires at least a twofold molar excess of follistatin over activin (Cataldo et al., 1994b). Similar antagonism between follistatin and activin was observed in theca cells (Shukovski et al., 1993). Inhibin binds to follistatin with 1,000-fold weaker affinity than does activin (Schneyer et al., 1994), consistent with its actions as primarily an antagonist of activin, not inhibin. Follistatin is produced by granulosa cells of every species examined. Follistatin expression is greater in healthy, dominant human follicles than in smaller antral follicles, and it is also expressed in the corpus luteum (Roberts et al., 1993). In cultured granulosa and granulosa-luteal cells, follistatin mRNA expression and protein secretion are stimulated by FSH or LH (Shimasaki et al., 1989; Saito et al., 1991; Tuuri et al., 1994). Serum follistatin levels are increased in women undergoing ovarian hyperstimulation but do not differ in the follicular and luteal phases of a spontaneous menstrual cycle, in postmenopausal women, and in men, suggesting that the ovary is not a significant source of circulating follistatin (Sugawara et al., 1990; Gilfillan and Robertson, 1994). Follistatin may serve as a local modulator of activin action in the healthy, gonadotropin-responsive follicle and corpus luteum, preventing down-modulation of thecal androgen and granulosa-luteal progesterone production, while failing to block the stimulatory effect of inhibin to increase thecal production of aromatizable androgen (Hillier and Miró, 1993).

Müllerian inhibiting substance. Müllerian inhibiting substance (MIS; also called anti-Müllerian hormone) is another member of the TGF-β superfamily. MIS was described initially as the testicular Sertoli cell factor responsible for regression of the Müllerian duct in the male fetus (reviewed in Lee and Donahoe, 1993). Subsequent studies have localized MIS to rat and human ovarian granulosa cells in follicles from the preantral through preovulatory stages, but not in primordial or atretic follicles or corpora lutea (Whitman and Pantazis, 1991; Hirobe et al., 1994). MIS has been found in human follicular fluid and shown to act as an oocyte meiosis inhibitor, and possibly a granulosa cell mitosis inhibitor as well (Lee and Donahoe, 1993; Kim et al., 1992). A receptor for MIS has been cloned and localized to granulosa cells in the rabbit (diClemente et al., 1994). The role of MIS in the adult ovary remains speculative.

Nerve growth factor family. Nerve growth factor (NGF), brain-derived neurotrophic factor (BDNF), neurotrophin-3 or hippocampus-derived neurotrophic factor (NT-3/HDNF), and neurotrophin-4/5 (NT-4/5) are members of the neurotrophin family of growth factors (Maness et al., 1994). These polypeptides promote the differentiation and survival of select neurons of the central and peripheral nervous systems (Barde, 1989; Thoenen, 1991). NGF synthesis has been demonstrated in immature rat ovary (Lara et al., 1990a), primarily in the thecal-interstitial cells of the prepubertal ovary (Ojeda and Dissen, 1995). Development of ovarian sympathetic innervation is NGF dependent (Lara et al., 1990b), and immunoneutralization of NGF actions results in retardation of follicular growth, blunting of gonadotropin-mediated steroidogenesis, and irregular estrous cycles (Lara et al., 1990b). NT-3 mRNA-expressing cells have been detected in the epithelium of secondary and tertiary but not primary follicles in the adult rat ovary leading to the speculation that NT-3 may also be involved in oocyte development and maturation (Ernfors et al., 1990). BDNF mRNA, however, has not been detected in the ovary (Ernfors et al., 1990).

With regard to receptors for this class of growth factors, a low affinity receptor known as p75-low affinity NGF (p75 NGF) receptor and high affinity trk receptors bind NGF and the other neurotrophins (Bothwell, 1991). Expression of the mRNA species for the low affinity NGF receptor has been documented in the developing rat ovary (Dissen et al., 1991). Immunoreactive NGF receptor is detected in thecal cells of antral and preantral follicles, as well as granulosa cells, albeit at levels lower than those found in theca. A significant decrease in receptor mRNA levels following the luteinization of thecal cells after the first ovulation is observed and suggests that NGF receptor expression may be linked to gonadotropin-mediated follicle differentiation (Dissen et al., 1991). In fact, more recent experiments positively correlate NGF receptor content loss with the degree of follicular atresia (Ojeda and Dissen, 1995). The presence of NGF receptors in these steroid-producing cells and their differential expression lend support to the concept of a direct involvement of neurotrophins in the regulation of ovarian follicular development. While the complementary patterns of neurotrophin and receptor expression described herein suggests a possible inductive role for neurotrophins in development of the rat ovary, their roles in human ovary remain to be elucidated.

Cytokines. Recent studies have contributed to an understanding of the role of cytokines as regulators of ovarian function. Cytokines comprise a group of proteins ranging in molecular weight from 6,000 to 60,000 that modulate a variety of cellular processes. In this section the roles of interleukins, interferons, and TNF-α in the human ovary are reviewed. These cytokines have been identified in cellular components of the ovary and in follicular fluid. The human corpus luteum is a rich source of cytokine-producing cells, including macrophages, T lymphocytes, and monocytes (Wang et al., 1992a). Messenger RNAs encoding interleukin-1 (IL-1), its receptor, and the IL-1 receptor antagonist are expressed in human ovary (Hurwitz et al., 1992). Interleukins-1, -2, and -6, and TNF-α are present in human follicular fluid (FF) obtained from gonadotropin-stimulated cycles (Barak et al., 1992a; 1992b; Buyalos et al., 1992a; Punnonen et al., 1992; Wang et al., 1992b). Production of IL-1 and -6 by normal ovarian epithelium has also been demonstrated (Lidor et al., 1993, Ziltener et al., 1993). Detectable levels of these cytokines, as well as the correlation of cytokine levels with steroid levels, support a regulatory role for these factors in ovarian function. Interleukin-6 levels measured in human FF are up to 30-fold higher than in serum (Buyalos et al., 1992a). Terranova et al., (1991) reported that atretic follicles contain greater levels of TNF-α than do estrogenic follicles. Similarly, lower serum levels of estradiol and progesterone are measured in the presence of detectable levels of TNF-α (Punnonen et al., 1992). Contrasting views exist about correlations between the interleukins and steroid levels. Positive correlations between human FF IL-2 and testosterone levels (Barak at el., 1992a) and between IL-1 and progesterone levels (Barak et al., 1992b) have been reported, whereas others (Wang and Norman 1992) have found no correlation between levels of interleukin-1, interleukin-2, progesterone, and estradiol in FF.

The potential roles for cytokines in the ovary, based on the data currently available, are presented in Table 5. Cytodifferentiative effects of IL-1 have been reported in human ovarian cells (Sjogren et al., 1991; Wang et al., 1991; Fukuoka et al., 1992), as well as effects on steroidogenesis. For example, in human granulosa cells, interferon-γ inhibits hCG-stimulated progesterone and FSH-stimulated estradiol secretion (Ijzermans and Marquet, 1989). Similarly, IL-1 and TNF-α inhibit FSH- or hCG-stimulated estradiol secretion by granulosa-luteal cells (Fukuoka et al., 1992; Barak et al., 1992c). TNF-α also inhibits hCG-stimulated progesterone production by granulosa-luteal cells from IVF patients (Best et al., 1994). These studies cumulatively suggest an inhibitory role for cytokines in gonadotropin-stimulated steroidogenesis and suggest that they may act as counter-regulatory

Table 5. Potential Roles for Cytokines in Human Ovary Based on *In Vitro* Studies with Human Granulosa Cells

- Inhibition of gonadotropin-stimulated steroidogenesis
- Modulators of ovulation via effects on prostaglandins and plasminogen activator/inhibitor system

factors to other growth factors that act as co-gonadotropins (*vide supra*). A role for IL-1β and TNF-α in the process of ovulation has been proposed based on their effects on granulosa prostaglandin (PG) production. Treatment of granulosa-luteal cells with interleukin-1β or recombinant TNF-α results in dose-dependent increased production of both PGE_2 and $PGF_{2\alpha}$ (Watanabe et al., 1993; Wang et al., 1992b), suggesting that these cytokines play a modulatory role in the process of ovulation. In addition, Polan and colleagues (1994) have observed that late follicular phase granulosa cells do not express IL-1 protein, whereas granulosa-luteal cells from IVF patients do. Interleukin-1β increases expression of mRNAs encoding IL-1β, IL-1 receptor type 1, plasminogen activator (PA) inhibitor-1, and PA inhibitor-2 in granulosa-luteal cells (Piquette et al., 1994). These data support a role for interleukin-1β in human ovarian remodeling that occurs during the periovulatory period. Studies in species such as the rat, rabbit, cow, and pig also support a modulatory role for cytokines in ovarian folliculogenesis, cellular differentiation, steroidogenesis, ovulation, and corpus luteum regression (reviewed in Ben-Rafael and Orvieto, 1992)

B. Intraovarian Peptides and Abnormal Ovarian Physiology

Polycystic Ovarian Syndrome (PCOS)

Ever since the first description by Stein and Cohen, 1935), polycystic ovary syndrome (PCOS) has received wide-spread attention as the most common cause of anovulation in women (Hull, 1989). It is a heterogeneous syndrome characterized by hyperandrogenism, persistent anovulation, oligomenorrhea or amenorrhea, hirsutism, elevated LH levels, and the accumulation of small cysts in the ovaries (Yen, 1980; Franks et al., 1988; Fauser, 1994). Many women with PCOS are obese and insulin-resistant, and hyperinsulinism has also been found in lean PCOS patients (Burghen et al., 1980; Chang et al., 1983; Dunaif et al., 1989, 1992). There is also a subset of rare syndromes of extreme insulin-resistance which all have in common hyperandrogenism, insulin-resistance, and polycystic ovaries (Poretsky, 1991; and Poretsky and Peiper, 1994 for reviews). At the level of the ovarian follicle, the initial stages of development, that is, recruitment and growth to the small antral stage, are not impaired in PCOS patients, although selection of a dominant preovulatory follicle does not occur. This leads to the accumulation of many small antral follicles in the ovary (Fauser, 1994). Ovulation can be induced in some patients with clomiphene citrate, although about 30% of PCOS women are clomiphene-resistant and require gonadotropins for ovulation induction. Resistance to exogenously administered gonadotropins is also often observed clinically. Although the subject of extensive research, basic underlying mechanisms of PCOS are still largely unknown. Since follicle growth is regulated by feedback mechanisms, the primary disturbance may be either at the hypothalamic-pituitary or ovarian level. Initial theories favored a primary ovarian abnormality as the cause of disturbed follicle growth in PCOS, based upon histologic evaluation and early reports of aromatase abnormalities in this disorder (Axelrod and Goldzieher, 1962; Goldzieher and Axelrod, 1963). However, subsequent observations, in particular, the presence of elevated circulating LH (high LH:FSH ratios) in many PCOS patients suggested a hypothalamic-pituitary dysfunction in the pathogenesis of this disorder (Yen, 1980). Recently, the ovary has again become the primary focus of investigation, as more information has been gleaned regarding intraovarian growth factors that co-regulate or mediate gonadotropin actions. Serum levels of bioactive FSH (Fauser et al., 1991) and follicular fluid levels of FSH in anovulatory women with PCOS are within the normal, preovulatory range (Fauser, 1994; Erickson et al., 1992). Also, granulosa cells of PCOS patients show normal (Erickson, et al., 1979, 1990; Mason et al., 1990) or even elevated (Mason et al., 1994) FSH-induced aromatase activity *in vitro*. Since FSH stimulation and granulosa aromatase activity are normal when PCOS granulosa are removed from the PCOS ovarian environment, it has been postulated that locally produced growth factors or inhibitors are involved in arrested follicular growth in women with PCOS. Such factors are likely to lead to a higher follicular FSH threshold (Brown, 1978), with the consequence of relative gonadotropin resistance and anovulation. Both elevated LH (acting on theca and increasing ovarian androgen production) and hyperinsulinism are believed to contribute to PCOS, and the potential contributions of several intraovarian peptides acting as co-gonadotropins in the pathogenesis of PCOS are reviewed herein.

Insulin-like peptides. IGFs stimulate granulosa cell proliferation and aromatase activity (*vide supra*), processes which are characteristically absent in the PCOS follicle *in vivo*. The lack of stimulation of aromatase may be due to intrinsic defects in PCOS granulosa aromatase, inadequate levels of intrafollicular IGFs or FSH, or inhibition of IGF and/or FSH action within the PCOS follicle. Follicular fluid in PCOS follicles contains physiological levels of both FSH (Fauser, 1994; Erickson et al., 1992) and IGF-I (Eden et al., 1988, 1990). Although PCOS granulosa cells produce little aromatase *in vivo*, they respond to FSH and/or IGF-I *in vitro* in a normal (Erickson, et al., 1979; Mason et al., 1990) or even elevated (Mason et al., 1994) fashion. Thus, the defect in PCOS is not a lack of stimulating hormones per se or a decreased responsiveness of the granulosa to IGFs (or FSH) with regard to aromatase production. Therefore, it has been postulated that locally active inhibitors of IGFs and/or FSH in follicular fluid of small antral PCOS follicles block stimulation of aromatase. IGFBPs are likely candidates. Indeed, IGFBP profiles in PCOS follicles are very similar to androgen-dominant, small antral (atretic) follicles obtained from regularly cycling women. There are high levels of IGFBP-2 and IGFBP-4 (Cataldo and Giudice, 1992b; San Roman and Magoffin, 1992; Schuller et al., 1993), in marked contrast to the near absence of these IGFBPs in estrogen-dominant follicles, detected by ligand binding techniques. *In situ* hybridization studies support these observations (El-Roeiy et al., 1994). In addition, the IGFBP-4 protease is not detectable in the PCOS follicle (Yap et al., 1995), similar to small, androgen-dominant follicles in normally cycling women and in marked contrast to healthy, estrogen-dominant follicles (Chandrasekher et al., 1995). In view of the mostly inhibitory actions of IGFBPs on IGF action, these observations are consistent with this class of molecules being one of several groups of inhibitors of ovarian follicular development in the pathogenesis of PCOS (Figure 5).

An association between hyperinsulinism and hyperandrogenism is well established (Conway et al., 1993; Dunaif et al., 1989; Jialal et al., 1987; Poretsky and PEiper, 1994). *In vitro* theca-interstitial cells obtained from women with hyperandrogenism release significantly more androgen per mg tissue than do those from non-hyperandrogenic women (Barbieri et al., 1986). Since the insulin and Type I IGF receptors are structurally similar (Poretsky et al., 1985) and insulin cross-reacts with the Type I IGF receptor (Nissley and Lopaczynski, 1991), it has been postulated that excessive insulin action on the theca in insulin-resistant women with PCOS, resulting in hyperandrogenism, occurs via the Type-I IGF receptor in this compartment of the ovary (reviewed in Poretsky, 1991). However, levels of insulin needed to cross-react at the Type I IGF receptor in theca would have to be several fold higher than those observed clinically in women with known hyperinsulinism.

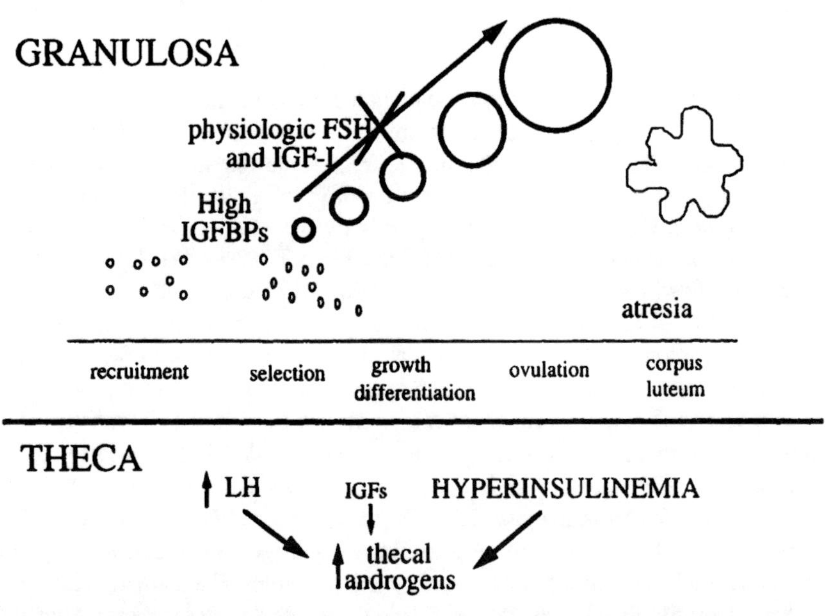

Figure 5. Models of IGF system in the pathogenesis of arrested follicular development observed in patients with PCOS and role of LH and insulin-like peptides in the pathogenesis of hyperandrogenemia associated with this syndrome (see text).

Several possibilities exist to explain the paradox of the ovary remaining sensitive to insulin, while other tissues are resistant (for reviews see Poretsky, 1991; Giudice, 1992). It is possible that there is an altered form of the insulin receptor or that a hybrid insulin/type I IGF receptor exists in the ovary, or alternatively, ovarian insulin receptors may be differently regulated compared to insulin receptors in other tissues (Poretsky et al., 1988, 1990). Also, post-receptor defects may also explain the different effects of insulin on glucose metabolism and steroidogenesis (Dunaif et al., 1992; Ciaraldi et al., 1992; Sun et al., 1991). Why there is not excessive stimulation of the granulosa, which are richly endowed with the Type I IGF receptor, have abundant substrate for aromatase, and bathe in physiologic FSH and IGF levels, is not certain at this time, although restriction of IGF bioavailability by high levels of intrafollicular inhibitory IGFBPs may contribute in part to this resistance to aromatase stimulation (vide supra). Thus, both elevated LH and hyperinsulinism and perhaps IGFs lead to hyperandrogenism observed in this disorder (Figure 5). In addition to direct action on the ovary, altered levels of SHBG can contribute to increased bioavailable androgens in the circulation.

With regard to the IGF system and its potential role in the pathogenesis of hyperandrogenism in women with PCOS, elevated circulating levels of IGF-I may have effects similar to those proposed for hyperinsulinemia. Most reports, however, demonstrate that serum IGF-I levels in PCOS patients are in the normal range (Lanzone et al., 1990; Homburg et al., 1992). An alternative approach to elevate IGF bioactivity is through changes in IGFBP concentrations. Since IGFs in serum are bound to a large extent to IGFBPs, and IGFBPs mostly have inhibitory effects on IGF actions, decreased IGFBP serum concentrations may lead to increased bioavailability of IGF within serum and thus in the ovary. Lower IGFBP-1 levels have indeed been reported in PCOS patients compared to controls (Pekonen et al., 1989; Suikkari et al., 1989b; Lanzone et al., 1990; Homburg et al., 1992), likely a consequence of hyperinsulinism. However, serum levels of the main circulating carrier of IGFs, IGFBP-3, are the same in normo-ovulatory women and women with PCOS (Cataldo and Giudice, 1992b; El-Roeiy et al., 1991). In addition, total and free circulating levels of IGFs have been reported to be higher in PCOS than in normo-ovulatory women in one study (Iwashita et al., 1990) but not others (Conway et al., 1990; Suikkari et al., 1991; Homburg et al., 1992). Higher levels of bioavailable IGFs (in serum or locally in the theca and stroma) may synergize with LH in stimulating thecal androgen production, resulting in hyperandrogenism, characteristic of this syndrome. High levels of free IGFs, however, may not synergize with FSH in stimulating granulosa aromatase activity, due to high intrafollicular levels of IGFBPs in the androgen-dominant PCOS follicles (Cataldo and Giudice, 1992b). Whether higher levels of free IGFs are present in the circulation or in the ovarian microenvironment (theca and stroma) of PCOS patients, compared to normo-ovulatory women, remains controversial and is worthy of further investigation.

Other Growth Factor Families

EGF, FGF, PDGF, and TGFβ. Several studies suggest that EGF may be one of several potential inhibitors in PCOS follicles responsible for the blunted granulosa response to gonadotropins (Franks and Mason, 1991). EGF levels in follicular fluid are higher in PCOS patients compared to normally cycling women (Volpe et al., 1991; Franks et al., 1988), although similar levels have also been reported (Eden et al., 1990). Also, EGF has been shown to antagonize FSH-induced aromatase activity in granulosa cells obtained from PCOS patients (Mason et al., 1990). However, since EGF levels in follicular fluid of regularly cycling controls correlate linearly with serum levels despite steroid level gradients, passive diffusion into the follicle from the serum rather than local production seems to occur (Westergaard and Andersen, 1989). Although these findings have not been confirmed in PCOS follicles, they support a counter-regulatory role for EGF to the IGFs and other co-gonadotropins in both normal and abnormal follicular development in humans.

Although experimental data in animals and *in vitro* studies using human granulosa cells show that FGF and PDGF can modulate FSH-induced LH receptor synthesis and aromatase activity (Esch et al., 1985; Heldin, et al., 1985), no data on the role of these peptides in ovarian pathophysiology are available. TGF-β has been considered a possible candidate as inhibitor of local FSH action (Franks and Mason, 1991). TGF-β increases IGFBP-3 production by granulosa cells of PCOS follicles (San Roman and Magoffin, 1992). Addition of FSH, together with TGF-β, causes an attenuation of the TGF-β stimulation to levels similar to untreated PCOS granulosa cells, suggesting that the production of IGFBP might be regulated by FSH and TGF-β. However, naturally occurring levels of TGF-β in PCOS follicles, compared to controls have not been reported, and the pathophysiological significance of TGF-β for abnormal follicle development remains to be determined and presents an exciting and challenging area for future research.

Inhibin. Because of the low FSH-to-LH ratios in serum of PCOS patients, an overproduction of inhibin by ovarian follicles has been suggested in this syndrome (Tanabe et al., 1983). Increased inhibin production in PCOS follicles has been reported (Tanabe et al., 1983), although others found no difference in inhibin levels between PCOS and small preantral cohort follicles (Pache et al., 1992). An increased overall production of inhibin due to a higher number of follicles in PCOS subjects has also been implicated (Tanabe et al., 1990). However, inhibin levels in PCOS are found to be in the early follicular range (Falcone et al., 1991; Lambert-Messerlian et al., 1994). A study into basal and gonadotropin-stimulated levels of inhibin in PCOS patients showed no evidence of a primary defect in ovarian inhibin physiology in PCOS patients (Buckler et al., 1988). One potential explanation for the equivocal data is that the inhibin immunoassay used most widely in the human is a heterologous assay with an antiserum that exclusively recognizes the inhibin subunit. The available data to date suggest a limited role of inhibin in disturbed follicle growth in PCOS.

Cytokines. Little is known about the relation between cytokines and PCOS. Treatment of PCOS patients with dexamethasone and controlled ovarian stimulation results in lower follicular fluid levels of TNF-α compared to those found in PCOS patients not treated with dexamethasone (Zolti et al., 1992). In view of the observations that greater levels of TNF-α are present in atretic compared to estrogenic follicles (*vide supra*), dexamethasone may in part overcome atresia by decreasing intrafollicular TNF-α levels. While cytokines likely play a role in the process of follicular growth, steroidogenesis, and corpus luteum function, little is known about their potential roles, specifically in the pathophysiology of PCOS.

Treatment. In PCOS, treatment of hyperinsulinemia often results in decreased hyperandrogenemia, whereas the opposite does not usually occur (for review see Poretsky and Peiper, 1994). These observations suggest that insulin-like peptides play a role, albeit unknown, in the pathogenesis of hyperandrogenism observed in PCOS. With regard to anovulation, treatment with antiestrogens (e.g., clomiphene citrate or tamoxifen) or exogenously administered gonadotropins are usually effective in inducing ovulation. Clinical problems associated with the use of gonadotropin therapy include multiple follicular development, ovarian hyperstimulation syndrome, and multiple gestation (Fauser, 1994). Treatment regimens have been devised that include a slow, low-dose regimen (Hamilton-Fairley et al., 1991; Shoham et al., 1991) and also a decremental dosing regimen (Fauser, 1995; van Santbrink et al., 1994, 1995), both of which have been used with success in achieving monofollicular development. When mechanisms underlying normal follicular development, atresia, and corpus luteum function and regression are understood, it is likely that the absence of selection in the face of apparently normal early follicular recruitment and development in PCOS will be understood and amenable to alternative therapies.

Other ovulatory disorders. Premature ovarian failure occurs in about 2-5% of women (Cohen and Speroff, 1991). It is characterized by elevated gonadotropins and low estradiol levels in the circulation and by gonadotropin resistance at the level of the ovary. While accelerated rates of atresia account for a small percentage of women with premature ovarian failure (as in Turner's mosaicism), it is most often associated with autoimmune disorders or is idiopathic. Perhaps included in the etiology of premature ovarian failure is an FSH receptor antagonist, similar to the precursor of the inhibin α subunit (Schneyer et al., 1991; Sluss and Schneyer, 1992). Potential roles of cytokines and growth factors in the pathogenesis of premature ovarian failure are largely unexplored, although their roles in follicular atresia in animal systems has recently been extensively reviewed (Hsueh et al., 1994).

II. OVIDUCT

Cyclic changes in the oviduct occur presumably to assist gamete and embryo trafficking at the appropriate time, as well as for embryo nourishment. The critical role of this tissue, beyond that of a conduit, is underscored by higher pregnancy rates in advanced reproductive technologies that include embryo or gamete transfer into the oviduct, compared to direct transfer into the uterine cavity (Martin, 1993). This section reviews the limited data that are available on growth factors in the oviduct, which may fulfill paracrine roles in gamete trafficking and pre-implantation embryonic development.

In human and animal models, relatively few studies have examined growth factor and growth factor receptor expression in oviductal epithelium, and functional studies are wanting. With regard to IGFs, IGF-I and IGF-II are present in oviductal fluid and in media conditioned by oviductal cells (Wiseman et al., 1992; Lee et al., 1992; Diehl et al., 1994). Oviductal cells express mRNAs encoding IGF-I, IGF-II, and both IGF receptors, as well as IGFBPs (Williams and Butler, 1991; Kincy et al., 1992; Andrews et al., 1994, Giudice et al., 1992b, 1993a, Carlsson et al., 1993; Watson et al., 1994). Immunohistochemical studies have confirmed the presence of IGF-I and the Type I IGF receptor in human, mouse, and rat oviductal epithelium (Andrews et al., 1994; Carlsson et al., 1993; Pfeifer and Chegini, 1994), and IGF-I mRNA and protein are differentially expressed in the late proliferative phase of the cycle (Giudice et al., 1993a; Pfeifer and Chegini, 1994). The data suggest that IGF-I may mediate estradiol action in this tissue, as it appears to do in uterine endometrium (*vide infra*).

EGF, TGF-α, their common receptors, and their respective mRNAs are expressed in oviductal epithelium in several species, including humans (Lei and Rao, 1992, Morishige et al., 1993; Andrews et al., 1994; Iguchi et al., 1993; El-Danasouri et al., 1993; Kennedy et al., 1994; Chegini, et al., 1994a). Other members of the EGF family have not been reported in oviduct. With regard to other growth factors, TGF-β1,β2, and β3 and the TGF-β type II receptor and their mRNAs are expressed in mouse, porcine, and human oviduct (Andrews et al., 1994; Zhou and Chegini, 1994, Buhi et al., 1994). Basic fibroblast growth factor (bFGF) has also been immunolocalized in the mouse oviduct (Wordinger and Palmer, 1994). Whether the growth factors that have been detected in oviductal tissue have any effect on synchronous beating of oviductal epithelial cilia, for example, to effect gamete trafficking is unknown. In addition, potential paracrine interactions between the cleaving pre-embryo during its journey through a growth factor-rich oviductal environment are largely unexplored. However, *in vitro* co-culture systems using mouse or human embryos and oviductal epithelium as substratum (reviewed in Giudice and Saleh, 1995; Bongso et al., 1993) suggest the importance of maternal-embryonic interactions in preimplantation embryonic development. Whether these interactions are mediated via growth factors remains a challenge for future investigation and will potentially enhance development of human embryos cultured *in vitro* for treating refractory infertility, since currently most synthetic media are devoid of growth factors and development is accelerated in co-culture systems.

III. ENDOMETRIUM

A. Normal Endometrial Physiology and Function

Uterine endometrium (Figure 1) is a dynamic tissue whose cellular components proliferate during the aptly named proliferative phase and undergo differentiation and some proliferation in the secretory phase of the menstrual cycle. It is comprised of a variety of cell types, including glandular and surface epithelium, stroma, vascular endothelium and smooth muscle cells, and a variety of immune cells (reviewed in Giudice and Ferenczy, 1995). The endometrium's mission is to provide a receptive environment to the conceptus. This includes appropriate apposition of the embryo to the surface epithelium, implantation through the epithelium, and invasion into the endometrial stroma and maternal vasculature. It must protect the conceptus from rejection by the maternal immune system and provide a nourishing environment. In the absence of implantation, a series of events is initiated, prompted by waning corpus luteum function, that permits orderly tissue shedding, efficient hemostasis, and tissue regeneration. These changes in endometrium are regulated by steroid hormones and growth factors and related peptides which coregulate or mediate the effects of the gonadal steroids on this tissue. While growth factors are found in many tissues, and endometrium is no exception, what is striking about endometrium is variation in levels of several growth factors and their receptors in this tissue throughout the menstrual cycle. Recently, growth factors in human endometrium have been extensively reviewed (Giudice, 1994), and this information is summarized below and in Figure 6 and Table 6.

Insulin-like Growth Factors

The IGFs are believed to be one of several growth factor systems that stimulate endometrial cellular mitosis and differentiation during the menstrual cycle and in early pregnancy. In rats estradiol increases uterine IGF-I mRNA expression by about 20-fold (Murphy et al., 1987a, 1987b; Norstedt et al., 1989; for review see Giudice, 1994). Further support for estrogen regulation of IGF-I in rat uterus comes from studies with estrogen receptor (ER) antagonists. For example, tamoxifen, a partial ER antagonist, up-regulates IGF-I gene expression in the rat uterus, whereas the complete ER antagonist, ICI 182780, down-regulates it (Huynh and Pollak, 1993). In the mouse, estradiol induces IGF-I gene expression primarily in endometrial epithelium, whereas progesterone induces it in endometrial stroma (Kapur et al., 1992). In humans, IGF-I and IGF-II are mitogenic to decidualized endometrial stromal cells. IGF-I mRNAs and protein are expressed exclusively in the stroma, primarily in late proliferative and early secretory phases of the cycle (Giudice et al., 1993b, Zhou et al., 1994; Tang et al., 1994). In view of the temporal expression of IGF-I (Figure 6a), it has been hypothesized that it may mediate the mitotic actions of estradiol in this tissue (Giudice et al., 1993b; Zhou et al., 1994). The IGF-II gene is expressed abundantly in mid-late secretory endometrium and early pregnancy decidua, and is expressed exclusively in stroma (Giudice et al., 1993b; Zhou et al., 1994), suggesting a role for it in tissue remodeling at the time of menstrual shedding or during early pregnancy. Messenger RNAs encoding the Type I and Type II IGF receptors are more abundantly expressed in the secretory phase and during early pregnancy, than in the proliferative phase of the cycle, and the Type I receptor is primarily expressed in glandular epithelium. Thus, with IGF peptides expressed in stroma and the primary signal-transducing receptor expressed primarily in the glands, it has been postulated that IGFs act by autocrine and paracrine interactions as mitogens and differentiative peptides in this tissue (Giudice et al., 1993b; Zhou et al., 1994).

IGFBP-1 is a *major* secretory protein of decidualized endometrium and is known by several names, including placental protein 12 (PP12) and α_1-progestin-associated endometrial globulin (α_1-PEG; reviewed in Seppala et al., 1991). Its mRNA is exclusively expressed in secretory compared to proliferative endometrium, and it is abundant in decidua. It is regulated in secretory endometrium and decidualizing endometrial stromal cells by progesterone, insulin-like peptides, and other growth factors (reviewed in Giudice, 1994). It is likely that IGFBP-1 regulates IGF action in endometrium and perhaps the trophoblast, since it inhibits IGF-I binding to

Table 6. Potential Roles for Growth Factors and Cytokines in Endometrium

proliferation		implantation	
	IGF-I		IGFBP-1
	EGF family		HB-EGF
	PDGF		TGF-β
	FGF		IL-1β
	INF-γ		CSF-1
			LIF
differentiation		vasoconstriction/dilatation	
	IGF-II		ETs
			Prostanoids
			NO
angiogenesis		hemostasis	
	FGF		tissue factor
	VEGF		

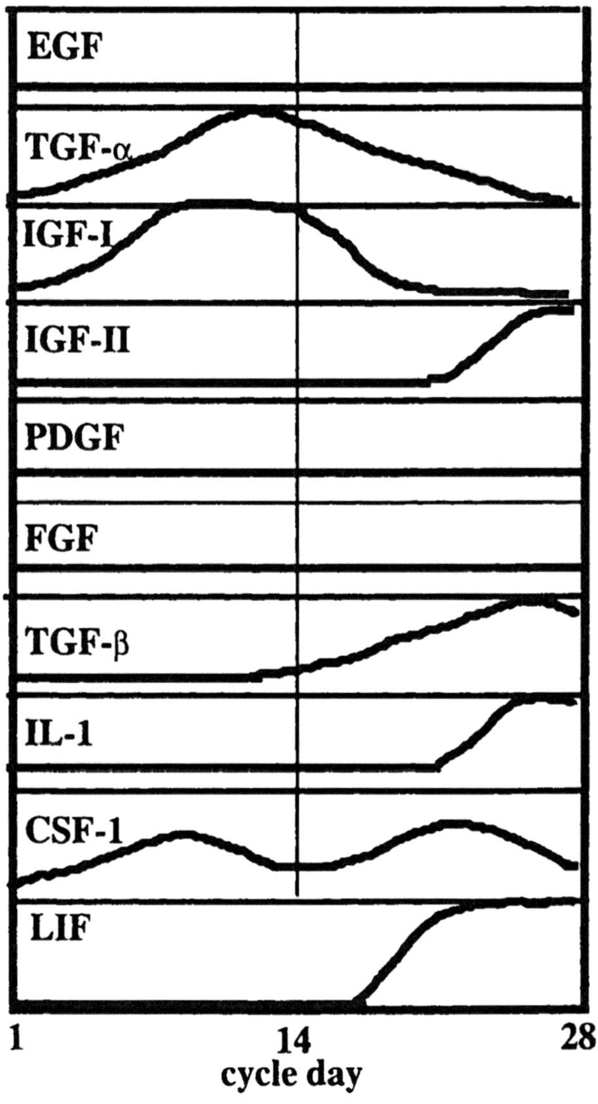
EGF
TGF-α
IGF-I
IGF-II
PDGF
FGF
TGF-β
IL-1
CSF-1
LIF
1
14
28
cycle day

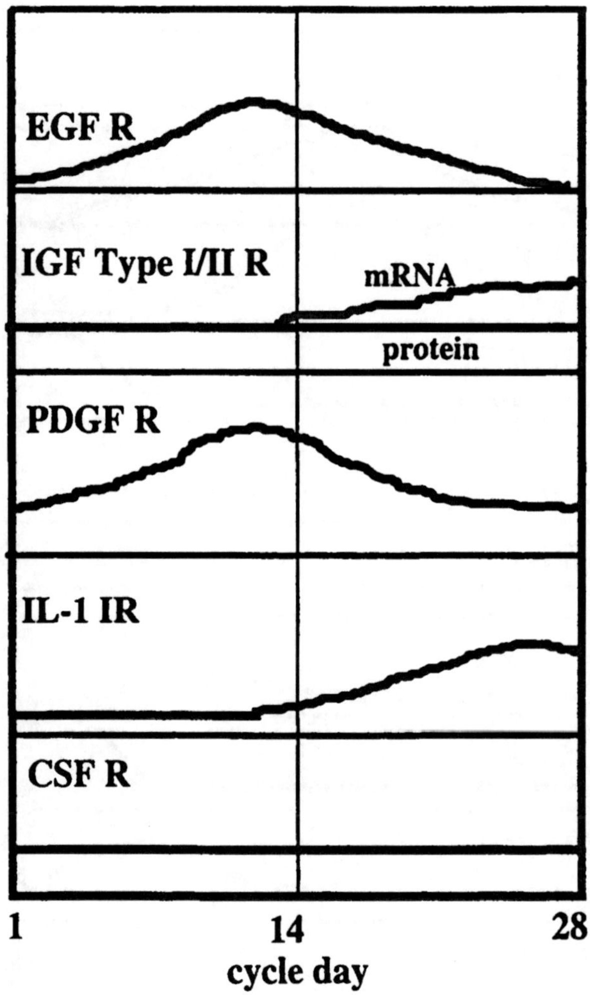

Figure 6. Relative amounts of growth factor (6a) and growth factor receptor (6b) mRNA and protein levels in human endometrium during the menstrual cycle. Levels illustrated are relative to levels during the early proliferative phase. (Reprinted with permission from Giudice, 1994.)

endometrial and trophoblast type I IGF receptors (Rutanen et al., 1988; Ritvos et al., 1989). Also, IGFBP-1 stimulates cell migration and binds to the fibronectin receptor (Jones et al., 1993), suggesting that this protein may influence intra-endometrial cell-cell and endometrial-trophoblast interactions, as well as cell-extracellular matrix interactions. Recent studies demonstrate that human cytotrophoblast invasion into decidualized endometrial stromal cultures *in vitro* is inhibited in the presence of IGFBP-1 (Irwin and Giudice, 1996). These observations suggest an important role for this major decidual product in implantation. Serum levels of IGFBP-1 do not change during normal menstrual cycles (Suikkari et al., 1987), although they increase in gonadotropin-stimulated cycles (Pekonen et al., 1992). Whether the ovary and/or the endometrium contributes to circulating IGFBP-1 (in addition to the liver) is not certain, and IGFBP-1 levels may prove useful as a marker of endometrial receptivity in gonadotropin-stimulated cycles. Further studies to evaluate this potential application are needed. During pregnancy, circulating levels of IGFBP-1 increase about 100-fold, and in amniotic fluid are about 1,000-fold higher in concentration than in pregnancy serum (for review see Seppala et al., 1991). Of clinical relevance is the observation that fetal birthweight is inversely proportional to IGFBP-1 levels in maternal serum (Hall et al., 1984), suggesting that high levels of this protein may bind available IGF peptide and decrease its availability for placental and fetal growth. It is likely that maternal serum or amniotic fluid IGFBP-1 levels may serve as adjunctive markers for fetal growth assessment in the future.

Epidermal Growth Factor Family

In humans, TGF-α is primarily present in the proliferative phase of the cycle, with levels diminishing after ovulation, whereas EGF mRNA and peptide are expressed in endometrium and do not exhibit cyclicity (Figure 6a). Binding studies reveal that the number of EGF/TGF-a receptor sites peaks at the end of the proliferative phase (just before ovulation), abruptly decreases thereafter, and reaches a nadir just before the next menses (Figure 6b). EGF actions on cultured human endometrial epithelial and stromal cells suggest that EGF, either alone or with estradiol, stimulates proliferation of both cell types *in vitro*. In addition, it is likely that EGF/TGF-a synergizes with other growth factors in its stimulation of endometrial cellular proliferation and possibly differentiation (for review see Giudice, 1994). Heparin-binding EGF-like growth factor (HB-EGF), amphiregulin, and betacellulin are recently identified members of the EGF family that also mediate their effects via the EGF receptor (Das et al., 1994; Johnson et al., 1993; Watanabe et al., 1994). Since TGF-α deficient mice exhibit normal implantation (Bruce-Mann et al., 1993; Luetteke et al., 1993), and since EGF mRNA is not detectable in the mouse uterus on the day prior to implantation (Huet-Hudson et al., 1990), the endogenous ligands that interact with the EGF receptor during normal endometrial development and implantation are, as yet, undefined. Recent evidence, however, indicates that HB-EGF may play a significant role in embryo-uterine interactions. The HB-EGF gene is expressed in the mouse uterine luminal epithelium just prior to implantation, and HB-EGF also promotes blastocyst development, hatching from the zona pellucida, and trophoblast outgrowth *in vitro* (Das et al., 1994). It has been postulated that HB-EGF of epithelial origin promotes blastocyst hatching *in vivo*, and while this hypothesis is attractive, it has yet to be proven. The expression and role of other members of the EGF family in the female reproductive tract and implantation await further clarification.

Platelet-derived Growth Factor

PDGF-AB has been localized in human endometrium to stromal cells and does not change with menstrual cycle stage (Figure 6a; for review see Giudice, 1994) In contrast, PDGF receptors are present in greater amounts in the proliferative compared to the secretory phase (Figure 6b). *In vitro* PDGF stimulates endometrial stromal cell proliferation. Since endometrial stromal cells are a source of PDGF, it is likely that it acts by autocrine mechanisms on the stroma and/or by paracrine mechanisms on neighboring epithelium.

Fibroblast Growth Factor

Basic fibroblast growth factor (FGF) is an angiogenic protein that is highly mitogenic for capillary endothelial cells *in vitro* and can induce angiogenesis *in vivo* (reviewed in Giudice, 1994). While it exhibits steroid-dependence in endometrial adenocarcinomas, in normal endometrium of cycling women FGF is present in high concentrations, but no steroid (cycle)-dependence has been observed (see Figure 6). Expression of other members of the FGF family have not been explored. In addition, little is known about FGF receptors in endometrium, although FGF stimulates endometrial stromal mitosis. Thus, FGF may be one of several growth factors acting by autocrine mechanisms on stroma and/or paracrine mechanisms between glandular epithelium, stroma, and vascular endothelium.

Transforming Growth Factor-β

The temporal expression of various TGF-β isoforms in specific uterine cells suggests that these growth factors have specific roles during the peri-implantation period. In the mouse uterus, TGF-β1 is primarily expressed in the luminal and glandular epithelium in the preimplantation period and to the uterine decidua in the postimplantation period (Tamada et al., 1990). TGF-β2 is synthesized by epithelial, myometrial, and decidual cells and its secretion may be regulated by estradiol, whereas TGF-β3 is primarily synthesized by myometrial cells, independent of ovarian steroids (Das et al., 1992). In humans, TGF-β mRNA is equally distributed in endometrial glands and stroma, and levels are lowest in the proliferative phase, increase twofold in the secretory phase, and about fivefold in decidua. The cycle dependence of TGF-β mRNA expression suggests that it may contribute to inhibition of cellular proliferation during periods when endometrial cellular differentiation predominates. In addition, a role for TGF-β in implantation has also been proposed (reviewed in detail in Lala and Graham, 1990; Strickland and Richards, 1992). Since TGF-β promotes differentiation of human cytotrophoblasts into noninvasive syncytiotrophoblasts, TGF-β of decidual origin may be an inhibitor of trophoblast invasion. The invading trophoblast secretes a variety of proteases (Lala and Graham 1990), including plasminogen activators, and since TGF-β induces plasminogen activator inhibitor (PAI) mRNAs, it has been postulated that TGF-β may act to control trophoblast invasion by regulating proteases and their inhibitors in the embryo and maternal decidua. Also, since the latent form of TGF-β is activated by plasmin, it is possible that maternal decidua stores latent TGF-β in the extracellular matrix, waiting for the invading trophoblast and its production of plasmin. By these mechanisms TGF-β could then increase PAIs and limit trophoblast invasion.

Cytokines

Cytokines are multifunctional proteins that promote cellular proliferation and differentiation, induce cell adhesion molecules and chemotactic agents, and exert their effects on cellular morphology. Their presence, regulation, and potential role(s) in human endometrium have recently been reviewed (Tabibzadeh, 1991; Simon and Polan, 1995).

Interleukins. Interleukin-1α (IL-1α) is expressed in human endometrial epithelial, endothelial, lymphoid, and stromal cells (reviewed in Giudice, 1994). IL-1β mRNA is expressed in the late secretory phase of the menstrual cycle. Plasma levels of IL-1β vary during the cycle, peaking in the secretory phase, but the source of elevated plasma levels remains unclear. IL-1 stimulates PGE_2 production by endometrial epithelial cells, and it has been suggested that this cytokine may play a role in menstruation. IL-1 also induces endometrial stromal cell secretion of IL-6, another cytokine. A recent study has shown that administration of the IL-1 type I receptor antagonist to mice prevents embryo attachment, and therefore, implantation (Simon et al., 1994). The precise mechanism for the antagonist's action is uncertain, although it is likely that the roles of the interleukins in endometrial cellular proliferation, induction of other cytokines, and their roles in implantation will unfold in the near future.

Colony stimulating factor-1. Colony stimulating factor-1 (CSF-1) is expressed in endometrial epithelium, primarily in mid-proliferative and mid-secretory phases compared to the rest of the cycle (Figure 6; reviewed in Giudice, 1994). With the onset of pregnancy, CSF-1 mRNA, and protein increase several fold in the decidua and are expressed in glandular epithelium and endothelium The receptor for CSF-1, encoded by the proto-oncogene c-*fms*, is not cycle-dependent in endometrium.

However, during pregnancy, its protein and mRNA levels in decidual glandular epithelium are highest during the first trimester for review see Giudice, 1994). In addition, placenta also expresses mRNAs encoding the growth factor and its receptor, which increase during gestation. It is likely that CSF-1 plays a central role in endometrial cellular function, as well as in regulating the endometrial macrophage population and placental growth and function during the first trimester of pregnancy.

Interferon-γ. Interferon-γ (INF-γ) is produced primarily by activated T lymphocytes. It is present in human endometrium, and INF receptors are present in endometrial epithelium and are not cycle-dependent. INF-γ induces HLA-DR molecules on endometrial epithelium which are most abundantly adjacent to lymphoid aggregates (Tabibzadeh, 1988). In addition to these effects, INF-γ inhibits proliferation of endometrial epithelium. Since activated T cells in lymphoid aggregates are adjacent to the glands in the basalis layer, and since activated T cells produce INF-γ, it has been proposed that the low proliferative activity of the basal epithelium may be due to the paracrine action of this cytokine (Tabibzadeh, 1988). Thus, INF-γ may be involved in regulating epithelial proliferation as well as epithelial-lymphocyte interactions.

Leukemia inhibitory factor. Leukemia inhibitory factor (LIF) has been extensively studied in the rodent. In gene knock-out studies, homozygous female mice deficient in LIF have normal ovulation and fertilization of oocytes, but the embryos remain in the oviduct, unable to implant (Stewart, 1994). The endometrium remains nondecidualized, and transfer of these embryos to a wild type pseudopregnant female results in implantation and pregnancy. Administration of recombinant LIF to LIF-deficient mice restores fertility and implantation (reviewed in Stewart, 1994). In human endometrium LIF expression is cycle dependent (Kojima et al., 1994; Stewart, 1994; Figure 6). However, unlike in the mouse where expression is limited to day 4 of pregnancy and is co-incident with the pre-implantation surge of estradiol, in humans LIF expression appears in endometrial epithelium at about the middle of the secretory phase of the cycle and remains expressed throughout the remainder of the phase (Figure 6). This is one of several growth factors that is believed to participate in endometrial cellular function and embryonic-decidual interactions, and precise mechanisms underlying these potential functions will likely unfold in the near future.

Angiogenic, Vasoactive, and Hemostatic Factors

Growth of the vascular elements in the endometrium is a major part of the proliferative phase and especially the mid-secretory phase of the menstrual cycle, in preparation for contact with an implanting blastocyst. In addition, vasoconstriction is an important process to prevent hemorrhage at the time of the menses, and if pregnancy ensues, vasodilatation is necessary to maintain adequate blood flow to the uterus and endometrium. In addition, endometrial hemostasis is important during early invasion of the maternal vasculature by the fetal trophoblast. Vascular epithelium and surrounding cells produce a variety of factors that likely participate in these mechanisms (Table 6).

Angiogenic factors. Basic (b) FGF is an angiogenic protein that is highly mitogenic *in vitro* for capillary endothelial cells and can induce angiogenesis *in vivo*. Its potential role as an angiogenic growth factor in endometrium has been discussed above. Vascular endothelial cell growth factor (VEGF) is also an angiogenic factor which is also a potent mitogen for these cells (Charnock-Jones et al., 1993). Endometrial glands and stroma express VEGF, and estradiol stimulates VEGF transformed endometrial glandular epithelium (Charnock-Jones et al., 1993). These data have led to the hypothesis that gonadal steroids regulate VEGF in non-endothelial cells in endometrium which can then act on resident endothelial cells by paracrine actions. Recent data show that VEGF is cell- and cycle-dependent in human endometrium and is regulated by estradiol (Shifren et al., 1996).

Vasoconstrictors/Vasodilators

Endothelins. Endothelins (ET-1, ET-2, and ET-3), potent vasoconstrictors derived from endothelial cells, interact with two unique receptors on vascular smooth muscle and endothelium. PreproET-1 mRNA is differentially expressed in menstrual compared to proliferative or secretory endometrium (Economos et al., 1992; O'Reilly et al., 1992), and specific binding sites for ETs are expressed on endometrial vascular endothelium and glandular epithelium (Davenport et al., 1991). It has been suggested that ET-1, derived from endometrial stromal and/or epithelial cells, act on the adventitial surface of neighboring spiral arterioles in the endometrium and promote vasoconstriction, required for menstrual shedding. This potent vasoconstrictor is tightly regulated at the levels of synthesis and degradation. ET-1 is degraded by enkephalinase, a membrane metalloendopeptidase, that is regulated by progesterone in endometrium (Head et al., 1993; Casey et al., 1991). High levels of this enzyme during the time of implantation suggests that low levels of bioactive ET-1 will be present during this time of the cycle when vasoconstriction is ill advised. Prostaglandins have both vasoconstrictive and vasodilatory properties and are highly regulated in endometrium. They have been extensively discussed elsewhere (Giudice and Ferenczy, 1995).

Nitric oxide. Endothelium-derived relaxing factor is released by vascular endothelial cells and has been identified as nitric oxide (NO). This gas is a potent vasodilator, perhaps more important than prostacyclin in maintaining vascular tone. Preliminary studies have shown the presence of a vascular relaxant in rabbit uterine perfusate, which is believed to be NO (Chaudhuri and Nathan, 1992). Identification of NO in human endometrium and the potential role of NO in the maintenance of vascular tone in the nonpregnant and pregnant uterus await further investigation.

Tissue factor. Tissue factor is a membrane-bound protein on non-endothelial cells that initiates coagulation upon direct contact with blood. In human endometrium during the secretory phase and in early pregnancy, tissue factor is present in stroma and is stimulated by progestins (Lockwood et al., 1993). It has been postulated that tissue factor from perivascular decidual cells may promote endometrial hemostasis during trophoblast invasion of the maternal endometrial vasculature (Lockwood and Schata, 1996).

B. Potential Roles of Growth Modulators in Abnormal Endometrial Physiology

Table 6 summarizes the potential roles of growth factors and cytokines in cyclic development and shedding of the endometrium and implantation and potential interactions with the developing preimplantation embryo. The endometrium (and the oviduct) produces a variety of growth modulators that may not only be self-serving for physiologic events during the menstrual cycle, but may also be important in nourishing an embryo as it traverses the oviduct and play roles in embryo-maternal interactions (Figure 7). Most studies conducted on growth modu-

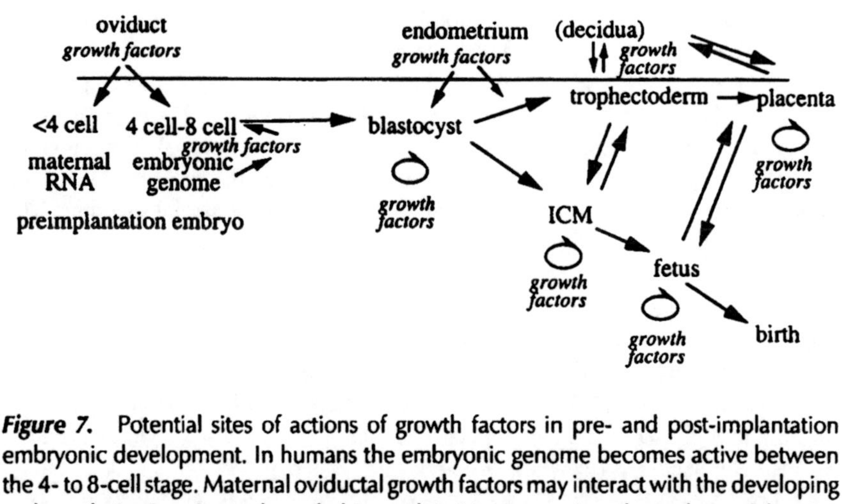

Figure 7. Potential sites of actions of growth factors in pre- and post-implantation embryonic development. In humans the embryonic genome becomes active between the 4- to 8-cell stage. Maternal oviductal growth factors may interact with the developing embryo during its sojourn through this conduit. Upon entry into the endometrial cavity, the blastocyst may respond to growth factors by autocrine and/or paracrine mechanisms (from the maternal endometrium [decidua]). After differentiation into the trophectoderm and the inner cell mass (ICM), subsequent placental and fetal development may occur by autocrine and paracrine growth factors. (Reprinted from Giudice and Saleh, 1995, with permission.)

lators have focused on normal or neoplastic endometrial and oviductal tissue and have used animal models for embryonic research. It is anticipated that when mechanisms underlying normal embryonic and endometrial/oviductal development are elucidated, then reproductive disorders reflecting abnormal development, including blighted ova, repetitive miscarriage, dysfunctional uterine bleeding, luteal phase deficiency, and unexplained infertility, for example, will be effectively treated with rationally designed therapies.

IV. ENDOMETRIOSIS

A. Clinical Significance

Human endometrium contains a variety of growth factors, growth factor receptors, and, in some instances, growth factor binding proteins, whose presence and cycle-dependence suggest potential roles for them as cellular mitogens and mediators of differentiation throughout the menstrual cycle and during implantation (*vide supra*). Growth factors and cytokines may also play a role in the pathogenesis and progression of growth of ectopic endometrium in endometriosis (Figure 8), a

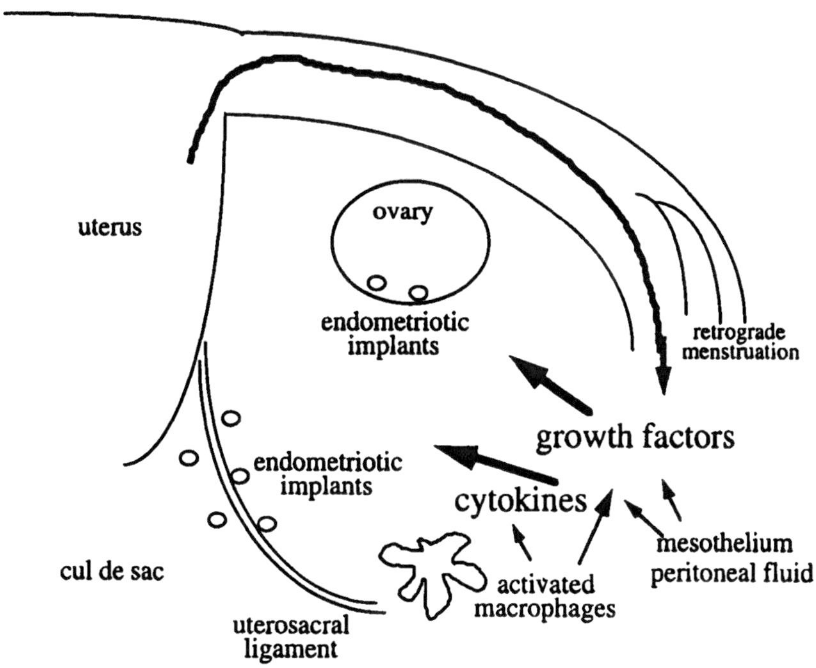

Figure 8. Model of growth factors and cytokines in growth of ectopic endometrium in the disorder of endometriosis.

benign gynecologic condition associated with often incapacitating pelvic pain and refractory infertility. This common gynecologic disorder affects up to 10% of women in their reproductive years and up to 50% of women with infertility (see Mahmood and Templeton, 1990, for review). There is a genetic predisposition in some women (Simpson et al., 1980). Although it has been intensively studied, the pathogenesis of this disorder is not fully understood. In women with endometriosis, endometrial cells attach to and proliferate and differentiate on the pelvic peritoneum and other pelvic organs and are believed to derive from retrograde menstruation (Sampson, 1924), metastasis via lymphatics and vascular channels from endometrium to distant organs (Halban, 1925), or from transformation of peritoneal mesothelium by undefined stimuli (Novak, 1931). Immune dysfunctions have been invoked to explain the development of endometriosis, with deficient cell immunity reflected in tolerance of growth of endometrium in ectopic sites (Dmowski et al., 1981). Also, autoimmunity to endometrial and other antigens has been observed (reviewed in Mahmood and Templeton, 1990), probably as part of the inflammatory response to retrograde menstruation debris.

B. Growth Factors and Cytokines

Activated macrophages are present in higher levels in peritoneal fluid in women with endometriosis than in those without the conditition (Haney et al., 1981; Halme et al., 1987). Macrophages release growth factors that stimulate proliferation of endometriosis implants (Halme et al., 1987; Olive et al., 1991) and cytokine embryotoxins (Fakih et al., 1987; Eisermann et al., 1988), the latter of which may adversely affect fertility. Growth factors in peritoneal fluid (PF), which bathes the pelvic and abdominal cavity, are likely to be stimuli to ectopic endometriosis implant growth (reviewed in Ramey and Archer, 1993) and are derived from several sources in addition to macrophages, including sloughed endometrium (by retrograde menstruation), ovarian follicular fluid, peritoneal cells, and endometriosis tissue itself (Figure 8). Several growth factors and cytokines have been found in PF and are listed in Table 7. These include EGF (Mcleon et al., 1986), IGF-I and IGF-II (Giudice et al., 1994), unidentified mitogens that bind heparin but are not fibroblast growth factor by molecular size (Koutsilieris et al., 1991, 1993), and the cytokines IL-1, IL-6 , and -8, TNF-α , M-CSF, and RANTES (Fakih et al., 1987; Hill and Anderson, 1989; Buyalos et al., 1992b; Arici et al., 1994a; Taketania et al., 1992; Awadalla et al., 1987; Eisermann et al., 1988; Khorram et al., 1993). RANTES is an 8 kD T cell-specific cytokine of the platelet factor 4 gene superfamily (Schall, 1991) and is a chemotactic factor for human blood monocytes and memory T lymphocytes (Schall et al., 1990). In addition, a chemotactic activity, not yet identified, has been found to be present in higher levels in women with, compared to those without, endometriosis (Lyttle et al., 1994). Also, monocyte chemotactic protein (MCP-1) has been identified in peritoneal fluid of women with endometriosis (Kliman, 1994) and in endometrium from patients with en-

Table 7. Growth Factors and Cytokines in Peritoneal Fluid

EGF
IGF-I, IGF-II
TNF-α
CSF
RANTES
IL-1, -6, -8
Heparin-binding growth factors
Chemotactic factors
MCP-1
Angiogenic activity

dometriosis (Akoum et al., 1994a, 1994b). Angiogenic activity has also been reported in peritoneal fluid, with high levels found in women with compared to those without endometriosis (Oosterlynck et al., 1993).

Although the pathogenesis of this disorder is still not known, it is believed that, in select patients, sloughed endometrial tissue is able to attach to the pelvic peritoneum and ovarian surfaces (and other intra-pelvic and intra-abdominal organs) and to invade the surface epithelium (Spuijbroek et al., 1992). Growth of endometriotic implants is dependent on estrogen and growth factors acting synergistically with or as mediators of estrogen. The inflammatory response to the presence of the endometriotic tissue is believed to contribute additional growth factors and cytokines which further stimulate growth and differentiation of the endometriotic foci. While the presence of a variety of growth factors and cytokines in peritoneal fluid suggests their involvement in the pathogenesis and growth of endometriosis, definite roles have not been proven. Therapy can be either medical or surgical, and both are temporizing. Definitive surgical treatment by hysterectomy and bilateral oophorectomy has been the gold standard in curing the disorder, although it is untenable for many women. Understanding basic mechanisms that predispose some women to endometriosis and the events that occur in the attachment, invasion, and growth of these implants will hopefully lead to more effective and better accepted therapies.

V. MYOMETRIUM AND UTERINE LEIOMYOMATA

A. Clinical Significance

The myometrial layer of the uterus (Figure 1) consists of interlacing bundles of smooth muscle cells, richly vascularized and supported by extracellular elements. Uterine leiomyomata ("fibroids") are benign smooth muscle cell tumors of myometrial origin, and are the most common female reproductive tract neoplasms.

Rare malignant neoplasms, leiomyosarcomas, are often confused with their benign counterpart, and whether leiomyosarcomas develop from preexisting leiomyomas or evolve independently has not yet been determined. Benign leiomyomas occur in approximately 25% of Caucasian women and 50% of Black women during their reproductive years, especially during the perimenopause, and many grow during pregnancy (Weingold, 1983; Cramer and Patel, 1990; Cramer, 1992). These observations, epidemiologic data, and the decrease in size that accompanies hypoestrogenic states (menopause or treatment with GnRH analogues) all suggest that these tumors are steroid-hormone-dependent, and traditional thinking has implicated estrogen specifically. Clinically, these benign tumors can cause abnormal uterine bleeding, lower abdominal and pelvic discomfort, and spontaneous abortion, depending on their size, location, and number. After surgical removal (myomectomy) additional leiomyomata often grow, and hysterectomy is a common treatment for these tumors. Of the 650,000 hysterectomies performed in the United.States in 1985, 27% were conducted because of symptomatic leiomyomata (Pokras and Jufnagel, 1987). It is projected that by the year 2005, 854,000 hysterectomies will be performed for leiomyomata in that year in the United.States, with annual costs totaling more than $1.15 billion in today's dollars (Pokras and Jufnagel, 1988). The mortality rate from hysterectomies, excluding those performed during pregnancy or for cancer-related conditions, is 8.6/10,000 (Wingo et al., 1985), and the morbidity associated with hysterectomy or myomectomy is several fold higher than the mortality rate. Clearly these tumors and the surgical therapy traditionally used to treat them pose significant health risks to women. Despite their common occurrence, risks, and cost, precise mechanisms for growth of these benign tumors and explanations for racial differences have not been well defined. While steroid hormones have been long implicated in normal and tumorous myometrial growth (reviewed by Cramer, 1992; Friedman, 1993; Rein and Nowak, 1992), roles for growth factors and related peptides in the pathogenesis of the growth of these tumors is beginning to be appreciated. This section reviews current literature on peptide growth factors in myometrial and myoma biology. Table 8 summarizes the major growth factor systems found in myometrium, leiomyoma, and leiomyosarcoma.

B. Growth Factors in Leiomyomata and Myometrium

Insulin-like Growth Factor Family

Myometrium, leiomyoma, and leiomyosarcoma express IGF-I and IGF-II mRNAs (Hoppener, et al., 1988; Boehm et al., 1990; Gloudemans et al., 1990; Vollenhoven et al., 1993). Independent investigators disagree on relative IGF-I gene expression in these tissues. Two studies report increased IGF-I expression in leiomyomas, compared to myometrium from the same uterus (Hoppener et al., 1988, Boehm et al., 1990), while two others suggest equivalent expression (Gloudemans et al., 1990, Vollenhoven et al., 1993). One report identifies reduced IGF-I mRNA in leiomyosarcoma (Gloudemans et al., 1990). IGF-I mRNA expression in leiomyomas was found to be dependent on the *in vivo* estrogen-status in one (Giudice et al., 1993c), but not another (Vollenhoven et al., 1993) study. However, in support of estrogen-dependence of IGF-I in leiomyomas are the observations that explants of leiomyomas (and myometrium) obtained at surgery from subjects pretreated with a GnRH agonist (resulting in hypoestrogenism) secrete markedly lower amounts of IGF-I, compared to explants of tissues obtained from placebo treated controls (Rein et al., 1990). In contrast to IGF-I, there is general agreement that IGF-II mRNA expression is significantly increased in leiomyoma, compared to myometrium, and is even more abundantly expressed in leiomyosarcoma (Boehm et al., 1990; Gloudemans et al., 1990; Vollenhoven et al., 1993). The latter has led to elevated circulating IGF-II and subsequent severe hypoglycemia (Daughaday et al., 1988), due to cross-reactivity of IGF-II at the insulin receptor. Aberrant expression of IGF-II in sarcomas of the uterus has been linked to hypomethylation of the IGF-II gene (Gloudemans et al., 1992).

Myometrium, leiomyoma, and leiomyosarcoma contain binding sites for IGF-I and -II (Tommola et al., 1989; Chandrasekher et al., 1992; Nagamani et al., 1994). IGF-I binding is higher in leiomyoma than in myometrium (Tommola et al., 1989; Chandrasekher et al., 1992), due to increased receptor numbers (Tommola et al., 1989). Scatchard analysis suggests higher IGF-II binding in uterine leiomyosarcoma compared to either myometrium or benign leiomyomata (Nagamani et al., 1994). Messenger RNA levels of both the Type I and Type II IGF receptor in leiomyomas do not to fluctuate with menstrual cycle phase or *in vivo* estrogen status (Giudice et al., 1993c). With regard to IGFBPs, IGFBP-2 through -6 have been identified in myometrium and leiomyoma (Giudice et al., 1993c; Vollenhoven et al., 1993). One study reports lower expression of IGFBP-3 in leiomyoma as compared to myometrium (Vollenhoven et al., 1993), although the effect that this may have on bioavailable IGF levels in leiomyoma tissue has yet to be determined. Likewise, differential expression of IGFBP-4, -5, and -6 in these tissues, and the activity of an IGFBP-4 protease found in uterine and leiomyoma smooth muscle cells (Pannier et al., 1995) have not to date been explored with regard to a potential role in the pathogenesis in leiomyoma growth. While IGFs are potent mitogens to smooth muscle cells in vitro (Rossi et al., 1992) in vitro doubling times are much more rapid than the slow growth generally observed for leiomyomas *in vivo*. It is likely that growth promoters and growth inhibitors and other mechanisms of tumor growth (e.g., increased extracellular matrix production [vide infra]) are operating in the slow growth of these benign tumors.

Epidermal Growth Factor Family

The presence of EGF and its receptor has been documented at the mRNA and protein levels in human myometrium and leiomyoma (Hoffman et al., 1984; Yeh et al., 1991; Rossi et al., 1992). While earlier radioligand studies suggested that

EGF bound equally well to leiomyomas and normal myometrium (Hoffman et al., 1984; Lumsden et al., 1988), more recent, independent studies found substantially less EGF binding in myoma (Fayed et al., 1989, Tommola et al., 1989). Binding differences are attributed to different receptor concentrations rather than altered receptor affinity (Tommola et al., 1989). EGF binding in human uterine smooth muscle cells does not vary during the menstrual cycle (Hoffman et al., 1984, Lumsden et al., 1988). Although EGF protein production has not been directly measured, comparable mRNA levels suggest that endogenous production of EGF by myometrial cells is constant throughout the menstrual cycle, whereas leiomyoma smooth muscle cells contain notably more EGF mRNA during the secretory phase. During the proliferative phase the two tissues have comparable levels of EGF mRNA expression (Harrison-Woolrych et al., 1994). These observations suggest that EGF may be regulated by progesterone, which increases in serum during the secretory phase of the cycle. While smooth muscle production of EGF in response to treatment with progesterone has not been verified *in vitro*, EGF is a potent mitogen to leiomyoma (and myometrial) smooth muscle cells in culture (Bhargava et al., 1979; Rossi et al., 1992). EGF has been classically thought of as an "estromedin" (i.e., as a mediator of estrogen actions), although it may act as a estrogen/progesterone mediator in leiomyomas. In support of this is the clinical observation that RU-486, a progesterone receptor antagonist, is useful in decreasing the size of uterine leiomyomata (Murphy et al., 1993). The estrogen-dependence of EGF, however, cannot be ignored. Lower levels of EGF and EGF receptor mRNAs and proteins are expressed in fibroids from GnRH agonist treated patients, compared to controls (Harrison-Woolrych et al., 1994; Chegini et al., 1994b). Furthermore, EGF binding in leiomyoma is significantly lower in tissue obtained from GnRH agonist treated patients, compared to untreated controls (Lumsden et al., 1988). Myometrial EGF binding is not altered by GnRH agonist therapy, suggesting increased EGF receptor sensitivity to estrogen in leiomyomatous tissue (Lumsden et al., 1988). While the evidence for estrogen/EGF-dependent growth of leiomyomata is clear, the clinical community would do well to reevaluate the concept that leiomyomas are directly estrogen-dependent tumors and consider the possibility that they are both estrogen- and progesterone-regulated. States associated with leiomyomata regression (GnRH analogue therapy and menopause) are not only hypoestrogenic conditions, but are also associated with low circulating progesterone, and during pregnancy, when progesterone (and estrogen) levels are high, about 50% of leiomyoma increase in size. It is possible that estrogen is required for the expression of progesterone receptors, and thus a combination of both hormones is necessary for leiomyoma growth. Progesterone receptor antagonists may, in the future, enter the repertoire of surgical adjuvants in treating leiomyomas.

Platelet-derived Growth Factor

Two mRNA transcripts for PDGF-B and one for PDGF-A are expressed in myometrium and leiomyoma (Mendoza et al., 1990; Boehm et al 1990). Expression of PDGF-B mRNAs is equivalent in the two tissues (Boehm et al., 1990). In gravid uteri, myometrial PDGF-A mRNA expression is increased relative to the nongravid state (Mendoza et al., 1990). Immunohistochemistry has identified expression of mature PDGF-AB protein in myometrial tissue at constant rates throughout the menstrual cycle (Rossi et al., 1992). Receptor sites for PDGF have been identified in myometrium and leiomyoma, with the latter tissue displaying more receptor sites but lower receptor affinity (Fayed et al., 1989). Immunohistochemical studies have shown PDGF-β (but not -α) receptor expression in myometrial cells, with no cycle dependence observed (Rossi et al., 1992). To date, no comparisons have been made between normal myometrium and leiomyosarcoma with regard to the PDGF system.

Transforming Growth Factor-β

Messenger RNAs encoding TGF-α, TGF-β1, -β2, and β3, as well as types I, II, and III TGF-β receptor (TGF-βRI, -βRII, -βRIII) are expressed in both myometrium and leiomyoma (Murphy et al., 1994; Chegini et al., 1994b). Two differences in TGF expression between myometrium and leiomyoma may play a factor in tumor growth. First, TGF-β3 expression, which acts through the TGF-βRI to mediate extracellular matrix (ECM) production, has been found to be several times higher in leiomyoma than in matched myometrium (Murphy et al., 1994; Arici et al 1994b). Secondly, TGF-βRII, which is postulated to mediate TGF-induced growth inhibition, is reported to be significantly reduced in leiomyoma, compared to myometrium (Murphy et al., 1994). Recent studies suggest that TGF-β may play an important role in leiomyoma growth by increasing ECM production. Collagen Type I, collagen Type II, and fibronectin mRNAs and proteins are increased in the proliferative phase of the menstrual cycle (Stewart et al., 1994). TGF-β3 expression in myometrial tissue depends on the estrogen status *in vivo* (Murphy et al., 1994; Arici et al., 1994b), and myometrial and leiomyoma cell cultures show estrogen sensitivity by exhibiting dose-dependent increases in TGF-β3 mRNA levels in response to estradiol (Arici et al., 1994b). It is possible that TGF-β isoforms via the TGFβRI receptor increase leiomyoma smooth muscle ECM production, contributing to increased volume of leiomyomas in estrogen-dominant states. In addition, TGF-β isoform interactions via the TGF-βRII receptor may promote growth inhibition, balancing the growth promoting mitogenic activities of other growth factors in this tissue.

Other Growth Factor Systems

Specific binding sites for insulin have been found in myometrium, leiomyoma, and leiomyosarcoma (Fayed et al., 1989; Nagamani and Stuart, 1992; Nagamani et al., 1994). Fayed et al., (1989) reported similar insulin binding in myometrium and leiomyoma, while Nagamani and Stuart (1992) found greater binding in leiomyoma due to increased receptor affinity. Growth hormone receptor (GHR) mRNA expression in myometrium and leiomyoma has been recently documented.

GHR mRNAs are more abundantly expressed in leiomyomas, compared to adjacent myometrium (Sharara and Neiman, 1994), perhaps explaining the greater prevalence of fibroids in acromegalic women (reviewed in Sharara and Neiman, 1994).

With regard to other growth factors, the mRNA encoding keratinocyte growth factor (KGF, also known as FGF-7) has been identified in myometrium (Pekonen et al., 1993). Although KGF receptor mRNA was not found in this tissue, the closely related FGF receptor-2 is expressed and may be the receptor for the endogenous KGF ligand. Four mRNA transcripts for vascular endothelial growth factor have been identified in myometrial tissue (Charnock-Jones et al., 1993), and this growth factor may play a role in angiogenesis in myometrium and leiomyomata. Myometrial cells specifically bind interleukin-1 (IL-1), and in culture are stimulated by IL-1 and also TNF-α (Hertelendy et al., 1993; Molnar et al., 1993).

Table 8. Summary of Major Growth Factor Systems Identified in Myometrium and Its Tumors

Growth Factor System in myometrium	*Alteration in leiomyoma (Reference)*	*Alteration in Leiomyosarcoma (Reference)*
IGF system		
IGF-I (Boehm, 1990)	↑ Expression (Boehm, 1990)	↓ Expression (Gloudemans, 1990)
IGF-II (Boehm, 1990)	↑ Expression (Gloudemans, 1990)	↑↑ Expression (Gloudemans, 1990)
IGFBP-2 thru -6 (Giudice, 1993)	↓ IGFBP-3 (Vollenhoven, 1993)	
IGF-I binding (Tommola, 1989)	↑ Binding (Tommola, 1989)	
IGF-II binding (Chandrasekher, 1992)		↑ Binding (Nagamani, 1994)
EGF system		
EGF (Yeh, 1991)	↑ mRNA in secretory phase only (Harrison-Woolrych, 1994)	
EGF binding (Yeh, 1991)	↓ Binding (Tommola 1989)	
PDGF system		
PDGF-A (Mendoza, 1990)		
PDGF-B (Boehm, 1990)		
PDGF-β Receptor (Rossi, 1992)		
PDGF binding (Fayed, 1989)	↑ Binding (Fayed, 1989)	
TGF system		
TGF-α (Chegini, 1994)		
TGF-β1 (Murphy, 1994)		
TGF-β2 (Chegini, 1994)		
TGF-β3 (Murphy, 1994)	↑ Expression (Murphy, 1994)	
TGF-β,II,&I 1994)	↓ βRII expression (Murphy, 1994)	

C. Summary

Figure 9 depicts the potential roles, discussed herein, of various growth factors as steroid-hormone modulators of leiomyoma growth. Other growth-promoting substances extracted from uterine and leiomyoma tissues have been described, although not yet characterized (Koutsilieris et al., 1993). The precise roles of all of these growth modulators within the context of the steroid hormone environment and the original smooth muscle cell precursor that clonally expands into a benign leiomyoma are not well understood. However, with further research in this area, it is hoped that alternative therapies will be devised to circumvent the traditional surgical therapies that are commonly the only definitive solution to symptomatic leiomyomas.

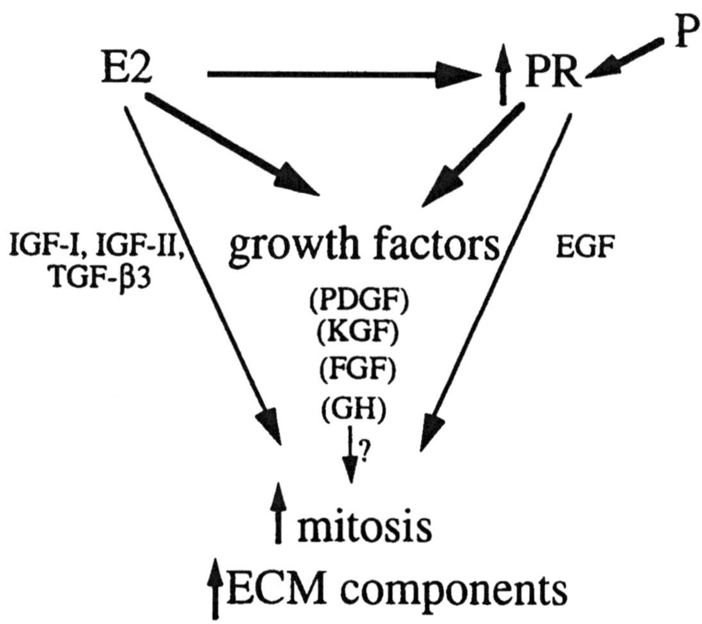

Figure 9. Model of steroid hormones (E_2 = estradiol; P = progesterone; PR = progesterone receptor) and growth modulators in mitosis and production of extracellular matrix (ECM) in leiomyomas.

VI. A LOOK TO THE FUTURE

The diverse processes that occur in the female reproductive tract culminate in coordinated and repetitive events that are necessary for the continuation of the species. In disorders associated with dyssynchronous events, abnormal secretion of humoral factors, or distortions of the normal anatomy, normal implantation can be threatened or inhibited. In addition, these abnormalities can and do contribute to morbidity of women of reproductive and post-reproductive age. As we approach the twenty-first century and molecular and genetic techniques permit further evaluation and understanding of growth factors and cytokines in health and disease of the female reproductive tract, it is likely that we shall be able to provide alternative therapies for disorders commonly observed today. These therapies may involve gene targeting, deleivery of antibodies to leiomyoma growth modulators or anti-sense RNA into the leiomyomas by unique delivery systems, coupled with more sophisticated invasive radiographic approaches and/or laparoscopic therapies, obviating the need for hysterectomy. In addition to improving the health of women, improvements in achieving fertility and fertility regulation will be within our grasp.

ACKNOWLEDGMENTS

This work was supported in part by NIH HD31579 (Linda C. Giudice), Stanford University School of Medicine Cancer Biology Program (Gary A. Ulaner), Fulbright Fellowship (Thierry van Dessel), Lalor Foundation Fellowship (Yasmin A.Chandrasekhar), American Fertility Society/Serono Award (Nicholas A. Cataldo), and Stanford University School of Medicine Medical Scholars Program (O. W. Stephanie Yap and Gary A. Ulaner).

REFERENCES

Adashi, E. Y. (1991). In: Reproductive Endocrinology (Yen, S.S.C., & Jaffe, R.B., Eds.), pp. 181-237. Saunders, Philadelphia.

Adashi, E. Y., & Resnick, C. E. (1986). Antagonistic interactions of transforming growth factors in the regulation of granulosa cell differentiation. Endocrinology 119, 1879-1881.

Adashi, E. Y., Resnick, C. E., & Twardzik, D. R. (1987). Transforming growth factor-a attenuates the acquisition of aromatase activity by cultured rat granulosa cells. J. Cell. Biochem. 33, 1-13.

Akoum, A.,. Lemay, A., Brunet, C., Hebert, J., & "le groupe d'investigation en gynecologie." (1994a). Secretion of monocyte chemotactic protein-1 (MCP-1) by cytokine-stimulated endometrial cells of women having endometriosis. The Endocrine Society, 76th Annual Meeting, Anaheim, CA, Abstract 1793.

Akoum, A., Lemay, A., Brunet, C., Hebert, J., & "le groupe d'investigation en gynecologie." (1994b). Endometriotic cells produce monocyte chemotactic protein-1(MCP-1) upon stimulation with interleukin-1 beta (IL-1b) and tumor necrosis factor alpha (TNF-a) in culture. The Endocrine Society, 76th Annual Meeting, Anaheim, CA, Abstract 1514.

Andrews, G.K., Dalton, T., & Kover, K. (1994). Analysis of the expression of growth factor, interleukin-1, and lactoferrin gene and the distribution of inflammatory leukocytes in the preimplantation mouse oviduct. Biol. Reprod. 51, 597-606.

Angervo, M., Koistinen, R., Suikkari, A-M., & Seppala, M. (1991). Insulin-like growth factor binding protein-1 inhibits the DNA amplification induced by insulin-like growth factor I in human granulosa-luteal cells. Hum. Reprod. 6, 770-773.

Angervo, M., Koistinen, R., & Seppala, M. (1992). Epidermal growth factor stimulates production of insulin-like growth factor-binding protein-1 in human granulosa-luteal cells. J. Endocrinol. 134, 12-131.

Arici, A., Tazuke, S., Attar, E., Lkiman, H., & Olive, D. (1994a). Interleukin (IL-8) concentration in peritoneal fluid and modulation of interleukin-8 expression in human mesothelial cells. The American Fertility Society, San Antonio, TX, Abstract 0-057.

Arici, A., Sozen, I., & Olive, D. (1994b). Modulation of transforming growth factor-b3 (TGF-b3) expression in myometrium and leiomyoma. Annual Meeting of the American Fertility Society, November 1994, Abstract O-0063 (pp.S31-32).

Attisano, L., Wrana, J.L., Cheifetz, S., & Massagué, J. (1992). Novel activin receptors: Distinct genes and alternative mRNA splicing generate a repertoire of serine/threonine kinase receptors. Cell 68, 97-108.

Awadalla, S.G,. Friedman, C.I., Haq, A.U., Roh, S.I., Chin, N.W., & Kim, M.H. (1987). Local peritoneal factors: Their role in infertility associated with endometriosis. Am. J. Obstet. Gynecol. 157, 1207-1214.

Axelrod, L.R., & Goldzieher, J.W. (1962). The polycystic ovary. III. Steroid biosynthesis in normal and polycystic ovarian tissue. J. Clin. Endocrinol. Metab. 22, 431-440.

Baird, A., & Hsueh, A. J. W. (1986). Fibroblast growth factor as an intraovarian hormone: Differential regulation of steroidogenesis by an angiogenic factor. Regul. Pept. 16, 243-250.

Baird, A., & Klagsbrun, M. (1991). The fibroblast growth factor family. Cancer Cells 3, 239-244.

Balboni, G. C., Vannelli, G. B., Barni, T., Orlando, C., & Serio, M. (1987). Transferrin and somatomedin C receptors in human ovarian follicles. Fertil. Steril. 48, 706-710.

Barak, V., Mordel, N., Holzer, H., Zajicek, G., Treves, A. J., & Laufer, N. (1992a). The correlation of interleukin 1 and tumor necrosis factor to oestradiol, progesterone and testosterone levels in periovulatory follicular fluid of *in vitro* fertilization patients. Hum. Reprod. 7, 462-464.

Barak, V., Mordel, N., Zajicek, G., Kalichman, I., Treves, A. J., & Laufer, N. (1992b). The correlation between interleukin 2 and soluble interleukin 2 receptors to oestradiol, progesterone and testosterone levels in periovulatory follicles of in vitro fertilization patients. Hum. Reprod. 7, 926-929.

Barak, V., Yanai, P., Treves, A. J., Roisman, I., Simon, A., & Laufer, N. (1992c). Interleukin-1: Local production and modulation of human granulosa luteal cells steroidogenesis. Fertil. Steril. 58, 719-725.

Barbieri, R.L., Makris, S., Randall, R.W., Daniels, G., Kistner, R.W., & Ryan, K.J. (1986). Insulin stimulates androgen accumulation in incubations of ovarian stroma obtained from women with hyperandrogenism. J. Clin. Endocrinol. Metab. 62, 904-910.

Barde, Y-A. (1989). Trophic factors and neuronal survival. Neuron 2, 1525-1534.

Basilico, C., & Moscatelli, D. (1992). The FGF family of growth factors and oncogenes. Adv. Cancer Res. 59, 115-165.

Baxter, R., & Martin, J. (1986). Radioimmunoassay of growth hormone-dependent insulin-like growth factor binding protein in human plasma. J. Clin. Invest. 78, 1504-1512.

Ben-Rafael, Z., & Orvieto, R. (1992). Cytokines-involvement in reproduction. Fertil. Steril. 58, 1093-1099.

Bergh, C., Olsson, J-H., & Hillensjo, T. (1991). Effect of insulin-like growth factor-I on steroidogenesis in cultured human granulosa cells. Acta Endocrinol. (Copenh). 125, 177-181.

Best, C. L., Pudney, J., Anderson, D. J., & Hill, J. A. (1994). Modulation of human granulosa cell steroid production in vitro by tumor necrosis factor alpha: Implications of white blood cells in culture. Obstet. Gynecol. 84, 121-127.

Bhargava, G., Rifas, L., & Makman, M.H. (1979). Presence of epidermal growth factor receptors and influence of epidermal growth factor on proliferation and aging in cultured smooth muscle cells. J. Cellular Physiology 100, 365-374.

Boehm, K.D., Daimon, M., Gorodeski, I.G., Sheean, L.A., Utian, W.H., & Ilan, J. (1990). Expression of the insulin-like and platelet-derived growth factor genes in human uterine tissues. Mol. Reprod. Dev. 27, 93-101.

Bongso, A., Fong, C-Y., Ng, S-C., & Ratnam, S. (1993). The search for improved *in vitro* systems should not be ignored: Embryo co-culture may be one of them. Human Reprod. 8, 1155-1162.

Bothwell, M. (1991). Keeping track of neurotrophin receptors. Cell 65, 915-918.

Brannian, J.D,. Woodruff, T.K., Mather, J.P., & Stouffer, R.L. (1992). Activin-A inhibits progesterone production by macaque luteal cells in culture. J. Clin. Endocrinol. Metab. 75, 756-761.

Brown, J.B. (1978). Pituitary control of ovarian function—concepts derived from gonadotrophin therapy. Aust. N. Z. J. Obstet. Gynaecol. 18, 47-54.

Bruce-Mann, G., Fowler, K.J., Gabriel, A., Nice, E.C., Linsay, W.R., & Dunn, A.R. (1993). Mice with a null mutation for the TGF-a gene have abnormal skin architecture, wavy hair and curly whiskers and often develop corneal inflammation. Cell 73, 249-261.

Buckler, H.M., McLachlan, R.I., MacLachlan, V.B., Healy, D.L., & Burger, H.G. (1988). Serum inhibin levels in polycystic ovary syndrome: Basal levels and response to luteinizing hormone-releasing hormone agonist and exogenous gonadotropin administration. J. Clin. Endocrinol. Metab. 66, 798-803.

Buhi, W., Chegini, N., Rong, H., & Alvarez, I. (1994). Immunolocalization of transforming growth factor (TGF-b) TGF-b_1, TGF-b_2, and TGF-b type II receptor protein in the porcine oviduct. [Abstract 256] 41st Annual Meeting of the Society for Gynecologic investigation, Chicago, IL.

Burghen, G.A., Givens, J. R., & Kitabchi, A. E. (1980). Correlation of hyperandrogenism with hyperinsulinism in polycystic ovarian disease. J. Clin. Endocrinol. Metab. 50, 113-116.

Buyalos, R. P., Watson, J. M., & Martinez-Maza, O. (1992a). Detection of interleukin-6 in human follicular fluid. Fertil. Steril. 57, 1230-1234.

Buyalos, R.P., Funari, V.A., Azziz, R., Watson, J.M., & Martinez-Maza, O. (1992b). Elevated interleukin-6 levels in peritoneal fluid of patients with pelvic pathology. Fertil. Steril. 58, 302-306.

Cameron, V.A., Nishimura, E., Mathews, L.S., Lewis, K.A., Sawchenko, P.E., & Vale, W.W. (1994). Hybridization histochemical localization of activin receptor subtypes in rat brain, pituitary, ovary, and testis. Endocrinology 134, 799-808.

Carlsson, B., Bergh, C., Bentham, J., Olsson, J-H., Norman, M.R., Billig, H., Roos, P., & Hillensjo, T. (1992). Expression of functional growth hormone receptors in human granulosa cells. Human Reprod. 7, 1205-1209.

Carlsson, B., Hillensjo, T., Nilsson, A., Tornell, J., & Billig, H. (1993). Expression of insulin-like growth factor (IGF-I) in the rat fallopian tube: Possible autocrine and paracrine action of fallopian tube-derived IGF-I on the fallopian tube and on the preimplantation embryo. Endocrinology 133, 2031-2039.

Carpenter, G. (1987). Receptor for epidermal growth factor and other polypeptide mitogens. Annu. Rev. Biochem. 56, 881-914.

Carpenter, G., & Cohen, S. (1979). Epidermal growth factor. Ann. Rev. Biochem. 48, 193-216.

Carpenter, G., & Zendegui, J. G. (1986). Epidermal growth factor, its receptor, and related proteins. Exp. Cell. Res. 164, 1-10.

Casey, M.L., Smith, J.W., Nagai, K., Hersh, L.B., & MacDonald, P.C. (1991). Progesterone-regulated cyclic modulation of membrane metalloendopeptidase (enkephalinase) in human endometrium. J. Biol. Chem. 266, 23041-23047.

Cataldo, N. A., & Giudice, L. C. (1992a). Insulin-like growth factor binding protein profiles in human ovarian follicular fluid correlate with follicular functional status. J. Clin. Endocrinol. Metab. 74, 821-829.

Cataldo, N.A., & Giudice, L.C. (1992b). Follicular fluid insulin-like growth factor binding protein profiles in polycystic ovary syndrome. J. Clin. Endocrinol. Metab. 74, 695-697.

Cataldo, N.A., Woodruff, T. K., & Giudice, L. C. (1993). Regulation of insulin-like growth factor binding protein production by human luteinizing granulosa cells cultured in defined medium. J. Clin. Endocrinol. Metab. 76, 207-215.

Cataldo, N.A., Fujimoto, V.Y., & Jaffe, R.B. (1994a). Activin-A stimulates the production of insulin-like growth factor binding proteins by human luteinizing granulosa cells. 41st Annual Meeting of the Society for Gynecologic Investigation, Chicago, March 1994, Abstract No.054

Cataldo, N.A., Fujimoto, V.Y., & Jaffe, R.B. (1994b). Follistatin antagonizes the effects of activin-A on steroidogenesis in human luteinizing granulosa cells. J. Clin. Endocrinol. Metab. 79, 272-277.

Caubo, B., DeVinna, R. S., & Tonetta, S. A. (1989). Regulation of steroidogenesis in cultured porcine theca cells. Endocrinology 125, 321-326.

Chandrasekher, Y.A., Heiner, J., Osuamkpe, M.S., & Nagamani, M. (1992). Insulin-like growth factor I and II binding in human myometrium and leiomyomas. Am. J. Obstet. Gynecol. 166, 64-69.

Chandrasekher, Y. A., & L.C. Giudice. (1994). Characterization of an insulin-like growth factor binding protein-4 protease in estrogenic follicular fluid. Abstract P284 presented at the 41st Annual Meeting of the Society for Gynecologic Investigation, Chicago, IL, March 22-26.

Chandrasekher, Y. A., Giudice, L.C., Bai, C., Liu, H-L., & Poretsky, L. (1994). Insulin and insulin-like growth factor I (IGF-I) inhibit IGF binding protein 1 (IGFBP-1) production by human granulosa-luteal cells (hGCs). Abstract P-159 presented at the 50th Annual Meeting of the American Fertility Society, San Antonio, TX, November 5-10.

Chang, R. J., Nakamura, R. M., Judd, H. L., & Kaplan, S. A. (1983) Insulin-resistance in non-obese patients with polycystic ovarian disease. J. Clin. Endocrinol. Metab. 57, 356-360.

Charnock-Jones, D.S., Sharkey, A.M., Rajput-Williams, J,. Burch, D., Schofield, J.P., Fountain, S.A., Boocock, C.A., & Smith, S.K. (1993). Identification and localization of alternately spliced mRNAs for vascular endothelial growth factor in human uterus and estrogen regulation in endometrial carcinoma cell lines. Biol. Reprod. 48, 1120-1128.

Chaudhuri, G., & Nathan, L. (1992). Nitric oxide and endometrial blood flow. In: Steroid Hormones and Uterine Bleeding (Alexander, N.J., & d'Arcangues, C., Eds.), pp. 241-244. American Association for the Advancement of Science Press, Washington, DC.

Chegini, N., & Williams, R. S. (1992). Immunocytochemical localization of transforming growth factors (TGFs) TGF-a and TGF-b in human ovarian tissues. J. Clin. Endocrinol. Metab. 74, 973-980.

Chegini, N., Zhao, Y., & Mclean, F.W. (1994a). Expression of messenger ribonucleic acid and presence of immunoreactive proteins for epidermal growth factor (EGF), transforming growth factor alpha (TGFa) and EGF/TGFa receptors and ^{125}I-EGF binding sites in human fallopian tube. Biol. Reprod. 50, 1049-1058.

Chegini, N., Rong, H., Zhao, Y., Williams, R.S. (1994b). Gonadotropin hormone releasing hormone agonist (GnRH) suppresses the expression of mRNA and proteins for growth factors and their receptors in uterine leiomyomata. Society for Gynecologic Investigation 41st Annual meeting, March, 1994, Chicago, Abstract P260, p. 324.

Christman, G. M., Randolph, J. G., Peegel, H., & Menon, K. M. (1991). Differential responsiveness of luteinized human granulosa cells to gonadotropins and insulin-like growth factor-I for induction of aromatase activity. Fertil. Steril. 55, 1099-1105.

Chun, S.-Y., Billig, H., Tilly, J.L., Furuta, I., Tsafriri, A., & Hsueh, A.J.W. (1994). Gonadotropin suppression of apoptosis in cultured preovulatory follicles: Mediatory role of endogenous insulin-like growth factor-I. Endocrinology 135, 1845-1853.

Ciaraldi, T., El-Roeiy, A., Madar, Z., Reichart, D., Olefsky, J., & Yen, S.S.C. (1992). Cellular mechanisms of insulin resistance in polycystic ovarian syndrome. J. Clin. Endocrinol. Metab. 75, 577-583.

Clemmons, D.R. (1993). IGF binding proteins and their functions. Mol. Reprod. Dev. 35,368-375.

Cohen, I., & Speroff, L. (1991). Premature ovarian failure: Update. Obstet. Gynecol. Survey 156-162.

Conway, G.S., Jacobs, H.S., Holly, J.M.P., & Wass, J.A.H. (1990). Effects of leutinizing hormone, insulin, insulin-like growth factor-1 and insulin-like growth factor small binding protein 1 in the polycystic ovary syndrome. Clin. Endocrinol. 33, 593-603.

Conway, G.S., Clark, P.M., & Wong, D. (1993). Hyperinsulinaemia in the polycystic ovary syndrome confirmed with a specific immunoradiometric assay for insulin. Clin. Endocrinol., 38, 219-222.

Cramer, D.W. (1992). Epidemiology of myomas. Semin. Reprod. Endocrinol. 10, 320-324.

Cramer, S.F., & Patel, D. (1990). The frequency of uterine myomas. Am. J. Clin. Pathol. 94, 435-438.

Das, S.K., Flanders, K.C., Andrews, G.K., & Dey, S.K. (1992). Expression of transforming growth factor-β isoforms (β2 and β3) in the mouse uterus: Analysis of the periimplantation period and effects of ovarian steroids. Endocrinology 130, 3459-3466.

Das, S.K., Wang, X,. Paria, B.C., Damm, D., Abraham, J.A., Klagsbrun, M., Andrews, G.K., & Dey, S.K. (1994). Heparin-binding EGF-like growth factor gene is induced in the mouse uterus temporally by the blastocyst solely at the site of its apposition: A possible ligand for interaction with blastocyst EGF-receptor in implantation. Development 120, 1071-1083.

Daughaday, W.H., Emanuele, M.A., Brooks, M.H., Barbato, A.L,. Kapadia, M., & Rotwein, P. (1988). Synthesis and secretion of insulin-like growth factor II by a leiomyosarcoma with associated hypoglycemia. N. Engl. J. Med. 319, 1434-1440.

Davenport, A.P., Cameron, I.T., Smith, S.K., & Brown, M.J. (1991). Binding sites for iodinated endothelin-1, endothelin-2, and endothelin-3 demonstrated on uterine glandular epithelial cells by quantitative high-resolution autoradiography. J. Endocrinol. 129, 149-154.

Davoren, J.B., & Hsueh A.J.W. (1986). Growth hormone increases ovarian levels of immuno-reactive somatomedinC/insulin-like growth factor I *in vivo*. Endocrinology 118, 888-890.

Derynck, R. (1986). Transforming growth factor-alpha: Structure and biological activities. J. Cell. Biochem. 32, 293-295.

Di Blasio, A. M., Vigano, P., & Ferrari A. (1994). Insulin-like growth factor-II stimulates human granulosa-luteal cell proliferation *in vitro*. Fertil. Steril. 61, 483-487.

diClemente, N., Wilson, C,. Faure, E., Boussin, L., Carmillo, P,. Tizard, R., Picard, J-Y., Vigier, B., Josso, N., & Cate, R. (1994). Cloning, expression, and alternative splicing of the receptor for anti-MŸllerian hormone. Molec. Endocrinol. 8, 1006-1020.

Diehl, J.R., Henricks, D.M., & Gray, S.L. (1994). EGF and IGF-I in the uterine and oviductal fluids of pregnant and nonpregnant pigs from day 10 to day 18. [Abstract] Biol. Reprod. 50 (Suppl. 1), 272.

Dissen, G. A., Hill, D. F., Costa, M. E., Ma, Y. J., & Ojeda, S. R. (1991). Nerve growth factor receptors in the peripubertal rat ovary. Mol. Endocrinol. 5, 1642-1650.

Dissen, G. A., Lara, H. E., Fahrenbach, W. H., Costa, M. E., & Ojeda, S. R. (1994). Immature rat ovaries become revascularized rapidly after autotransplantation and show a gonadotropin-dependent increase in angiogenic factor gene expression. Endocrinology 134, 1146-1154.

Dmowski, W.P., Steele, R.W., & Baker, G.F. (1981). Deficient cellular immunity in endometriosis. Am. J. Obstet. Gynecol. 141, 377-383.

Dor, J., Ben-Shlomo, I., Lunenfeld, B., Pariente, C., Levran, D., Karasik, A., Seppala, M., & Mashiach, S. (1992). Insulin-like growth factor -I (IGF-I) may not be essential for ovarian follicular development: Evidence from IGF-I deficiency. J. Clin. Endocrinol. Metab. 74, 539-542.

Dorrington, J., Chuma, A. V., & Bendell, J. J. (1988). Transforming growth factor b and follicle-stimulating hormone promote rat granulosa cell proliferation. Endocrinology 123, 353-359.

Duffy, D. M., Lindheim, S. R., Ary, B. A., Saad, M., Vijad, M.A., Chang, L., & Lobo, R.A. (1993). The effects of growth hormone releasing factor (GRF) and insulin-like growth factor -I IGF-I) on folliculogenesis in normal women. Program of the American Fertility Society. Abstract. O-155.

Dunaif, A., Segal, K.R., Futterweit, W., & Dobrjansky, A. (1989). Profound peripheral insulin resistance, independent of obesity, in polycystic ovary syndrome. Diabetes 38, 1165-74.

Dunaif, A., Segal, K.R., Shelley, D., Green, G., Dobrjansky, A., & Licholai, T. (1992). Evidence for distinctive and intrinsic defects in insulin action in polycystic ovary syndrome. Diabetes 41, 1257-1260.

Dunkel, L., Tilly, J. L., Shikone, T., Nishimori, K., & Hsueh, A. J. W. (1994). Follicle-stimulating hormone receptor expression in the rat ovary: Increases during prepubertal development and regulation by the opposing actions of transforming growth factors b and a. Biol. Reprod. 50, 940-948.

Economos, K., MacDonald, P.C., & Casey, M.L. (1992). Endothelin-1 gene expression and protein biosynthesis in human endometrium: Potential modulator of endometrial blood flow. J. Clin. Endocrinol. Metab. 74, 14-19.

Eden, J. A., Jones J., Carter, G.D., & Alagnband-Zadeh, J. (1988). A comparison of follicular fluid levels of insulin-like growth factor-I in normal dominant and cohort follicles, polycystic and multicystic ovaries. Clin. Endocrinol. 29, 327-336.

Eden, J. A., Jones, J., Carter, G. D., & Alaghband-Zadeh, J. (1990). Follicular fluid concentrations of insulin-like growth factor 1, epidermal growth factor, transforming growth factor-alpha and sex-steroids in volume matched normal and polycystic human follicles. Clin. Endocrinol. 32, 395-405.

Eisermann, J., Gast, M.G., Pineda, J., Odem, R.R., & Collins, J.L. (1988). Tumor necrosis factor in peritoneal fluid of women undergoing laparoscopic surgery. Fertil. Steril. 50, 573-579.

El-Danasouri, F.A., & Westphal, L.M. (1993). Immunocytochemical localization of transforming growth factor-a and epidermal growth factor receptor in human fallopian tubes and cumulus cells. Am. J. Reprod. Immunol. 30, 82-87.

El-Roeiy, A., Baxter, R.C., Chen, X., Laughlin, G., & Yen, S.S.C. (1991). The role of Growth hormone, insulin growth factors (IGF-I and IGF-II) and their binding proteins (IGFBP-3 and IGFBP-1) in PCO syndrome. Society for Gynecologic Investigation, San Antonio, TX, March, 1991, Abstract, No.517.

El-Roeiy, A., Chen, X., Roberts, V. J., LeRoith, D., Roberts, C. T., & Yen, S. S. C. (1993). Expression of insulin-like growth factor-I (IGF-I) and IGF-II and the IGF-I, IGF-II, and insulin receptor genes and localization of the gene products in the human ovary. J. Clin. Endocrinol. Metab. 77, 1411-1418.

El-Roeiy, A., Chen, X., Roberts, V. J., Shimasaki, S., Ling, N., LeRoith, D., Roberts, C. T., & Yen, S. S. C. (1994). Expression of the genes encoding the insulin-like growth factors, the IGF and insulin receptors, and IGF-binding proteins-1-6 and the localization of their gene products in normal and polycystic ovary syndrome ovaries. J. Clin. Endocrinol. Metab. 78, 1488-1496.

Erickson, G.F., Hsueh, A.J.W., Quigley, M.E., Rebar, R.W., & Yen, S.S.C. (1979). Functional studies of aromatase activity in human granulosa cells from normal and polycystic ovaries. J. Clin. Endocrinol. Metab. 49, 514-519.

Erickson, G. F., Garzo, V. G., & Magoffin, D.A. (1989). Insulin-like growth factor-I regulates aromatase activity in human granulosa and granulosa-luteal cells. J. Clin. Endocrinol. 69, 716-720.

Erickson, G. F., Magoffin, D. A., Cragun, R. J., & Chang R. J. (1990). The effects of insulin and insulin-like growth factors-I and -II on estradiol production by granulosa cells of polycystic ovaries. J. Clin. Endocrinol. Metab. 70, 894-902.

Erickson, G. F., Garzo, V. G., & Magoffin, D. A. (1991). Progesterone production by human granulosa cells cultured in serum free medium: Effects of gonadotrophins and insulin-like growth factor-I (IGF-I). Hum. Reprod. 6, 1074-1081.

Erickson, G.F., Magoffin, D.A., Garzo, V.G., Cheung, A.P., & Chang, R.J. (1992). Granulosa cells of polycystic ovaries: Are they normal or abnormal? Hum. Reprod. 7, 293-299.

Erickson, G.F., Li, D., Sadrkhanloo, R., Liu, X-J., Shimasaki, S., & Ling, N. (1994). Extrapituitary actions of gonadotropin-releasing hormone: Stimulation of insulin-like growth factor binding protein-4 and atresia. Endocrinology 134, 1365-1372.

Ernfors, P., Wetmore, C., Olson, L., & Persson, H. (1990). Identification of cells in rat brain and peripheral tissues expressing mRNA for members of the nerve growth factor family. Neuron 5, 511-526.

Esch, F., Baird, A., Ling, N., Ueno, N., Hill, F., & Denoroy, L. (1985). Primary structure of bovine pituitary basic fibroblast growth factor (FGF) and comparison with the amino-terminal sequence of bovine brain acidic FGF. Proc. Natl. Acad. Sci. USA 82, 6507-6511.

Fakih, H., Baggett, B., Holtz, G., Tsang, K.-Y., Lee, J.C., & Williamson, H.O. (1987). Interleukin-1: A possible role in the infertility associated with endometriosis. Fertil. Steril. 47, 213-217.

Falcone, T., Billiar, R., & Morris, D. (1991). Serum inhibin levels in polycystic ovary syndrome: Effect of insulin resistance and insulin secretion. Obstet. Gynecol. 78, 171-176.

Fauser, B.C.J.M. (1994). Observations in favor of normal early follicle development and disturbed dominant follicle selection in polycystic ovary syndrome. Gynecol. Endocrinol. 8, 1-8.

Fauser, B.C.J.M. (1995). Step down follicle stimulating hormone regimens in Polycystic Ovary Syndrome. In: Ovulation Induction: Basic Science and Clinical Advances (Filicori, M., & Flamigni, M., Eds.). Elsevier, Int. Congres. Series, Amsterdam, The Netherlands.

Fauser, B.C.J.M., Pache, T.D., Lamberts, S.W., Hop, W.C., De Jong, F.H., & Dahl, K.D. (1991). Serum bioactive and immunoreactive luteinizing hormone and follicle-stimulating hormone levels in women with cycle abnormalities, with or without polycystic ovarian disease. J. Clin. Endocrinol. Metab. 73, 811-817.

Fayed, Y.M., Tsibris, J.C., Langenberg, P.W., & Robertson, A.L., Jr. (1989). Human uterine leiomyoma cells: Binding and growth responses to epidermal growth factor, platelet-derived growth factor, and insulin. Lab. Invest. 60, 30-37.

Fielder, P.J., Pham, H., Adashi, E.Y., & Rosenfeld, R.G. (1993). Insulin-like growth factors (IGFs) block FSH-induced proteolysis of IGF-binding protein-5 (BP-5) in cultured rat granulosa cells. Endocrinology 133, 415-418.

Findlay, J.K. (1993). An update on the roles of inhibin, activin, and follistatin as local regulators of folliculogenesis. Biol. Reprod. 48, 15-23.

Franks, S., & Mason, H.D. (1991). Polycystic ovary syndrome: Interaction of follicle stimulating hormone and polypeptide growth factors in oestradiol production by human granulosa cells. J. Ster. Biochem. Mol. Biol. 40, 405-409.

Franks, S., Mason, H.D., Polson, D.W., Winston, R.M., Margara, R., & Reed, M.J. (1988). Mechanism and management of ovulatory failure in women with polycystic ovary syndrome. Hum. Reprod. 3, 531-534.

Friedman, A.J. (1993). Treatment of uterine myomas with GnRH agonists. Seminars in Reproductive Endocrinology 11, 154-161.

Fukuoka, M., Yasuda, K., Emi, N., Fujiwara, H., Iwai, M., Takakura, K., Kanzaki, H., & Mori, T. (1992). Cytokine modulation of progesterone and estradiol secretion in cultures of luteinized human granulosa cells. J. Clin. Endocrinol. Metab. 75, 254-258.

Gangrade, B. K., Davis, J. S., & May, J. V. (1991). A novel mechanism for the induction of aromatase in ovarian cells *in vitro*: Role of transforming growth factor alpha-induced protein tyrosine kinase. Endocrinology 129, 2790-2792.

Gates, G. S., Bayer, S., Seibel, M., Poretsky, L., Flier, J. S., & Moses, A. C. (1987). Characterization of insulin-like growth factor binding to human granulosa cells obtained during *in vitro* fertilization. J. Receptor Res. 7, 885-889.

Geisthovel, F., Moretti-Rojas, I. M., Asch, R. H., & Rojas, F. J. (1989a). Expression of insulin-like growth factor-II (IGF-II) messenger ribonucleic acid (mRNA), but not IGF-I mRNA in human preovulatory granulosa cells. Hum. Reprod. 4, 899-903.

Geisthovel, F., Moretti-Rojas, I. M., Rojas, F. J., & Asch, R. H. (1989b). Immunoreactive insulin-like growth factor I in human follicular fluid. Hum. Reprod. 4, 35-40.

Gilfillan, C. P., & Robertson, D. M. (1994). Development and validation of a radioimmunoassay for follistatin in human serum. Clin. Endocrinol. 41, 453-461.

Giudice, L. C. (1992). Insulin-like growth factors and ovarian follicular development. Endocr. Rev. 13, 641-669.

Giudice, L. C. (1994). Growth factors and growth modulators in human uterine endometrium: Their potential relevance to reproductive medicine. Fertil. Steril. 61, 1-17.

Giudice, L.C. & Ferenczy, A. (1995). In: Reproductive Endocrinology, Surgery, and Technology (Adashi, E.Y., Rock, J.A., & Rosenwaks, Z., Eds). Raven Press, New York.

Giudice, L.C., & Saleh, W. (1995). Growth factors in reproduction. Trends Endocrinol. Metab. 6, 60-69.

Giudice, L. C., Farrell, E. M., Pham, H., & Rosenfeld R.G. (1990a). Identification of insulin-like growth factor binding protein-3 (IGFBP-3) and IGFBP-2 in human follicular fluid. J. Clin. Endocrinol. Metab. 71, 1330-1338.

Giudice, L. C., Farrell, E. M., Pham, H., Lamson, G. L., & Rosenfeld, R. G. (1990b). Insulin-like growth factor binding proteins in maternal serum throughout gestation and in the puerperium: Effects of a pregnancy-associated serum protease activity. J. Clin. Endocrinol. Metab. 71, 806-816.

Giudice, L. C., Milki, A.M., Milkowski, D. A., & El-Danasouri, I. (1991). Human granulosa contain messenger ribonucleic acids (mRNAs) encoding insulin-like growth factor binding proteins (IGFBPs) and in culture secrete IGFBPs. Fertil. Steril. 56, 475-480.

Giudice, L. C., Chandrasekher, Y. Aladin, & Cataldo, N.A. (1992a). The potential roles of intraovarian peptides in normal and abnormal mechanisms of reproductive physiology. Cur. Opin. Obstet. Gynecol. 5, 350-359.

Giudice, L.C., Dsupin, B.A., Irwin, J.C., & Eckert, R.L. (1992b). Identification of insulin-like growth factor binding proteins in human oviduct. Fertil. Steril. 57, 294-301.

Giudice, L.C., Dsupin, B.A., Jin, J.H., Vu, T.H., & Hoffman, A.R. (1993a). Insulin-like growth factor (IGF), IGF binding protein, and IGF receptor gene expression in human oviductal fimbria and ampulla during the menstrual cycle. [Abstract 318] 40th Annual Meeting of the Society for Gynecologic Investigation, Toronto.

Giudice, L.C., Dsupin, B.A., Vu, T.H., Lin, I.L., & Hoffman, A.R. (1993b). Differential expression of mRNAs encoding insulin-like growth factors and their receptors in human endometrium during the menstrual cycle and in early pregnancy. J. Clin. Endocrinol. Metab. 76, 1115-1122.

Giudice, L.C., Irwin, J.C., Dsupin, B.A., Pannier, E.M., Jin, I.H., Vu, T.H., & Hoffman, A.R. (1993c). Insulin-like growth factor (IGF), IGF binding protein (IGFBP), and IGF receptor gene expression and IGFBP synthesis in human uterine leiomyomata. Hum. Reprod. 8, 1796-1806.

Giudice, L.C., Dsupin, B.A., Gargosky, S.E., Rosenfeld, R.G., & Irwin, J.C. (1994). The insulin-like growth factor system in human peritoneal fluid: Its effects on endometrial stromal cells and its potential relevance to endometriosis. J. Clin. Endocrinol. Metab. 79, 1284-1293.

Gloudemans, T., Prinsen, I., Van Unnik, J.A., Lips, C.J., Den Otter, W., & Sussenbach, J.S. (1990). Insulin-like growth factor gene espression in human smooth muscle tumors. Cancer Res. 50, 6689-6695.

Gloudemans, T., Pospiech, I., Van Der Ven, L.T., Lips, C.J., Schneid, H., Den Otter, W., & Sussenbach, J.S. (1992). Expression and CpG methylation of the insulin-like growth factor II gene in human smooth muscle tumors. Cancer Res. 52, 6516-6521.

Gloudemans, T., Pospiech, I., Van der Ven, L.T., Lips, C.J,. Den Otter, W., & Sussenbach, J.S. (1993). An avaII restriction fragment length polymorphism in the insulin-like growth factor II gene and the occurrence of smooth muscle tumors. Cancer Res. 53, 5754-5758.

Godowski, P.J,. Leung, D.W., & Meacham, L.R. (1989). Characterization of the human growth hormone receptor gene and demonstration of a partial gene deletion in two patients with Laron-type dwarfism. Proc. Natl. Acad. Sci. USA 86, 8083.

Goldzieher, J.W., & Axelrod, L.R. (1963). Clinical and biochemical features of polycystic ovarian disease. Fertil. Steril. 14, 631-653.

G6mez, E., Tarin, J. J., & Pellicer, M. D. (1993). Oocyte maturation in humans: The role of gonadotropins and growth factors. Fertil. Steril. 60, 40-46.

Gospodarowicz, D., & Bialecki, H. (1978). The effects of epidermal and fibroblast growth factor on the replicative lifespan of bovine granulosa cells in culture. Endocrinology 103, 854-858.

Gospodarowicz, D., & Bialecki, H. (1979). Fibroblast and epidermal growth factors are mitogenic agents for cultured granulosa cells of rodent, porcine, and human origin. Endocrinology 104, 757-764.

Gospodarowicz, D., Plouet, J., & Fujii, D. K. (1989). Ovarian germinal epithelial cells respond to basic fibroblast growth factor and express its gene: Implications for early folliculogenesis. Endocrinology 125, 1266-1276.

Gougeon, A. (1996). Regulation of ovarian follicular development in primates: Facts and hypotheses. Endocr. Rev. 17, 121-155.

Grimes, R.W., & Hammond, J.M. (1994). Proteolytic degradation of insulin-like growth factor (IGF)-binding protein-3 by porcine ovarian granulosa cells in culture: Regulation by IGF-I. Endocrinology 134, 337-343.

Groome, N.P., Illingworth, P.J., O'Brien, M., Cooke, I,. Ganesan, T.S,. Baird, D.T., & McNeilly, A.S. (1994). Detection of dimeric inhibin throughout the human menstrual cycle by two-site enzyme immunoassay. Clin Endocrinol. 40,717-723.

Halban, J. (1925). Hysteroadenosis metastatica. Die lymphogenese der sog. Adenofibromatosis heterotopica. Arch. Gynak. 124, 1205-1210.

Hall, K., Hansson, U., Lundin, G., Luthman, M., Persson, B., Povoa, G., Stangenberg, M., & Ofverholm, U. (1984). Serum levels of somatomedins and somatomedin-binding protein in pregnant women with Type I or gestational diabetes and their infants. J. Clin. Endocrinol. Metab. 63, 1300-1406.

Halme, J., Becker, S., & Haskill, S. (1987). Altered maturation and function of peritoneal macrophages: Possible role in the pathogenesis of endometriosis. Am. J. Obstet. Gynecol. 156, 783-786.

Hamilton-Fairley, D., Kiddy, D., Watson, H., Sagle, M., & Franks, S. (1991). Low-dose gonadotropin therapy for induction of ovulation in 100 women with PCOS. Human. Reprod. 6, 1095-1099.

Hammond, J. M., Samaras, S. E., Grimes, R., Leighton, J., Barber, J., Canning, S. F., & Guthrie, H. D. (1993). The role of insulin-like growth factors and epidermal growth factor-related peptides in intraovarian regulation in the pig ovary. J. Reprod. Fertil. Suppl. (England) 48, 117-125.

Hamori, M., Blum, W. F., Torok, A., Stehle, R., waibel, E., Cledon, P., & Ranke, M.B. (1991a). Insulin-like growth factors and their binding proteins in human follicular fluid. Hum. Reprod. 6, 313-318.

Hamori, M., Blum, W. F., Torok, A., Stehle, R., waibel, E., Cledon, P., & Ranke, M.B. (1991b). Immunoreactive insulin-like growth factor binding protein-3 in the culture of human luteinized granulosa cells. Acta. Endocrinol. (Copenh). 124, 685-690.

Haney, A.F., Muscato, J. J., & Weinberg, J.B. (1981). Peritoneal fluid cell populations in infertility patients. Fertil. Steril. 25, 696-700.

Harrison-Woolrych, M.L., Charnock-Jones, D.S., & Smith, S.K. (1994). Quantification of messenger ribonucleic acid for epidermal growth factor in human myometrium and leiomyomata using reverse transcriptase polymerase chain reaction. J. Clin. Endocrinol. Metabol. 78, 1179-1184.

Hartsthorne, G. M., Bell, S. C., & Waites, G. T. (1990). Binding proteins for insulin-like growth factors in the human ovary: Identification, follicular fluid levels and immunohistological localization of the 29-32 kd type 1 binding protein, IGF-bp1. Hum. Reprod. 5, 649-660.

Head, J.R., MacDonald, P.C., & Casey M.L. (1993). Cellular localization of membrane metalloendo-peptidase (enkepahlinase) in human endometrium during the ovarian cycle. J. Clin. Endocrinol. Metab. 76, 769-776.

Heldin, C. H., Wateson, A., & Westermark, B. (1985). Platelet-derived growth factor. Mol. Cell. Endocrinol. 39, 169-187.

Henriksen, R., Funa, K., Wilander, E., Backstrom, T., Ridderheim, M., & Oberg, K. (1993). Expression and prognostic significance of platelet-derived growth factor and its receptors in epithelial ovarian neoplasms. Cancer Res. USA 53, 4550-4554.

Hernandez, E. R., Hurwitz, A., Payne, D. W., Dharmarajan, A. M., Purchio, A., & Adashi, E. Y. (1990). Transforming growth factor-b1 inhibits ovarian androgen production: Gene expression, cellular localization, mechanism(s), and site(s) of action. Endocrinology 127, 2804-2811.

Hernandez, E. R., Hurwitz, A., Vera, A., Pellicer, A., Adashi, E.Y., LeRoith, D., & Roberts, C.T., Jr. (1992). Expression of the genes encoding the insulin-like growth factors and their receptors in the human ovary. J. Clin. Endocrinol. Metab. 74, 419-425.

Hertelendy, F., Romero, R., Molnar, M., Todd, H., & Baldassare, J.J. (1993). Cytokine-initiated signal transduction in human myometrial cells. Am. J. Reprod. Immunol. 30, 49-57.

Hill, J.A., & Anderson, D.J. (1989). Lymphocyte activity in the presence of peritoneal fluid from fertile women and infertile women with and without endometriosis. Am. J. Obstet. Gynecol. 161, 861-864.

Hillier, S.G., & Miró, F. (1993). Inhibin, activin, and follistatin: Potential roles in ovarian physiology. Ann. N.Y. Acad. Sci. 687, 29-38.

Hillier, S. G., Yong, E. L., Illingworth, P. J., Baird, D.T., Schwall, R. H., & Mason, A. J. (1991). Effect of recombinant activin on androgen synthesis in cultured human thecal cells. J. Clin. Endocrinol. Metab. 72, 1206-1211.

Hirobe, S., He, W.W., Gustafson, M.L., MacLaughlin, D.T., & Donahoe, P.K. (1994). Müllerian inhibiting substance gene expression in the cycling rat ovary correlates with recruited or graafian follicle selection. Biol. Reprod. 50, 1238-1243.

Hofman, G.E., Rao, C.V., Barrows, G.H., Schultz, G.S., & Sanfilippo, J.S (1984). Binding sites for epidermal growth factor in human uterine tissues and leiomyomas. J. Clin. Endocrinol. Metabol. 58, 880-884.

Hofmann, G. E., Scott, R. T., Jr., Bryzski, R. G., & Jones, H. W., Jr. (1990). Immunoreactive epidermal growth factor concentrations in follicular fluid obtained from *in vitro* fertilization. Fertil. Steril. 54, 303-307.

Holly, J. M. P., Eden, J. A., Alaghband-Zadeh, J., Carter, G.D., Jemmott, R.C., Cianfarani, S., Chard, T., & Wass, J.A. (1990). Insulin-like growth factor binding proteins in follicular fluid from normal dominant and cohort follicles, polycystic and multicystic ovaries. Clin. Endocrinol. (Oxf). 33, 53-57.

Homburg, R., Pariente, C., Lunenfeld, B., & Jacobs, H.S. (1992). The role of insulin-like growth factor-1 (IGF-1) and IGF binding protein-1 (IGFBP-1) in the pathogenesis of polycystic ovary syndrome. Hum. Reprod. 7, 1379-1383.

Hoppener, J.W.M., Mosselman, S., Roholl, P.J.M., Lambrechts, C., Slebos, R.J., de Pagter-Holthuizen, P., Lips, C.J., Jansz, H.S., & Sussenbach, J.S. (1988). Expression of insulin-like growth factor-I and -II genes in human smooth muscle tumors. EMBO J. 7, 1379-1385.

Hossenlopp, P., Segovia, B., Lassarre, C., Roghani, M., Bedon, M., & Binoux, M. (1990). Evidence of enzymatic degradation of insulin-like growth factor binding proteins in the 150K complex during pregnancy. J. Clin. Endocrinol. Metab. 71, 759-805.

Hsueh, A. J. W., Welsh, T. H., & Jones, P. B. C. (1981). Inhibition of ovarian and testicular steroidogenesis by epidermal growth factors. Endocrinology 108, 2002-2006.

Hsueh, A. J. W., Adashi, E. Y., Jones, P. B. S., & Welsh, T. H. (1984). Hormonal regulation of the differentiation of cultured ovarian granulosa cells. Endocr. Rev. 5, 76-127.

Hsueh, A.J.W., Dahl, K.D., Vaughan, J., Tucker, E., Rivier, J., Bardin, C.W., & Vale, W. (1987). Heterodimers and homodimers of inhibin subunits have different paracrine action in the modulation of luteinizing hormone-stimulated androgen biosynthesis. Proc. Nat. Acad. Sci. 84, 5082-5086.

Hsueh, A.J.W., Billig, H., & Tsafriri, A. (1994) Ovarian follicle atresia: A hormonally controlled apoptotic process. Endocr. Rev. 15, 707-724.

Huet-Hudson, Y.M., Andrews, G.K., & Dey, S.K. (1989). Cell type-specific localization of c-myc protein in the mouse uterus: Modulation by steroid hormones and analysis of the periimplantation period. Endocrinology 125, 1683-1690

Hull, M.G.R. (1989). Polycystic ovarian disease: Clinical aspects and prevalence. Res. Clin. Forum 11, 21-33.

Hurst, B. S., Zacur, H. A., Schlaff, W. D., Damewood, M. D., & Berkowitz, G. D. (1993). The effect of epidermal growth factor in human granulosa cells varies with follicle size. J. Endocrinol. Invest. 16, 143-145.

Hurwitz, A., Hernandez, E. R., Resnick, C. E., Packman, J. N., Payne, D. W., & Adashi, E. Y. (1990). Basic fibroblast growth factor inhibits gonadotropin-supported ovarian androgen biosynthesis: Mechanism(s) and site(s) of action. Endocrinology 126, 3089-3095.

Hurwitz, A., Loukides, J., Ricciarelli, E., Botero, L., Katz, E., McAllister, J. M., Garcia, J. E., Rohan, R., Adashi, E.Y., & Hernandez, E.R. (1992). The human intraovarian interleukin-1 (IL-1) system: Highly compartmentalized and hormonally-dependent regulation of the genes encoding IL-1, its receptor, and its receptor antagonist. J. Clin. Invest. 89, 1746-1754.

Huynh, H., & Pollak, M. (1993). Insulin-like growth factor I gene expression in the uterus is stimulated by tamoxifen and inhibited by the pure antiestrogen ICI 182780. Cancer Res. 53, 5585-5590.

Iguchi, T., Edery, M., Tsai, P., Ozawa, S., Sato, T., & Bern, H.A. (1993). Epidermal growth factor receptor levels in reproductive organs of female mice exposed neonatally to diethylstilbestrol. Proc. Soc. Exp. Biol. Med. 204, 110-116.

Ijzermans, J. N. M., & Marquet, R. L. (1989). Interferon gamma: A review. Immunobiology. 179, 456-464.

Irwin, J.C., & Giudice, L.C. (1996). Insulin-like growth factor binding protein-1 binds to placental cytotrophoblast $\alpha5\beta1$ integrin and inhibits cytotrophoblast invasion into decidualized endometrial stromal cultures. J. Clin. Endocrinol. Metab. In press.

Iwashita, M., Mimuro, T., Watanabe, M., Setoyama, T., Matsuo, A., Adachi, T., Takeda, Y., & Sakamoto. (1990). Plasma levels of insulin-like growth factor-1 and its binding protein in prolycystic ovary syndrome. Horm. Res. 33, (Suppl. 2), 21-26.

Jalkanen, J., Suikkari, A.-M., Koistinen, R., Butzow, R., Ritvos, O., Seppala, M., & Ranta, T. (1989). Regulation of insulin-like growth factor binding protein-1 production in human granulosa-luteal cells. J. Clin. Endocrinol. Metab. 69, 1174-1180.

Jaye, M., Schlessinger, J., & Dionne, C. A. (1992). Fibroblast growth factor receptor tyrosine kinases: Molecular analysis and signal transduction. Biochim. Biophys. Acta. 1135, 185-189.

Jialal, I., Naiker, P., Reddi, K., Moodley, J., & Joubert, S.M. (1987). Evidence for insulin resistance in nonobese patients with polycystic ovarian disease. J. Clin. Endocrinol. Metab. 64, 1066-1069.

Johnson, G.R., Kannan, B., Shoyab, M., & Stromberg, K. (1993). Amphiregulin induces tyrosine phosphorylation of the epidermal growth factor receptor and $p185^{erb}$ B2. J. Biol. Chem. 268, 2924-2931.

Jones, J.I., & Clemmons, D. (1995). Insulin-like growth factors and their binding proteins: Biological actions. Endo. Rev. 16, 3-34.

Jones, J.I., Gockerman, A., Busby, W.H., Wright, G., & Clemmons, D.R. (1993). Insulin-like growth factor binding protein 1 stimulates cell migration and binds to the a5b1 integrin by means of its Arg-Gly-Asp sequence. Proc. Natl. Acad. Sci. USA 90, 10553-10557.

Jones, P. B. C., Welsh Jr., T. H., & Hsueh, A. J. W. (1982). Regulation of ovarian progestin production by EGF in cultured rat granulosa cells. J. Biol. Chem. 257, 11268-11273.

Kamada, S., Kubota, T., Taguchi, M., Wen-Rong, H., Sakamoto, S., & Aso, T. (1992). Effects of insulin-like growth factor-II on proliferation and differentiation of ovarian granulosa cells. Horm. Res. 37, 141-145.

Kapur, S., Tamada, H., Dey, S.K., & Andrews, G.K. (1992). Expression of insulin-like growth factor-I (IGF-I) and its receptor in the peri-implantation mouse uterus, and cell-specific regulation of IGF-I gene expression by estradiol and progesterone. Biol. Reprod. 46, 208-219.

Katz, Riciarelli, E., & Adashi, E.Y. (1993). The potential relevance of growth hormone to female reproductive physiology and pathophysiology. Fertil. Steril. 59, 8-34.

Kennedy, T. G., Brown, K. D., & Vaughan, T. J. (1993). Expression of the genes for the epidermal growth factor receptor and its ligands in porcine corpora lutea. Endocrinology 132, 1857-1859.

Kennedy, T.G., Brown, K.D,. & Vaughan, T.J. (1994). Expression of the genes for the epidermal growth factor receptor and its ligands in porcine oviduct and endometrium. Biol. Reprod. 50, 751-756.

Khorram, O., Taylor, R.N., Ryan, I.P., Schall, T.J., & Landers, D.V. (1993). Peritoneal fluid concentrations of the cytokine RANTES correlate with the severity of endometriosis. Am. J. Obstet. Gynecol. 169, 1545-1549.

Kim, J. H., Seibel, M. M., MacLaughlin, D. T., Donahoe, P. K., Ransil, B. J., Hametz, P. A., & Richards, C. J. (1992). The inhibitory effects of mullerian-inhibiting substance on epidermal growth factor induced proliferation and progesterone production of human granulosa-luteal cells. Endocrinology 75, 911-917.

Kincy, V.L., Williams, J.E., & Butler, J.E. (1992). Expression of the insulin-like growth factor -II (IGF-II) gene by ovine oviductal cells. Biol. Reprod. 46 (Suppl. 1),68 (Abstract).

Kliman H. (1994). Role of cytokines in the pathogenesis of endometriosis. Alternatives to Hysterectomy: Bench to Bedside, National Institutes of Health, May 1994, Abstract No. 10.

Knecht, M., & Catt, K. Y. (1983). Modulation of cAMP-mediated differentiation in ovarian granulosa cells by epidermal growth factor and platelet-derived growth factor. J. Biol. Chem. 258, 2789-2794.

Kojima, K., Kanzaki, H,. Iwai, M., Hatayama, H., Fujimoto, M., Ihoue, T., Hovie, K., Nakayama, H., Fujita, J., & Mori, T. (1994). Expression of leukemia inhibitory factor in human endometrium and placenta. Biol. Reprod. 50, 882-887.

Koos, R. D., & Olson, C. E. (1989). Expression of basic fibroblast growth factor in the rat ovary: Detection of mRNA using reverse transcription-polymerase chain reaction amplification. Mol. Endocrinol. 3, 2041-2048.

Koos, R., D., & Seidel, R. H. (1989). Detection of acidic fibroblast growth factor mRNA in the rat ovary using reverse transcription-polymerase chain reaction amplification. Biochem. Biophys. Res. Commun. 165, 82-88.

Koutsilieris, M., Allaire-Michaud, L., Fortier, M., & Lemay, A. (1991). Mitogen(s) for endometrial-like cells can be detected in human peritoneal fluid. Fertil. Steril. 56, 888-893.

Koutsilieris, M., Niklinski, W., Frenette, G., & Lemay, A. (1993). Heparin-sepharose binding growth factors in peritoneal fluid of women with endometriosis. Fertil. Steril. 59, 93-97.

Kubota, T., Kamada, S., Ohara, M., Taguchi, M., Sakamoto, S., Shimizu Y., & Aso, T. (1993). Insulin-like growth factor-II in follicular fluid of the patients with in vitro fertilization and embryo transfer. Fertil. Steril. 59, 844-849.

Lala, P.K., & Graham, C.H. (1990). Mechanisms of trophoblast invasiveness and their control: The role of proteases and protease inhibitors. Cancer Metastasis Rev. 9, 369-374.

Lambert-Messerlian, G.M., Hall, J.E., Sluss, P.M., Taylor, A.E., Martin, K.A., Groome, N.P., Crowley, W.F., Jr., & Schneyer, A.S. (1994). Relatively low levels of dimeric inhibin circulate in men and women with polycystic ovarian syndrome using a specific two-site enzyme-linked immunosorbent assay. J. Clin. Endocrinol. Metab. 79, 45-50.

Lamson, G., Giudice, L.C., Cohen, P., Liu, F., Gargosky, S., Muller, H.L., Oh, Y., Wilson, K.F., Hintz, R.L., & Rosenfeld, R.G. (1993). Proteolysis of IGFBP-3 may be a common regulatory mechanism of IGF action *in vivo*. Growth Regul. 3, 91-95.

Lanzone, A., Fulghesu, A.M., Pappalardo, S., Proto, C., Le Donne, M., Andreani, C.L., Muscatello, R., Caruso, A., & Mancuso, S. (1990). Growth hormone and somatomedin-C secretion in patients with polycystic ovarian disease. Their relationships with hyperinsulinism and hyperandrogenism. Gynecol. Obstet. Invest.

LaPolt, P.S., Soto, D., Su, J-G., Campen, C.A., Vaughan, J., Vale, W., & Hsueh, A.J.W. (1989). Activin stimulation of inhibin secretion and messenger RNA levels in cultured granulosa cells. Molec. Endocrinol. 3, 1666-1673.

LaPolt, P. S., Yamoto, M., Veljkovic, M., Sincich, C., Ny, T., Tsafriri, A., & Hsueh, A. J. W. (1990). Basic fibroblast growth factor induction of granulosa cell tissue-type plasminogen activator expression and oocyte maturation: Potential role as a paracrine ovarian hormone. Endocrinology 127, 2357-2363.

Lara, H. E., Hill, D. F., Katz, K. H., & Ojeda, S. R. (1990a). The gene encoding nerve growth factor is expressed in the immature rat ovary: Effect of denervation and hormonal treatment. Endocrinology 126, 357-363.

Lara, H. E., McDonald, J. K., & Ojeda, S. R. (1990b). Involvement of nerve growth factor in female sexual development. Endocrinology 126, 364-375.

Laron, Z. (1984). In: Advances in Internal Medicine and Pediatrics (von P Frick, H., von Harnack, G.A., Kochsiek, K., Martini, G.A., & Prader, A., Eds.). Springer-Verlag, Berlin.

Laron, Z., Pertzelan, A., & Mannheimer, S. (1966). Genetic pituitary dwarfism with high serum concentrations of growth hormone–a new inborn error of metabolism? Isr. J. Med. Sci. 2, 152-155.

Lee, C.T., Mitchell, K.R., Diehl, J.R., & Henricks, D.M. (1992). Secretion and expression of IGF-I in cultured porcine oviductal cells. Biol. Reprod. 46 (Suppl. 1), 251 (Abstract).

Lee, M.M., & Donahoe, P.K (1993). Mullerian inhibiting substance: A gonadal hormone with multiple functions. Endocr. Rev. 14, 152-164.

Lei, Z.M., & Rao, ChV. (1992). Expression of epidermal growth factor (EGF) receptor and its ligands, EGF and transforming growth factor-a, in human fallopian tubes. Endocrinology 131, 947-957.

Li, R., Phillips, D.M., & Mather, J.P. (1995). Activin promotes ovarian follicle development *in vitro*. Endocrinol. 136, 849-856.

Li, W., Ho, Y. B., & Leung, P.C.K.. (1992). Inhibition of progestin accumulation by activin-A in human granulosa cells. J. Clin. Endocrinol. Metab. 75, 285-289.

Lidor, Y. J., Xu, F. J., Martinez-Maza, O., Olt, G. J., Marks, J. R., Berchuck, A., Ramakrishnan, S., Berek, J. S., & Bast, R. C. (1993). Constitutive production of macrophage colony-stimulating factor and interleukin-6 by human ovarian surface epithelial cells. Experimental Cell Res. 207, 332-339.

Lockwood, C.J., & Schatz, F. (1996). A biological model for the regulation of peri-implantational hemostasis and menstruation. J. Soc. Gynecol. Invest. 3, 159-165.

Luetteke, N.C., Qui, T.H., Peiffer, R.L., Oliver, P., Smithies, O., & Lee, D.C. (1993). TGF-a deficiency results in hair follicle and eye abnormalities in targeted and waved-1 mice. Cell 73, 263-278.

Lumsden, M.A., West, C.P., Bramley, T., Rumgay, L., & Baird, D.T. (1988). The binding of epidermal growth factor to the human uterus and leiomyomata in women rendered hypo-oestrogenic by continuous administration of an LHRH agonist. Br. J. Obstet. Gynecol. 95, 1299-1304.

Lunenfeld, B., Menashe, Y., Dor, Y., Pariente, C., & Insler, V. (1991). The effect of growth hormone, growth factors, and their binding globulins on ovarian responsiveness. Contracept. Fertil. Sex. 19, 133-136.

Lyttle, C.R., Levia, M., & Weil, S. (1994). Peritoneal macrophage migration. In: Alternatives to Hysterectomy: Bench to Bedside, National Institutes of Health, May 1994, Abstract No. 11.

Magoffin, D. A., Gancedo, B., & Erickson, G. F. (1989). Transforming growth factor-b promotes differentiation of ovarian thecal-interstitial cells but inhibits androgen production. Endocrinology 125, 1951-1958.

Mahmood, T.A., & Templeton, A. (1990). Pathophysiology of mild endometriosis: Review of literature. Human. Reprod. 5,756-760.

Maness, L. M., Kastin, A. J., Weber, J. T., Banks, W. A., Beckman, B. S., & Zadina, J. E. (1994). The neurotrophins and their receptors: Structure, function, and neuropathology. Neurosci. Biobehav. Rev. 18, 143-159.

Martin, MC. (1993). Collection and report of assisted reproductive technology outcome data: Problems and solutions. Fertil. Steril. 57, 762-763.

Maruo, T., Ladines-Llave, Cecilia A., Samoto, T., Matsuo, H., Manalo, A. S., Ito, H., & Mochizuki, M. (1993). Expression of epidermal growth factor and its receptor in the human ovary during follicular growth and regression. Endocrinology 132, 924-931.

Mason, H. D., Margara, R, Winstron, R. M. L., Beard, R. W., Reed, M. J., & Franks, S. (1990). Inhibition of oestradiol production by epidermal growth factor in human granulosa cells of normal and polycystic ovaries. Clin. Endocrinol. 33, 511-517.

Mason, H.D., Willis, D., Holly, J. M. P., Cwyfan-Hughes, S. C., Seppala, M., & Franks, S. (1992). Inhibitory effects of insulin-like growth factor binding proteins on steroidogenesis by human granulosa cells in culture. Mol. Cell. Endocrinol. 89, R1.

Mason, H., Margara, R., Winston, R. M. L., Seppšlš, M., Koistinen, R., & Franks, S. (1993a). Insulin-like growth factor-1 (IGF-1) inhibits production of IGF-binding protein-1 while stimulating estradiol secretion in granulosa cells from normal and polycystic human ovaries. J. Clin. Endocrinol. Metab. 76, 1275-1279.

Mason, H., Willis, D., Seppala, M., & Franks, S. (1993b). Insulin inhibits insulin-like growth factor binding protein-1 (IGFBP-1) production by granulosa, theca, and stroma from human ovaries. J . Endocrinol. 137, 179-185.

Mason, H.D., Willis, D.S., Beard, R.W., Winston, R.M.L., Margara, R., & Franks, S. (1994). Estradiol production by granulosa cells of normal and polycystic ovaries: Relationship to menstrual cycle history and concentrations of gonatropins and sex steroids in follicular fluid. J. Clin. Endocrinol. Metab. 79, 1355-1360.

Massague, J. (1990). Transforming growth factor-a. A model for membrane-anchored growth factors. J. Biol. Chem. 265, 21393-21396.

Massague, J., & Pandiella, A. (1993). Membrane-anchored growth factors. Annu. Rev. Biochem. 63, 515-541.

Mathews, L.S., & Vale, W.W. (1991). Expression cloning of an activin receptor, a predicted transmembrane serine kinase. Cell 65, 973-982.

May, J. V., Frost, J. P., & Schomberg, D. W. (1988). Differential effect of epidermal growth factor, somatomedin-C/insulin-like growth factor I, and transforming growth factor-b on porcine granulosa cell deoxyribonucleic acid synthesis and cell proliferation. Endocrinology 123, 168-179.

McLachlan, R.I., Robertson, D.M., Burger, H.G., & DeKretser, D.M. (1986). The radioimmuno-assay of bovine and human follicular fluid and serum inhibin. Mol. Cell. Endocrinol. 46, 175-185.

Mcleon, F.D., Vijayakumay, R., Brown, M., Rao, C. V., Yussman, M.A., & Schultz, G. (1986). Peritoneal fluid volume, estrogen, progesterone, prostaglandin, and epidermal growth factor concentration in patients with and without endometriosis. Obstet. Gynecol. 68, 189-196.

McNatty, K.P., Moore-Smith, D., Makris, A., Osathanondh, R., & Ryan, K.J. (1979). The micro-environment of the human antral follicle: Interrelationship among the steroid levels in antral fluid, the population of granulosa cells, and the status of the oocyte *in vitro* and *in vivo*. J. Clin. Endocrinol. Metab. 49, 851-860.

Menashe, Y., Sack, J., & Mashiach, S. (1991). Spontaneous pregnancies in two women with Laron-type dwarfism: Are growth hormone and circulating insulin-like growth factor mandatory for induction of ovulation? Human Reprod. 6, 670-671.

Mendoza, A.E., Young, R., Orkin, S.H., & Collins, T. (1990). Increased platelet-derived growth factor A-chain expression in human uterine smooth muscle cells during the physiologic hypertrophy of pregnancy. Proc. Natl. Acad. Sci. USA 87, 2177-2181.

Molnar, M., Romero, R., & Hertelendy, F. (1993). Interleukin-1 and tumor necrosis factor stimulate arachidonic acid release and phospholipid metabolism in human myometrial cells. Am. J. Obstet. Gynecol. 169, 825-829.

Mondschein, J., & Schomberg, D. W. (1981a). Growth factors modulate gonadotropin receptor induction in granulosa cell culture. Science 211, 1179-1180.

Mondschein, J., & Schomberg, D. W. (1981b). Platelet-derived growth factor enhances granulosa cell luteinizing hormone receptor induction by follicle-stimulating hormone and serum. Endocrinology 109, 325-327.

Morishige, K., Hirohisa, K., Amemiya, K., Adachi, H., Adachi, K., Sakoyam, Y., Mikake, A., & Tanizzawa, O. (1993). Menstrual stage-specific expression of epidermal growth factor and transforming growth factor-a in human oviduct epithelium and their role in early embryogenesis. Endocrinology 133, 199-207.

Mulheron, G. W., Bossert, N. L., Lapp., J.A., Walmer, D.K., & Schomberg, D. W. (1992). Human granulosa-luteal and cumulus cells express transforming growth factors-beta type 1 and type 2 mRNA. J. Clin. Endocrinol. Metab. 74, 458-460.

Murphy, A.A., Kettel, L.M,. Morales, A.J., Roberts, V.J., & Yen, S.S.C. (1993). Regression of uterine leiomyomata in response to the antiprogesterone RU 486. J. Clin. Endocrinol. Metab. 76, 513-517.

Murphy, J., Tsibris, J,. Tsibris, A., Pendleton, S., Mayer, J., Maroulis, G., & Spellacy, W. (1994). Regulation by estrogen (E) of the transforming growth factor-b (TGF-b) system in uterine leiomyomas (L). Society for Gynecologic Investigation 41st Annual meeting, March, 1994, Chicago, Abstract P21, p. 205.

Murphy, L.J., Murphy, L.C., & Friesen, H.G. (1987a). A role for the insulin-like growth factors as estromedins in the rat uterus. Trans. Assoc. Am. Physicians 99, 204-214.

Murphy, L.J., Murphy, L.C., & Friesen, H.G. (1987b). Estrogen induces insulin-like growth factor-I expression in the rat uterus. Mol. Endocrinol. 1, 445-450.

Nagamani, M., & Stuart, C.A. (1992). Specific binding sites for insulin in the human myometrium and leiomyomas of the uterus. Fertil. Steril. 58, 120-123.

Nagamani, M., Stuart, C.A., & Huang, H.Q. (1994). Specific binding and growth effect of insulin and insulin-like growth factors in a human leiomyosarcoma cell line. Society for Gynecologic Investigation 41st Annual meeting, March, 1994, Chicago, Abstract P250, p. 319.

Nakamura, T., Takio, K., Eto, Y., Shibai, H., Titani, K., & Sugino, H. (1990). Activin-binding protein from rat ovary is follistatin. Science 247, 836-838.

Neufeld, G., Napoleone, F., Schweigerer, L., Mitchell, R., & Gospodarowicz, D. (1987). Bovine granulosa cells produce basic fibroblast growth factor. Endocrinology 121, 597-603.

Nissley, P., & Lopaczynski, W. (1991). Insulin-like growth factor receptors. Growth Factors 5, 29-36.

Norstedt, G., Levinovitis, A., & Eriksson, H. (1989). Regulation of insulin-like growth factor -I mRNA and insulin-like growth factor -II mRNA by estrogen in the rat. Acta Endocrinol. 120, 466-472.

Novak, E. (1931). Pelvic endometriosis: Spontaneous rupture of endometrial cysts with a report of three cases. Am. J. Obstet. Gynecol. 22, 826-830.

Oh, W-S., Robertson, D.M., & deKretser, D.M. (1989). Inhibin as an oocyte maturation inhibitor. Molec. Cell. Endocrinol. 62, 307-311.

O'Reilly, G., Charnock-Jones, D.S., Davenport, A.P., Cameron, I.T., & Smith, S.K. (1992). Presence of messenger ribonucleic acid for endothelin-1, endothelin-2, and endothelin-3 in human endometrium and a change in the ratio of ET_A and ET_B receptor subtypes across the menstrual cycle. J. Clin. Endocrinol. Metab. 75, 1545-1549.

Ojeda, S.R., & Dissen, G.A. (1995). Developmental regulation of the ovary via growth factor tyrosine kinase receptors. Trends Endocrinol. Metab.

Olive, D.L., Montoya, I,. Riehl, R.M., & Schenken, R.S. (1991). Macrophage-conditioned media enhance endometrial stromal cell proliferation *in vitro*. Am. J. Obstet. Gynecol. 164, 953-958.

Olsson, J-H., Carlsson, B., & Hillensjo, T. (1990). Effect of insulin-like growth factor-I on deoxyribonucleic acid synthesis in cultured human granulosa cells. Fertil. Steril. 54, 1052-1057.

Oosterlynck, D.J., Meuleman, C., Sobis, H., Vandeputte, M., & Koninckx, P.R. (1993). Angiogenic activity of peritoneal fluid from women with endometriosis. Fertil. Steril. 59, 778-783.

Pache, T.D., Hop, W.C., De Jong, F.H., Leerentveld, R.A., van Geldorp, H., Van de Kamp, T.M., Gooren, L.J., & Fauser, B.C. (1992). 17 beta-Oestradiol, androstenedione and inhibin levels in fluid from individual follicles of normal and polycystic ovaries, and in ovaries from androgen treated female to male transsexuals. Clinical Endocrinology 36, 565-571.

Pannier, E.M., Irwin, J.C., Ulaner, G.A., Kim, E., & Giudice, L.C. (1995). Benign, but not malignant, human uterine smooth muscle cells secrete an IGFBP-4 protease that is regulated by IGFs. In preparation.

Pekonen, F., Nyman, T., & Rutanen, E.M. (1993). Differential expression of keratinocyte growth factor and its receptor in the human uterus. Mol. Cell. Endocrinol. 95, 43-49.

Pekonen, F., Laatikainen, T., Buyalos, R., & Rutanen, E.M. (1989). Decreased 34K insulin-like growth factor binding protein in polycystic ovarian disease. Fertil. Steril. 51, 972-975.

Pekonen, F., Rutanen, E.-M., Kurunmaki, H., & Hovatta, O. (1992) Ovulation induction increases serum levels of insulin-like growth factor binding protein-1. Int. J. Fertil. 37, 188-191.

Pfeifer, T., & Chegini, N. (1994). Immunohistochemical localization of insulin-like growth factor (IGF-I), IGF-I receptor, and IGF binding proteins 1-4 in human fallopian tube at various reproductive stages. Biol. Reprod. 50, 281-289.

Piquette, G. N., Crabtree, M. E., El-Danasouri, I., Milki, A., & Polan, M. L. (1993). Regulation of plasminogen activator inhibitor-1 and -2 messenger ribonucleic acid levels in human cumulus and granulosa-luteal cells. Endocrinology 76, 518-523.

Piquette, G. N., Simon, C., el Danasouri, I., Frances, A., & Polan, M. L. (1994). Gene regulation of interleukin-1 beta, interleukin-1 receptor type I, and plasminogen activator inhibitor-1 and -2 in human granulosa-luteal cells. Fertil. Steril. 62, 760-770.

Pokras, R., & Jufnagel, V.G. (1987). National Center for Health Statistics, 1965-1984. Washington, D.C.: U.S. Government Printing Office; DHHS Publications no (PHS) 87-1753. (Vital and health statistics; series 13; No 92).

Pokras, R., & Jufnagel, V.G. (1988). Hysterectomy in the U.S. 1965-1984. AJPH 78, 852-854.

Polan, M. L., Loukides, J. A., & Honig, J. (1994). Interleukin-1 in human ovarian cells and in peripheral blood monocytes increases during the luteal phase: Evidence for a midcycle surge in the human. Am. J. Obstet. Gynecol. 170, 1000-1006.

Poretsky, L. (1991). On the paradox of insulin-induced hyperandrogenism in insulin-resistant states. Endo. Rev. 12, 3-10.

Poretsky, L., & Peiper, B. (1994). Insulin resistance, hypersecretion of LH, and a dual-defect hypothesis for the pathogenesis of polycystic ovary syndrome. Obst. Gynecol. 84, 613-621.

Poretsky, L., Grigorescu, F., Seibel, M., Moses, A.C., & Flier, J.S. (1985). Distribution and characterization of insulin and insulin-like growth factor I receptors in normal human ovary. J. Clin. Endocrinol. Metab. 61, 728-734.

Poretsky, L., Bhargava, G., & Kalin, M. (1988). Regulation of the insulin receptors in the human ovary (*in vitro*). J. Clin. Endocrinol. Metab. 67, 774-780.

Poretsky, L., Bhargava, G., Saketos, M., & Dunaif, A. (1990). Regulation of human ovarian insulin receptors *in vivo*. Metabolism 39, 161-165.

Punnonen, J., Heinonen, P. K., Teisala, K., Kujansuu, E., Jansen, C. T., & Punnonen, R. (1992). Demonstration of tumor necrosis factor-a in preovulatory follicular fluid: Its association with serum 17b- estradiol and progesterone. Gynecol. Obstet. Invest. 33, 80-84.

Rabinovici, J., Spencer, S.J,. Doldi, N., Goldsmith, P.C., Schwall, R., & Jaffe, R.B. (1992). Activin-A as an intraovarian modulator: Actions, localization, and regulation of the intact dimer in human ovarian cells. J. Clin. Invest. 89, 1528-1536.

Rabinovici, J., Dandekar, P., Angle, M., Rosenthal, S., & Martin, M. (1990a). Insulin-like growth factor I levels in follicular fluid from human preovulatory follicles: Correlation with serum IGF-I levels. Fertil. Steril. 54, 428-433.

Rabinovici, J., Spencer, S.J., & Jaffe, R.B. (1990b). Recombinant human activin-A promotes proliferation of human luteinized granulosa cells *in vitro*. J. Clin. Endocrinol. Metab. 71, 1396-1398.

Ramasharma, K., & Li, C. H. (1987). Human pituitary and placental hormones control human insulin-like growth factor II secretion in human granulosa cells. Proc. Natl. Acad. Sci. 84, 2643-2648.

Ramasharma, K., Cahrera, C. M., & Li, C. H. (1986). Identification of insulin-like growth factor II in human seminal and follicular fluids. Biochem. Biophys. Res. Commun. 140, 536-540.

Ramey, J.W., & Archer, D.F. (1993). Peritoneal fluid: Its relevance to the development of endometriosis. Fertil. Steril. 60, 1-14.

Rein, M.S., & Nowak, R.A. (1992). Biology of uterine myomas and myometrium *in vitro*. Seminars in Reproductive Endocrinology 10, 310-319.

Rein, M.S,. Friedman, A.J., Pandian, M.R., & Heffner, L.J. (1990). The secretion of insulin-like growth factors I and II from explant cultures of fibroids and myometrium from women treated with a gonadotropin-releasing hormone agonist. Obstet. Gynecol. 76, 388-394.

Richards, J. (1994). Hormonal control of gene expression in the ovary. Endo. Rev. 15, 725-751.

Ritvos, O., Ranta, T., Jalkanen, J., Suikkari, A.-M., Voutilainen, R., Bohn, H., & Ratanen, E.M. (1988). Insulin-like growth factor (IGF) binding protein from human decidua inhibits the binding and biological action of IGF-I in cultured choriocarcinoma cells. Endocrinology 122, 2150-2157.

Roberts, A. J., & Skinner, M. K. (1991). Transforming growth factor-a and -b differentially regulate growth and steroidogenesis of bovine thecal cells during antral follicle development. Endocrinology 129, 2041-2048.

Roberts, V.J., Barth, S,. El-Roeiy, A., & Yen, S.C.C. (1993). Expression of inhibin/activin subunits and follistatin messenger ribonucleic acids and proteins in ovarian follicles and the corpus luteum during the human menstrual cycle. J. Clin. Endocrinol. Metab. 77, 1402-1410.

Roberts, V.J., Barth, S., El-Roeiy, A., & Yen, S.S.C. (1994). Expression of inhibin/activin system messenger ribonucleic acids and proteins in ovarian follicles from women with polycystic ovarian syndrome. J. Clin. Endocrinol. Metab. 79, 1434-1439.

Robertson, D.M., Foulds, L.M., deVos, F., Leversha, L., & DeKretser, D.M. (1990). Identification of inhibin and inhibin-related proteins in human follicular fluid. Reprod. Fertil. Dev. 2, 327-335.

Robinson, C. J. (1991). Multiple receptors found for the growing FGF family. Trends Pharmacol. Sci. 12, 12-124.

Rosenfeld, R.G., Lamson, G., Pham, H., Oh, Y., Conover, C., De Leon, D.D., Donovan, S.M., Ocrant, I., & Giudice, L.C. (1990). Insulin-like growth factor-binding proteins. Recent Prog. Horm. Res. 46, 99-159.

Ross, R., Raines, E. W., & Bowen-Pope, D. F. (1986). Biology of platelet-derived growth factor. Cell 46, 155-169.

Rossi, M.J., Chegini, N., & Masterson, B.J. (1992). Presence of epidermal growth factor, platelet-derived growth factor, and their receptors in human myometrial tissue and smooth muscle cells: Their action in smooth muscle cells *in vitro*. Endocrinology 130, 1716-1727.

Rotwein, P. (1991). Structure, evolution, expression, and regulation of insulin-like growth factors I and II. Growth Factors 5, 3-18.

Rutanen, E.-M., Pekonen, F., & Makinen, T. (1988). Soluble 34K binding protein inhibits the binding of insulin-like growth factor I to its receptors in human secretory phase endometrium: Evidence for autocrine/paracrine regulation of growth factor action. J. Clin. Endocrinol. Metab. 66, 173-180.

Sadatsuki, M., Tsutsumi, O., Yamada, R., Muramatsu, M., & Taketani, Y. (1993). Local regulatory effects of activin A and follistatin on meiotic maturation of rat oocytes. Biochem. Biophys. Res. Comm. 196, 388-395.

Saito, S., Nakamura, T., Titani, K., & Sugino, H. (1991). Production of activin-binding protein by rat granulosa cells *in vitro*. Biochem. Biophys. Res. Comm. 176, 413-422.

Salustri, A., Ulisse, S., Yanagishita, M., and Hascall V. C. (1990). Hyaluronic acid synthesis by mural granulosa cells and cumulus cells *in vitro* is selectively stimulated by a factor produced by oocytes and transforming growth factor-b. J. Biol. Chem. 265, 19517-19523.

Sampson, J.A. (1924). Benign and malignant endometrial implants in peritoneal cavity, and their relation to certain ovarian tumors. Surg. Gynecol. Obstet. 38, 287-293.

San Roman, G.A. & Magoffin, D.A. (1992). Insulin-like growth factor binding proteins in ovarian follicles from women with polycystic ovarian disease: Cellular source and levels in follicular fluid. J. Clin. Endocrinol. Metab. 75, 1010-1016.

San Roman, G. A., & Magoffin, D. A. (1993). Insulin-like growth factor binding proteins in healthy and atretic follicles during natural menstrual cycles. J. Clin. Endocrinol. Metab. 76, 625-632.

Schall, T.J. (1991). Biology of the RANTES/SIS cytokine family. Cytokine 3, 165-170.

Schall, T.J., Bacon, K., Toy, K.J., & Goeddel, D.V. (1990). Selective attraction of monocytes and T lymphocytes of the memory phenotype by cytokine RANTES. Nature 347, 669-675.

Schneyer, A.L., Mason, A.J., Burton, L.E., Ziegner, J.R., & Crowley, W.F., Jr. (1990). Immunoreactive inhibin alpha-subunit in human serum: Implications for radioimmunoassay. J. Clin. Endocrinol. Metab. 70, 1208-1212.

Schneyer, A.L., Sluss, P.M., Whitcomb, R.W., Martin, K.A., Sprengel, R., & Crowley, W.F., Jr. (1991). Precursors of alpha-inhibin modulate follicle-stimulating hormone receptor binding and biological activity. Endocrinology. 129, 1987-1999.

Schneyer, A.L., Rzucidlo, D.A., Sluss, P.M., & Crowley, W.F., Jr. (1994). Characterization of unique binding kinetics of follistatin and activin or inhibin in serum. Endocrinology. 135, 667-674.

Schuller, A.G.P., Lindenbergh-Kortleve, D.J., Pache, T.D., Zwarthoff, E.C., Fauser, B.C.J.M., & Drop, S.L.S. (1993). Insulin-like growth factor binding protein-2, 28 kDa and 24 kDa IGF-BP levels are decreased in fluid of dominant follicles, obtained from normal and polycystic ovaries. Reg. Pep. 48, 157-163.

Schwall, R.H., Mason, A.J,. Wilcox, J.N., Bassett, S.G., & Zeleznik, A.J. (1990). Localization of inhibin/activin subunit mRNAs within the primate ovary. Molec. Endocrinol. 4, 75-79.

Schwall, R.H., Nikolics, K., Szonyi, E., Gorman, C., & Mason, A.J. (1988). Recombinant expression and characterization of human activin A. Molec. Endocrinol. 2, 1237-1242.

Seppala, M., Julkunen, M., Riittinen, L., & Koistinen, R. (1992). Endometrial proteins: A reappraisal. Human. Reprod. 7, 31-40.

Seppala, M., Wahlstrom, T., Koskimes, A.I., Tenhunen, A., Rutanen, E.M., Koistinen, R., Huhtaniemi, I., Bohn, H., & Stenman, U.H. (1984). Human preovulatory follicular fluid, luteinized cells of hyperstimulated preovulatory follicles, and corpus luteum contain placental protein 12. J. Clin. Endocrinol. Metab. 58, 505-510.

Sharara, F., & Neiman, L. (1994). Growth hormone receptor mRNA expression in leiomyoma and surrounding myometrium. The American Fertility Society Annual Meeting, San Antonio, TX, Abstract O-0062, p. S31.

Shifren, J.L., Tseng, J.F., Zaloudek, C.J., Ryan, I.P., Meng, Y.G., Ferrara, N., Jaffe, R.B., & Taylor, R.N. (1996). Ovarian steroid regulation of vascular endothelial growth factor in human endometrium: Implications for angiogenesis during the menstrual cycle and in the pathogenesis of endometriosis. J. Clin. Endcrinol. Metab. 81, 3112-3118.

Shikone, T., Yamoto, M., & Nakano, R. (1992). Follicle-stimulating hormone induces functional receptors for basic fibroblast growth factor in rat granulosa cells. Endocrinology 131, 1063-1068.

Shimasaki, S., & Ling, N. (1992). Identification and molecular characterization of insulin-like growth factor binding proteins (IGFBP-1, -2, -3, -4, -5, and -6). Prog. Growth Factor Res. 3, 243-246.

Shimasaki, S., Koga, M., Buscaglia, M.L., Simmons, D.M., Bicsak, T.A., & Ling, N. (1989). Follistatin gene expression in the ovary and extragonadal tissues. Molec. Endocrinol. 3, 651-659.

Shimonaka, M., Inouye, S., Shimasaki, S,. & Ling, N. (1991). Follistatin binds to both activin and inhibin through the common beta-subunit. Endocrinology 128, 3313-3316.

Shoham, Z., Patel, A., & Jacobs, H.S. (1991). Polycystic ovary syndrome: Safety and effectiveness of a stepwise and low-dose administration of purified FSH. Fertil. Steril. 55, 1051-1056.

Shukovski, L., Dyson, M., & Findlay, J.K. (1993). The effects of follistatin, activin and inhibin on steroidogenesis by bovine thecal cells. Molec. Cell. Endocrinol. 97, 19-27.

Simon, C., & Polan, M.L. (1994). Cytokines and reproduction. Western J. Med. 160, 425-429.

Simon, C., Frances, A., Piquette, G.N., el Dahasouri, I., Zurawski, G., Dang, W., & Polan, M.L. (1994). Embryonic implantation in mice is blocked by interleukin-1 receptor antagonist. Endocrinology 134, 521-528.

Simpson, J.L., Elias, S., Malinak, L.R., & Buttram, V.C., Jr., (1980). Heritable aspects of endometriosis. I. Genetic studies. Am. J. Obstet. Gynecol. 137, 327-333.

Sjogren, A., Holmes, P. V., & Hillensjo, T. (1991). Interleukin-1 modulates luteinizing hormone stimulated cyclic AMP and progesterone release from human granulosa cells *in vitro*. Hum. Reprod. 6, 910-913.

Skinner, M. K., & Coffey, R. J., Jr. (1988). Regulation of ovarian cell growth through the local production of transforming growth factor-a by theca cells. Endocrinology 123, 2632-2638.

Sluss, P.M., & Schneyer, A.L. (1992). Low molecular weight follicle-stimulating hormone receptor binding inhibitor in sera from premature ovarian failure patients. J. Clin. Endocrinol. Metab. 74, 1242-1246.

Sporn, M. B., Roberts, A. B., Wakefield, L. M., & Assoian, R. K. (1985). Transforming growth factor-b: Biological function and chemical structure. Science 233, 532-534.

Spuijbroek, M.D.E.H., Dunselman, G.A.J., Menheere, P.P.C.A., & Evers, J.L.H. (1992). Early endometriosis invades the extracellular matrix. Fertil. Steril. 58, 929-934.

Stein, I.F., & Cohen, M.R. (1935). Surgical treatment of polycystic ovaries, amenorrhea and sterility. Am. J. Obstet. Gynecol. 38, 465-480.

Steinkampf, M. P., Mendelson, C. R., & Simpson, E. R. (1988). Effects of epidermal growth factor and insulin-like growth factor I on the levels of mRNA encoding aromatase cytochrome P-450 of human ovarian granulosa cells. Mol. Cell Endocrinol. 59, 93-98.

Stewart, C. (1994). The role of leukemia inhibitory factor (LIF) and other cytokines in regulating implantation in mammals. Annals N.Y. Acad. Sci. 734, 157-165.

Stewart, E.A., Friedman, A.J., Peck, K., & Nowak, R.A. (1994). Relative overexpression of collagen type I and Collagen type II messenger ribonucleic acids by uterine leiomyomas during the proliferative phase of the menstrual cycle. J. Clin. Endocrinol. Metab. 79, 900-906.

Stirling, D., Magness, R. R., Stone, F., Waterman, M. R., & Simpson, E.R. (1991). Expression of mRNA encoding basic fibroblast growth factor (bFGF) in bovine corpora lutea and cultured luteal cells. J. Reprod. Fertil. 91, 1-8.

Stouffer, R.L., Woodruff, T.K., Dahl, K.D., Hess, D.L., Mather, J.P., & Molskness, T.A. (1993). Human recombinant activin-A alters pituitary luteinizing hormone and follicle-stimulating hormone secretion, follicular development, and steroidogenesis, during the menstrual cycle in rhesus monkeys. J. Clin. Endocrinol. Metab. 77, 241-248.

Strickland, S., & Richards, W.G. (1992). Invasion of the trophoblasts. Cell 71, 355-361.

Sugawara, M., DePaolo, L., Nakatani, A., DiMarzo, S.J., & Ling, N. (1990). Radioimmunoassay of follistatin: Application for *in vitro* fertilization procedures. J. Clin. Endocrinol. Metab. 71, 1672-1674.

Suikkari, A-M., Rutanen, E-M., & Seppala, M. (1987). Circulating levels of immunoreactive insulin-like growth factor binding protein in non-pregnant women. Human Reprod. 2, 297-302.

Suikkari, A.-M., Jalkanen, J., Koistinen. R., Butzow, R., Rituos, O., Ranta, T., & Seppala, M. (1989a). Human granulosa cells synthesize low molecular weight insulin-like growth factor-binding protein. Endocrinology. 124, 1088-1092.

Suikkari, A.M., Ruutiainen, K., Erkkola, R., & Seppala, M. (1989b). Low levels of low molecular weight insulin-like growth factor-binding protein in patients with polycystic ovarian disease. Human Reprod. 4, 136-139.

Suikkari, A.-M., Tiitinen, A., Stenman, U.-H., Seppala, M., & Laatikainen, T. (1991). Oral contraceptives increase insulin-like growth factor binding protein-1 concentration in women with polycystic ovarian disease. Fertil. Steril. 55, 895-899.

Sun, X.J., Rothenberg, P., Kahn, C.R., Backer, J.M., Araki, E., Wilden, P.A., Cahill, D.A., Goldstein, B.J., & White, M.F. (1991). Structure of the insulin receptor substrate IRS-1 defines a unique signal transduction protein. Nature 352, 73-77.

Tabibzadeh, SS. (1991). Human endometrium: An active site of cytokine production and action. Endocr. Rev. 12, 272-290.

Tabibzadeh, S.S., Satyaswaroop, P.G., & Rao, P.N. (1988). Antiproliferative effect of interferon gamma in human endometrial epithelial cells *in vitro*: Potential local growth modulatory role in endometrium. J. Clin. Endocrinol. Metab. 67, 131-138.

Taketania, Y., Kui, T-M., & Mizuno, M. (1992). Comparison of cytokine levels and embryo toxicity in peritoneal fluid in infertile women with untreated or treated endometriosis. Am. J. Obstet. Gynecol. 167, 265-270.

Tamada, H,. McMaster, M.T,. Flanders, K.C., Andrews, G.K., & Dey, S.K. (1990). Cell type-specific expression of transforming growth factor-b1 in the mouse uterus during the periimplantation period. Mol. Endocrinol. 4, 965-972.

Tanabe, K., Gagliano, P., Channing, C.P., Nakamura, Y., Yoshimura, Y., Iizuka, R., Fortuny, A., Sulewski, J., & Rezai, N. (1983). Levels of inhibin-F activity and steroids in human follicular fluid from normal women and women with polycystic ovarian disease. J. Clin. Endocrinol. Metab. 57, 24-31.

Tanabe, K., Saijo, A., Park, J.Y., Kohriyama, S., Sano, Y., Nakamura, Y., & Iizuka, R. (1990). The role of inhibin in women with polycystic ovary syndrome (PCOS). Horm. Res. 33 Suppl. 2, 10-17.

Tang, X.-M., Rossi, M.J., Masterson, B.J,. & Chegini, N. (1994). Insulin-like growth factor (IGF)-I, IGF-I receptors, and IGF binding proteins 1-4 in human uterine tissue: Tissue localization and IGF-I action in endometrial stromal and myometrial smooth muscle cells *in vitro*. Biol. Reprod. 50, 1113-1125.

Tapanainen, J., Leiononen, P. J., Tapanainen, P., Yamamoto, M., & Jaffe, J. B. (1987). Regulation of human granulosa-luteal cell progesterone production and proliferation by gonadotropins and growth factors. Fertil. Steril. 48, 576-580.

Terranova, P. F. (1991). Regulation of the granulosa cell: Growth factor interactions. Semin. Reprod. Endocrinol. 9, 313-320.

Terranova, P. F., Roby, K.F., Sancho-Tello, M., Weed, J., & Lyles, R. (1991). Tumor necrosis factor α: Localization and actions within the preovulatory follicle. In: Growth Factors in Reproduction (Schomberg, D.W., Ed.), pp. 63-78. Springer-Verlag, New York.

Thoenen, H. (1991). The changing scene of neurotrophic factors. Trends Neurosci. 14, 165-170.

Tilly, J.L., Kowalski, K.I., Schomberg, D.W., & Hsueh, A.J.W. (1992a). Apoptosis in atretic ovarian follicles is associated with selective decreases in messenger ribonucleic acid transcripts for gonadotropin receptors and cytochrome P450 aromatase. Endocrinology 131, 1670-1675.

Tilly, J. L., Billig, H., Kowalski, K. I., & Hsueh, A. J. W. (1992b). Epidermal growth factor and basic fibroblast growth factor suppress the spontaneous onset of apoptosis in cultured rat ovarian granulosa cells and follicles by a tyrosine kinase-dependent mechanism. Mol. Endocrinol. USA 6, 1942-1950.

Tommola, P., Pekonen, F., & Rutanen, E-M. (1989). Binding of epidermal growth factor and insulin-like growth factor I in human myometrium and leiomyomata. Obstet. Gynecol. 74, 658-662.

Tsafriri, A., Vale, W., & Hsueh, A. J. W. (1989) Effects of transforming growth factors and inhibin-related proteins on rat preovulatory Graafian follicles *in vitro*. Endocrinology 125, 1857-1862.

Tuuri, T., Erämaa, M., Hildén, K., & Ritvos, O. (1994). Activin-binding protein follistatin messenger ribonucleic acid and secreted protein levels are induced by chorionic gonadotropin in cultured human granulosa-luteal cells. Endocrinology. 135, 2196-2203.

van Dessel, T.J.H.M., Faessen, G., Yap, O.W.S., Chandrasekher, Y.A., Fauser, B.C.J.M., & Giudice, L.C. (1996). Serum and ovarian follicular fluid levels of insulin-like growth factor (IGF)-I,

IGF-II, and IGF binding proteins -1 and -3 during the normal menstrual cycle. J. Clin. Endocrinol. Metab. 81, 1224-1231.

van Santbrink, E., van Dessel, T.J.H.M., & Fauser, B.C.J.M. (1994). Decremental FSH and follicle growth and selection during the normal menstrual cycle.

van Santbrink, E., Donderwinkel, P.F.J., van Dessel, T.J.H.M., & Fauser, B.C.J.M. (1995). Gonadotrophin induction of ovulation using a step-down dose regimen: Single-center clinical experience in 82 patients. Human Reprod.

Vollenhoven, B.J., Herington, A.C., & Healy, D.L. (1993). Messenger ribonucleic acid expression of the insulin-like growth factors and their binding proteins in uterine fibroids and myometrium. J. Clin. Endocrinol. Metabol. 76, 1106-1110.

Volpe, A., Coukos, G., DõAmbrogio, G., Artini, P. G., & Genazzani, A. R. (1991). Follicular fluid steroid and epidermal growth factor content, and in vitro estrogen release by granulosa-luteal cells from patients with polycystic ovaries in an IVF/ET program. Eur. J. Obstet. Gynecol. Reprod. Biol. 42, 195-199.

Voutilainen, R., & Miller, W. L. (1987). Coordinate trophic hormone regulation of mRNAs for insulin-like growth factor II and the cholesterol side-chain cleavage enzyme, P450ssc, in human steroidogenic tissues. Proc. Natl. Acad. Sci. 84, 1590-1595.

Wandji, S. A., Pelletier, G., & Sirard, M. A. (1992). Ontogeny and cellular localization of 125I-labeled basic fibroblast growth factor and 125I-labeled epidermal growth factor binding sites in ovaries from bovine fetuses and neonatal calves. Biol. Reprod. 47, 807-813.

Wang, L., Robertson, S., Seamark, R. F., & Norman, R. J. (1991). Lymphokines, including interleukin-1, alter gonadotropin-stimulated progesterone production and proliferation of human granulosa-luteal cells *in vitro*. J. Clin. Endocrinol. Metab. 72, 824-831.

Wang, L. J., & Norman, R. J. (1992). Concentrations of immunoreactive interleukin-1 and interleukin-2 in human preovulatory follicular fluid. Hum. Reprod. 7, 147-150.

Wang, L. J., Pascoe, V., Petrucco, O. M., & Norman, R. J. (1992a). Distribution of leukocyte subpopulations in the human corpus luteum. Hum. Reprod. 7, 197-202.

Wang, L. J., Brannstrom, M., Robertson, S. A., & Norman, R. J. (1992b). Tumor necrosis factor alpha in the human ovary: presence in follicular fluid and effects on cell proliferation and prostaglandin production. Fertil. Steril. 58, 934-940.

Watanabe, H., Nagai, K., Yamaguchi, M., Ikenoue, T., & Mori, N. (1993). Interleukin-1 beta stimulates prostaglandin E2 and F2 alpha synthesis in human ovarian granulosa cells in culture. Prostaglandins Leukotrienes and Essential Fatty Acids 49, 963-967.

Watanabe, T., Shintani A., Nakata, M., Shing, Y., Folkman, J., Igarashi, K., & Sasada, R. (1994). Recombinant human betacellulin. J. Biol. Chem. 269, 9966-9973.

Weingold, A.B. (1983). In: Principles and Practice of Clinical Gynecology (Kase, N.G., & Weingold, A.B., Eds.). John Wiley and Sons, New York.

Westergaard, L. G., & Andersen, C. Y. (1989). Epidermal growth factor (EGF) in human preovulatory follicles. Hum. Reprod. 4, 257-260.

Whitman, G.F., & Pantazis, C.G. (1991). Cellular localization of müllerian inhibiting substance messenger ribonucleic acid during human ovarian follicular development. Amer. J. Obstet. Gynecol. 165, 1881-1886.

Williams, J.E., & Butler, J.E. (1991). Expression of insulin-like growth factor-I (IGF-I) gene by ovine oviductal cells. J. Anim. Sci. 69 (Suppl. 1), 406 (Abstract).

Wingo, P.A., Huezo, C.M., Rubin, G.L., et al., (1985). The mortality risk associated with hysterectomy. Am. J. Obstet. Gynecol. 152, 803-808.

Wiseman, D.L., Henricks, D.M., Eberhardt, D.M., & Bridges, W.C. (1992). Identification and content of insulin-like growth factors in porcine oviductal fluid. Biol. Reprod. 47, 126-132.

Wong, W.L., Garg, S.J., Woodruff, T., Bald, L., Fendly, B., & Lofgren, J.A. (1993). Monoclonal antibody based ELISAs for measurement of activins in biological fluids. J. Immunol. Methods. 165, 1-10.

Wood, A. M., Lambert, A., Hooper, M. A. K., Mitchell, G. G., & Robertson, W. R. (1994). Exogenous steroids and the control of estradiol secretion by human granulosa-lutein cells by follicle stimulating hormone and insulin-like growth factor-I. Hum. Reprod. 9, 19-25.

Woodruff, T.K., Lyon, R.J., Hansen, S.E., Rice, G.C., & Mather, J.P. (1990). Inhibin and activin locally regulate rat ovarian folliculogenesis. Endocrinology 127, 3196-3205.

Woodruff, T.K., Meunier, H., Jones, P.B.C., Hsueh, A.J.W,. & Mayo, K.E. (1987). Rat inhibin: Molecular cloning of a and b-subunit complementary deoxyribonucleic acids and expression in the ovary. Molec. Endocrinol. 1, 561-568.

Woodruff, T.K., Krummen, L.A., Lyon, R.J., Stocks, D.L., & Mather, J.P. (1993a). Recombinant human inhibin A and recombinant human activin A regulate pituitary and ovarian function in the adult female rat. Endocrinology. 132, 2332-2341.

Woodruff, T.K., Krummen, L., Baly, D., Garg, S., Allison, D., Sadick, M., Wong, W., Mather, J., & Soules, M. (1993b). Quantitative two-site enzyme-linked immunosorbent assays for inhibin A, activin A, and activin B. Hum. Reprod. 8 (Suppl. 2), 133-137.

Wordinger, R., & Palmer, L. (1994). Immunolocalization of basic fibroblast growth factor within the mouse oviduct during early pregnancy. Biol. Reprod. 50 (Suppl. 1), 274 (Abstract).

Yamoto, M., Minami, S., Nakano, R., & Kobayashi, M. (1992). Immunohistochemical localization of inhibin/activin subunits in human ovarian follicles during the menstrual cycle. J. Clin. Endocrinol. Metab. 74, 989-993.

Yap, O.W.S., Chandrasekher, Y.A., & Giudice, L.C. (1994). Growth factor regulation of IGFBP regulation by cultured human granulosa-luteal cells. Abstract, Third International Symposium on Insulin-like Growth Factors, Sydney, February 1994.

Yap, O.W.S., van Dessel, T.J.H.M., Chandrasekher, Y.A., Fauser, B.C.J.M., & Giudice, L.C. (1996) Insulin-like growth factors (IGFs) and IGF binding protein proteases in follicles from normally cycling women and androgen-treated female-to-male transsexuals. Fertil. Steril. Submitted.

Yeh, J., Rein, M., & Nowak, R. (1991). Presence of messenger ribonucleic acid for epidermal growth factor (EGF) and EGF receptor demonstrable in monolayer cell cultures of myometria and leiomyomata. Fertil. Steril. 56, 997-1000.

Yen, S.S.C. (1980). The polycystic ovary syndrome. Clin. Endocrinol. 12, 177-182.

Ying, S.Y. (1988). Inhibins, activins, and follistatins: Gonadal proteins modulating the secretion of follicle-stimulating hormone. Endocr. Reviews 9, 267-293.

Yong, E. L., Baird, D. T., Yates, R., Reichert, L. E., & Hillier S. G. (1992). Hormonal regulation of the growth and steroidogenic function of human granulosa cells. J. Clin. Endocrinol. Metab. 74, 842-850.

Zhao, Y., & Chegini, N. (1994). The expression and cellular localization of TGF-b_1. TGF-b_2, TGF-b_3 and TGF-b type II receptor mRNA and protein and ^{125}I-TGF-b_1 binding sites in human Fallopian tube. 41st Annual Meeting of the Society for Gynecologic investigation, Chicago, IL. Abstract 209.

Zhou, J., & Bondy, C. (1993). Anatomy of the human ovarian insulin-like growth factor system. Biol. Reprod. 48, 467-470.

Zhou, J., Dsupin, B.A., Giudice, L.C., & Bondy, C. (1994). Insulin-like growth factor (IGF), IGF binding protein, and IGF receptor gene expression in human endometrium. J. Clin. Endocrinol. Metab. 79, 1723-1734.

Ziltener, H. J., Maines-Bandiera, S., Schrader, J. W., & Auersperg, N. (1993). Secretion of bioactive interleukin-1, interleukin-6, and colony-stimulating factors by human ovarian surface epithelium. Biol. Reprod. 49, 635-641.

Zolti, M., Bider, D., Seidman, D.S., Mashiach, S., & Ben-Rafael, Z. (1992). Cytokine levels in follicular fluid of polycystic ovaries in patients treated with dexamethasone. Fertil. Steril. 57, 501-504.

CPSIA information can be obtained at www.ICGtesting.com
Printed in the USA
LVOW05*0355020114

367657LV00008B/73/P

9 780762 301188